Cement Production Technology
Principles and Practice

Cement Production Technology

Principles and Practice

Anjan Kumar Chatterjee

CRC Press is an imprint of the
Taylor & Francis Group, an **informa** business

CRC Press
Taylor & Francis Group
6000 Broken Sound Parkway NW, Suite 300
Boca Raton, FL 33487-2742

© 2018 by Taylor & Francis Group, LLC
CRC Press is an imprint of Taylor & Francis Group, an Informa business

No claim to original U.S. Government works

Printed on acid-free paper

International Standard Book Number-13: 978-1-138-57066-5 (Hardback)

This book contains information obtained from authentic and highly regarded sources. Reasonable efforts have been made to publish reliable data and information, but the author and publisher cannot assume responsibility for the validity of all materials or the consequences of their use. The authors and publishers have attempted to trace the copyright holders of all material reproduced in this publication and apologize to copyright holders if permission to publish in this form has not been obtained. If any copyright material has not been acknowledged please write and let us know so we may rectify in any future reprint.

Except as permitted under U.S. Copyright Law, no part of this book may be reprinted, reproduced, transmitted, or utilized in any form by any electronic, mechanical, or other means, now known or hereafter invented, including photocopying, microfilming, and recording, or in any information storage or retrieval system, without written permission from the publishers.

For permission to photocopy or use material electronically from this work, please access www.copyright.com (http://www.copyright.com/) or contact the Copyright Clearance Center, Inc. (CCC), 222 Rosewood Drive, Danvers, MA 01923, 978-750-8400. CCC is a not-for-profit organization that provides licenses and registration for a variety of users. For organizations that have been granted a photocopy license by the CCC, a separate system of payment has been arranged.

Trademark Notice: Product or corporate names may be trademarks or registered trademarks, and are used only for identification and explanation without intent to infringe.

Visit the Taylor & Francis Web site at
http://www.taylorandfrancis.com

and the CRC Press Web site at
http://www.crcpress.com

To my parents, Sibani and Phani Bhusan, who are no more, but who, I have a strange feeling, are protecting me and my family from an unknown remoteness.

Contents

Preface xiii
Author xvii
Notation xix

1 Basics of mineral resources for cement production 1

1.1 Preamble 1
1.2 Characterization of minerals and rocks 2
1.3 Nature of limestone occurrence 6
1.4 Assessment of limestone deposits 8
1.5 Mineral composition and quality of limestone 18
1.6 Limestone mining 24
1.7 Quarry design and operational optimization 27
1.8 Argillaceous materials 29
1.9 Corrective materials 32
1.10 Natural gypsum 33
1.11 Influence of raw materials on unit operations 36
1.12 Summary 37
References 39

2 Raw mix proportioning, processing, and burnability assessment 41

2.1 Preamble 41
2.2 Stoichiometric requirements in raw mix computation 41
2.3 Raw mix computation 48
2.4 Preparation process for raw mix 53
2.5 Burnability features of raw meal 63
2.6 Use of mineralizers 68

2.7	Summary	69
	References	70

3 Fuels commonly in use for clinker production — 73

3.1	Preamble	73
3.2	Characteristics of fuels	74
3.3	Coal resources of the world	81
3.4	Basic chemistry and physics of combustion	83
3.5	Coal preparation and firing	87
3.6	Relation of process parameters with combustion	92
3.7	Petcoke as a substitute fuel	96
3.8	Summary	100
	References	101

4 Alternative fuels and raw materials — 103

4.1	Preamble	103
4.2	Broad classification	104
4.3	Feasibility of an AFR project	107
4.4	Inventory and material characteristics	109
4.5	Systematic quality assessment	118
4.6	Co-processing of alternative fuels	121
4.7	Systemic requirements for using alternative fuels	124
4.8	Gasification technology	133
4.9	Alternative raw materials	135
4.10	Environmental aspects	136
4.11	Summary	138
	References	139

5 Pyroprocessing and clinker cooling — 141

5.1	Preamble	141
5.2	Clinker formation process	141
5.3	Preheater-precalciner systems	146
5.4	Rotary kiln systems	152
5.5	Kiln burners and combustion	155
5.6	Clinker coolers	157
5.7	Volatiles cycle in preheater-precalciner kiln systems	161
5.8	Refractory lining materials in the kiln system	162
5.9	Energy consumption and kiln emissions	167
5.10	Kiln control strategies	169
5.11	Summary	171
	References	172

6 Clinker grinding and cement making — 175

6.1	Preamble	175
6.2	Clinker characteristics	176
6.3	Clinker grinding systems	185
6.4	Energy conservation and material characteristics	200
6.5	Grinding aids in cement manufacture	205
6.6	Storage, dispatch, and bagging of cement	209
6.7	Summary	211
	References	212

7 Composition and properties of Portland cements — 213

7.1	Preamble	213
7.2	Basic grades and varieties	214
7.3	Characteristics of Portland cements	219
7.4	Phase-modified Portland cements	225
7.5	Blended Portland cements	226
7.6	Characterization of cements and practical implications of properties	230
7.7	Overview of the hydration reactions	235
7.8	Cement for durable concrete	245
7.9	Summary	248
	References	249

8 Advances in plant-based quality control practice — 251

8.1	Preamble	251
8.2	Sampling guidelines	252
8.3	Sampling stations in cement plants	254
8.4	Computer-aided run-of-mine limestone quality control	256
8.5	Preblending operation	258
8.6	Raw mix control	260
8.7	Kiln operation monitoring	263
8.8	Cement grinding process	264
8.9	X-ray diffractometry for phase analysis	265
8.10	Online quality control in cement plants	267
8.11	Flue gas analysis	272
8.12	Process measurements	280
8.13	Total process control system	284
8.14	Summary	285
	References	286

9 Environmental mitigation and pollution control technologies — 287

    9.1 Preamble — 287
    9.2 Pollutants emitted into the atmosphere during manufacture of cement — 288
    9.3 Generation and broad characteristics of dust — 293
    9.4 Sulfur dioxide emissions — 313
    9.5 Nitrogen oxide emissions — 315
    9.6 Noise pollution — 319
    9.7 Selected monitoring techniques — 321
    9.8 Current environmental outlook — 323
    9.9 Summary — 324
    References — 325

10 Trends of research and development in cement manufacture and application — 327

    10.1 Preamble — 327
    10.2 Sustaining technologies in the growth of the cement industry — 328
    10.3 Status of potentially disruptive pyroprocessing technologies — 331
    10.4 Portland cement derivatives with niche application potential — 336
    10.5 Complex building products formulation with CAC — 352
    10.6 Research thrusts towards low-carbon cement industry — 354
    10.7 Technology options for converting Co_2 into fuel products — 358
    10.8 Low-carbon cements and concretes — 359
    10.9 Nanotechnology in cement research — 366
    10.10 Non-hydraulic cements — 367
    10.11 Summary — 368
    References — 369

11 Global and regional growth trends in cement production — 371

    11.1 Preamble — 371
    11.2 Capacity and production growth perspectives — 372
    11.3 National economy versus cement consumption — 377
    11.4 Change drivers of production and application of cement — 379

11.5	Future design of cement plants	382
11.6	Growth of the Indian cement industry: a case study	384
	References	391

12 Epilogue 393

Index 397

Preface

This book is borne out of my very long and intimate association with the science and technology of cement production. In the early 1960s, when I visited what was then the Soviet Union (now the Russian Federation) to study, I got the flavor of cement production technology from my teachers, who offered interesting lessons in materials science, phase equilibrium, minerals processing, and the petrography of synthetic materials with industrial relevance and illustrations. The production of cement became more fascinating to me when I was exposed to various plants that included cement manufacturing units. For the first time, I stood in front of a gigantic kiln, 7 m in diameter and 230 m long, with a production capacity of only about 1000 tons per day of clinker by the wet process of manufacture prevalent in those days.

Back in my own country, I had the opportunity of serving what was then the Cement Research Institute of India (now the National Council of Cement and Building Materials) and becoming a part of the changing canvas of the Indian cement industry, marked by the process transformation from wet to dry; from single-kiln plants to multiple dry preheater kilns operating in specific locations; the advent of kilns with a capacity of 1,000,000 tons or more through pre-calcination technology; from meagre off-line process control to a high level of instrumentation and computer-aided process control systems. Later, as Research Director of what was then the Associated Cement Companies Limited (now ACC Limited) I had the rare opportunity of integrating industrial research and development (R&D) with plant adoption and practices, transferring new product technologies and upgrading operating processes. Over roughly five decades I had innumerable occasions of organizing training and skill development programs for practising professionals and technicians, and delivering lectures and lessons to them. This book, in essence, is an updated compendium of what I taught in these classes and presented in the seminars and workshops.

The question that may logically appear is why one should look for yet another book on this subject when there are several available. The simple answer is that the published books and compendia are focused on specific themes of chemistry, chemical engineering, plant and equipment design, project engineering, plant maintenance, and so on. If one has to understand the multiple facets of cement production technology as a beginner, one may have to browse through multiple publications. The present book is an attempt to compress, in one place, the essential information of a diverse nature required to be familiar with the wide subject of cement production technology. A reader, however, requires an adequate background in chemistry and chemical engineering for making the best use of this book, and it is expected to be most useful to those endeavouring to have a career in the cement industry, or those who have freshly entered this profession, or—for that matter—those practising engineers and scientists who intend to refresh their fundamentals in the subject without entering the classroom.

Looking at the rapid growth and spread of the cement industry worldwide, I am all the more convinced that a textbook or a handbook of multidisciplinary nature—but comprehensible to young professionals with different academic backgrounds—is a need of the hour. World production of cement was more than 4.0 billion tons per annum in 2013, corresponding to about 600 kg per capita consumption of cement, which is, perhaps, higher than the per-capita food consumption of human beings. Further, this mass of cement is not used only as cement but produces cement-based building products and materials that amount to approximately seven times the quantity of cement produced. The consumption of such cement-based materials appears to be second only to mankind's consumption of water. I refer to these material volumes only in order to demonstrate the magnitude of the cement industry that many of us serve. It is essential that we understand all the intricacies of the science and technology governing such a colossal manufacturing activity. This has been another compelling reason for me to author this book.

I have been an ardent advocate of R&D in cement production and application. The cement industry, being a mature one, has always been dependent on sustaining technological developments, often borrowed from sister industries such as ceramics and electronics. Research on disruptive technologies has been rare and often stinted. Nevertheless, the Portland cement industry's almost 200-year history has seen a large number of spectacular technological advances, which, I believe, should make up part of the knowledge of those participating in the industry's growth. Hence, the more important developments of the past, and present trends of research, have been captured in this book to the extent a non-researcher should know.

This book is organized in 11 chapters. The first two chapters are devoted to the winning and processing of raw materials. Chapters 3, 4, and 5 deal with fuels and clinker making. Chapters 6 and 7 delve into clinker grinding technology and characterizing different varieties of Portland cement, respectively. The overall quality control and environmental technology aspects have been elaborated in Chapters 8 and 9, respectively. The past and present trends of R&D have been concisely narrated in Chapter 10. The global and regional growth perspectives of the industry have been outlined in Chapter 11.

It may be pertinent to mention that the book has drawn freely from previous literature as well as the author's personal experience and prior publications. The more relevant ones are listed as references at the end of each chapter.

I am grateful to the large number of friends and professional colleagues who have encouraged me to author this book. The list is too long to be presented, but I would be failing in my duties if I did not thank my erstwhile colleagues at Conmat Technologies Private Limited, Kolkata, for their unstinting support and assistance during the preparation of the initial version of the manuscript. I am also sincerely grateful to Professor Binay Dutta, an academician of great repute and former Chairman of the West Bengal Pollution Control Board in India, for having guided me in the last mile of the arduous task of making this book worthy of publishing.

It would be a matter of great satisfaction if readers at large benefited from this book.

Author

Anjan Kumar Chatterjee is presently the chairman of Conmat Technologies Private Limited, which is a research and consulting outfit in Kolkata, India, that is engaged in providing technical support services to the cement, concrete, and mineral industries in India and abroad. He is also director-in-charge of Dr. Fixit Institute of Structural Protection and Rehabilitation, Mumbai, which is a not-for-profit knowledge center specifically devoted to repair, restoration, and renewal engineering of concrete buildings. He is also associated with the major cement companies in the country in various advisory capacities. Prior to taking up the above assignments, Dr. Chatterjee was an employee of the Associated Cement Companies Limited (now ACC Limited) for over two decades, and retired as its Whole-time executive director. While at ACC, he was responsible for the company's R&D and Project Engineering departments, as well as several other business units.

Academically, Dr. Chatterjee is a postgraduate in Geology and holds a doctorate in Materials Science. He has researched extensively in the fields of electro-remelting slags, phase equilibrium studies on oxy-fluoride systems, and microstructural studies of cement, concrete, and ceramics at the Institute of Metallurgy in Moscow, Moscow State University, Russia, and the Building Research Establishment, United Kingdom. Dr. Chatterjee has been on various international assignments with UNIDO and IVAM Environmental Research, University of Amsterdam, The Netherlands.

He is a fellow of the Indian National Academy of Engineering, Indian Concrete Institute, and Indian Ceramic Institute and a founding member of the Asian Cement & Concrete Research Academy in Beijing, China. He has been conferred lifetime achievement awards by the Indian Concrete Institute, Association of Consulting Civil Engineers, Confederation of Indian Industries, and Cement Manufacturers' Association, and has many other awards and a large number of publications to his credit.

Notation

Standard cement chemistry notation is assumed throughout this book:

- $A=Al_2O_3$ $C=CaO$ $F=Fe_2O_3$ $H=H_2O$ $M=MgO$ $K=K_2O$ $N=Na_2O$ $S=SiO_2$ $\acute{S}=SO_3$ C-S-H denotes a variable composition.
- AFm denotes a solid solution range within the monosulfate-type structure.
- Aft denotes a solid solution range within the ettringite-type structure.
- "Ton" has been used for capacity/product mass without arithmetic correction of "tonne."

CHAPTER ONE

Basics of mineral resources for cement production

1.1 Preamble

It is said that cement manufacture begins in the quarry (Figure 1.1), which is the commonly used term for opencast mining of minerals and rocks including limestone, the primary raw material for cement production. Since there is hardly any alternative to using limestone for making cement, it is treated as one among the most essential resources of a country. However, given its importance in the economy of any nation, our understanding of the genesis, occurrence, composition, properties, mining, and application of limestone is rather poor. This is because limestone has long been regarded as a "common" rock, and past geological studies were limited in scope, focused essentially on mapping deposits, analyzing rocks, and—occasionally—evaluating aquifers and petroleum reservoirs. However, a different set of data and a deeper understanding are needed to make more efficient use of limestone as a cement raw material. Further, there are certain other naturally occurring raw materials like limestone that are also used in cement manufacture, although in smaller quantities and only when required, and they include clay, bauxite, iron ore, sandstone, and so on. These materials are also obtained from opencast mining operations. Hence, a basic knowledge of geology and mining is important for cement chemists, technologists, and plant engineers. Keeping this in view, this chapter presents the fundamentals of geology, chemistry, and the mining of raw materials as relevant to cement production.

FIGURE 1.1 A view of the limestone quarry at Cedar Creek, Virginia, USA. (Courtesy: Mr Robert Shenk; rob.shenk@gmail.com.)

1.2 Characterization of minerals and rocks

Any naturally occurring chemical compound making up a part of the Earth's crust is called a mineral. Minerals may occur in solid, liquid, or gaseous form. Natural gas is the best example of the occurrence of minerals in a gaseous state. Petroleum, mercury, and water are the commonest examples of minerals in liquid form. Solid minerals can either be native elements like gold, silver, etc., or compounds like calcium carbonate (calcite), calcium silicate (wollastonite), aluminum silicate hydrate (kaolinite), etc. The last group of solid minerals in the form of simple or complex compounds is of particular relevance for cement manufacture.

Aggregates of such minerals of more or less invariable composition forming independent geological bodies are known as rocks. Depending on their genesis, rocks are classified as igneous, sedimentary, or metamorphic. Igneous rocks are formed by the cooling of the molten mass, called magma, inside the Earth. Primary igneous rock types include granite, syenite, dunite, gabbro, basalt, diorite, etc. Based on their silica content, these rocks are further classified as acidic ($SiO_2 > 65\%$), intermediate (SiO_2 65–52%), basic (SiO_2 52–40%), and ultrabasic ($SiO_2 < 40\%$). Amongst the commonly known igneous rocks, granite is acidic and basalt is basic in composition.

Sedimentary rocks are formed by the deposition of the products of weathering of igneous and metamorphic rocks by mechanical, chemical, or organic means from water or air that acts as the carrier of weathering products. Common examples of sedimentary rocks include limestone, sandstone, coal, iron ore, etc. These rocks are the most relevant ones for cement manufacture.

Metamorphic rocks are derived from previously formed rocks of any kind through primarily in-situ action of high pressure and temperature, as well as of chemicals from hot liquids and gases. Quartzite, schist, gneiss, marble, etc. are typical examples of metamorphic rocks.

Minerals and their intrinsic properties

It is common knowledge that the properties of minerals are a function of their chemical composition and system of crystallization. It is possible to have 32 crystal classes belonging to six systems. All these systems and classes are distinguished from one another by their symmetry. The six main systems are shown in Figure 1.2 (2). While discussing the mineral constituents of different rocks, as well as the properties of different synthetic phases of clinker and cement, reference is often made to the crystal systems to which they belong.

The physical characters of minerals can either be intrinsic or extrinsic, as illustrated below:

a. Characters depending upon cohesion and elasticity, such as cleavage, fracture, hardness, elasticity, etc.
b. Specific gravity or density compared with that of water.
c. Characters depending upon light, such as color, luster, transparency, special optical properties, etc.
d. Characters depending upon heat, such as conductivity, change of form, fusibility, etc.
e. Characters depending upon electricity and magnetism.
f. Characters depending upon the action of senses such as taste, odor, feel, etc.

As already mentioned, all the properties of a mineral depend upon the nature of the chemical elements of which it is composed, and perhaps even more upon the way in which their atoms are arranged in the crystal structure. One of the common methods of identifying and characterizing

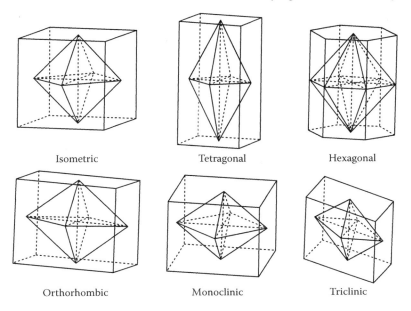

FIGURE 1.2 Different crystal systems based on their symmetries. (E. S. Dana and W. E. Ford: *A Textbook of Mineralogy*. 1995. Copyright Wiley-VCH Verlag GmbH & Co. KGaA. Reproduced with permission.)

a mineral in hand specimens is to apply the Mohs scale of hardness. In this scale, the degree of hardness is determined by observing the comparative ease or difficulty with which one mineral is scratched by another, or by a file or knife. In minerals, there are all grades of hardness, from that of talc, impressible by the fingernail, to that of diamond. For all practical purposes, the Mohs scale of hardness is followed:

1. Talc
2. Gypsum
3. Calcite
4. Fluorite
5. Apatite
6. Orthoclase
7. Quartz
8. Topaz
9. Corundum
10. Diamond

For example, if the mineral under examination is scratched by a knife-blade as easily as is calcite, its hardness is said to be 3; if less easily than calcite but more easily than fluorite, its hardness is between 3 and 4. It has generally been observed that compounds of heavy metals, such as silver, copper, mercury, lead, etc., are soft. Carbonates, sulfates, phosphates, sulfides, and hydrous salts are neither soft nor hard, whereas oxides and silicates containing aluminum are conspicuously hard.

Physical and technological properties of rocks

The general properties of rocks, as expected, depend primarily on their mineral composition and microstructure, that is, their structural and textural features. Texturally, rocks can be friable, where the grains are not interconnected as in sand, pebble, gravel, etc.; or coherent, where water-colloidal bonds hold the particles together through plasticity, such as in clay, loam, bauxite, etc.; or hard, where there are rigid elastic bonds between the mineral particles, as in limestone, granite, gneiss, marble, etc.

Depending on grain size and crystalline characteristics, rocks are classified as shown in Table 1.1.

Table 1.1 Classification of rocks based on grain size

Nomenclature	Ranges of individual grains
Coarse-grained	1–5 cm.
Medium-grained	1–10 mm.
Fine-grained	< 1 mm.
Aphanitic	Grains distinguishable under a lens.
Cryptocrystalline	Grains not identifiable under an optical microscope.
Glassy	Solid but non-crystalline.

Mining practice and theory utilize all the properties of rocks relating to their mechanics, thermodynamics, and electrodynamics, which are clubbed as physical properties. These are mostly coupled with another set of rock characteristics, known as technological properties, which refer to different conditions of the breaking of rocks under complex states of stressing. Each of the technological properties has narrow limits related to the application of certain techniques, machines, or technological media upon the rock. The technological indices reflect on a function not solely of the rock itself but also of the mechanisms acting on them. The physical and technological properties are summarized in Tables 1.2 and 1.3, respectively (3).

Table 1.2 Important physical properties of rocks

Property	Parameter	Symbol	Unit	Definitions
Density	Apparent specific gravity	Υ	g/cm^3	Weight of a unit volume of rock in its natural state with its pores, joints, etc.
	Porosity	P	Ratio	Relative volume of all pores in a unit weight of rock.
Mechanical	Ultimate compressive strength	σcom	kg/cm^2	Uniaxial compressive stress at which a rock breaks.
	Ultimate tensile strength	σten	kg/cm^2	Tensile stress at which a rock breaks.
	Young's modulus	E	Ratio	Ratio of acting longitudinal stress to the corresponding relative longitudinal strain.
	Poisson's ratio	υ	Ratio	Ratio of elastic longitudinal strain to elastic lateral strain under normal uniaxial stress.
Thermal	Thermal conductivity	λ	W/m.°C	Quantity of heat passing in unit time through unit section in a direction perpendicular to that section with a temperature difference of 1°C over unit distance.
	Specific heat	c	J/kg.°C	The amount of heat required to raise the temperature of 1 kg of a substance through 1°C.
	Coefficient of linear expansion	β	1/°C	Percentage increase in the dimensions of a body on heating through 1°C.
Electro magnetic	Electrical resistivity or Specific resistance	ρ	Ohm.m	Reciprocal of the intensity of current passing through unit area of a specimen at an electric field strength in the specimen equal to unity.
	Dielectric constant or Relative permittivity	ε	–	Coefficient showing how much the electric field intensity falls when a rock specimen is placed in the field.
	Magnetic permittivity	μ	–	Coefficient showing a change in the intensity of a magnetic field when a rock specimen is brought into it.

Table 1.3 Technological properties of rocks

Parameter	Symbol/Unit	Description
Hardness	H	Resistance to indentation.
Toughness		Relative index characterizing the resistance of rocks to forces tending to separate them.
Abrasiveness	k_{ab}, cm³/m.kgf	Capacity of rocks to wear away metals by friction estimated by the volume of the metal lost for a unit path of movement over the rock under 1 kg pressure.
Crushability (breakability)		Consumption of power for dynamic crushing of a unit volume of rock.
Drillability	m/min	Degree of resistance of rock to breaking by drilling tools expressed by the length of blast hole drilled in unit time.
Blastability	kg/m³	Degree of resistance of rock to breaking by blasting, expressed by the amount of explosives consumed to break 1 m³ of rock in a mass.
Unit amount of drilling	m/m³	Meters of blast holes needed to break 1 m³ of rock in a mass.
Natural moisture content	w	Quantity of water contained in rocks in natural conditions of occurrence, ratio of its weight in a rock to the weight of dry rock.
Moisture capacity	w′	Capacity of a rock to absorb water on exposure.
Lower limit of plasticity of clay		Minimum moisture content at which brittle failure of clay does not occur.
Upper limit of plasticity of clay		Minimum moisture content at which the clay flows.
Bulk weight	Υ_b, kg/m³	Weight of unit volume of rock in its heaped-up state.
Angle of repose	ϕ_0	Angle formed by the free surface of loose rock with the horizontal plane.
Looseness factor	k_l	Ratio of a volume of loosened rock to its volume in a pillar.

The physical and technological properties that have been briefly summarized in the above tables influence the geological prospecting and mining operations in various ways. The exploration of an ore deposit that is essential for delineation of the ore body demands the knowledge of apparent specific gravity, electrical resistance, thermal conductivity, behavior of elastic waves through rocks, magnetic susceptibility, etc. More particularly, operations like geophysical exploration, drilling, blasting, rock breaking, and mineral extraction depend upon the strength properties of rocks, their plasticity, porosity, and fracturing. The size reduction and transportation of rocks involve the knowledge of hardness, crushability, water saturation characteristics, abrasiveness, etc. The physics of rocks are as important as their chemistry in any mineral processing industry.

1.3 Nature of limestone occurrence

The term limestone is applied to any calcareous sedimentary rock consisting essentially of carbonates. Most limestone deposits are formed in shallow, calm, warm marine water, which provides an environment where organisms capable of forming calcium carbonate shells and skeletons can easily extract the required ingredients. When these organisms die, their shell and skeletal debris accumulate as sediment and with time consolidate and compact into limestone rock. This kind of limestone is known as biological or organic sedimentary rock. Their organic genesis is often reflected in the rocks by the presence of fossils. Limestone can also form by direct precipitation of calcium carbonate from marine or fresh water and is known as chemical or inorganic sedimentary rock. It has generally been observed that most commercially viable limestone deposits were formed by the organic route. Limestone altered by dynamic or contact metamorphism becomes coarsely crystalline and is referred to as "marble" and "crystalline limestone." It may be of theoretical interest that some carbonate bodies are of igneous origin—known as carbonatite—resembling marble, although geochemically much different. However, the occurrence of carbonatite is rare and is not important for cement production.

Distribution and exploitation of limestone occurrences

The distribution of limestone in space and time has been dealt with in (4). It makes up about 10% of the Earth's total land surface and is found in many countries. At the same time, limestone deposits are encountered in almost in all geological ages, spanning over 600 million years, although rocks of certain periods, such as the Ordovician or Cretaceous periods, show a predominance of carbonates. Build-up of limestone is more abundant between global latitudes of 30° North and 30° South, with such deposits particularly abundant in the Caribbean Sea, the Indian Ocean, the Persian Gulf, and the Gulf of Mexico. China, the USA, Russia, Japan, India, Brazil, Germany, Mexico, and Italy are some of the world's largest limestone producers. Bands of limestone emerge from the Earth's surface in often spectacular rocky outcrops and islands. The Swedish island of Gotland, the Niagara Escarpment in Canada and the USA, Notch Peak in Utah, the Ha Long Bay National Park in Vietnam, and the hills around the Lijiang River and Guilin City in China are some examples of such limestone occurrence. The Florida Keys, islands off the south coast of Florida, should also be mentioned in this context, as deposits there are composed of unusually textured "oolitic" limestone, described later in this chapter. So far as limestone quarries are concerned, the world's largest is reported to be at the Michigan Limestone and Chemical Company in Roger's City, Michigan, USA. Huge quarries are also found in Europe, such as the quarry system of Mount Saint Peter in Belgium and the Netherlands, which extends to over 100 kilometers.

Global production of limestone was estimated at 4.5 billion tons by Oates in 1998 (5). Since then, in about fifteen years, cement production itself has exceeded 4 billion metric tons. Hence, it would not be wrong to consider that the present consumption of limestone primarily for making cement and construction aggregates and also for other less voluminous industrial applications might be close to 6 billion tons. The USA consumes between 5 and 10% of the global production of industrial limestone and its production was 1.3 billion metric tons when calculated roughly a decade ago (6). For China, the world's largest producer of cement, specific data on limestone production were not readily available, but Chinese statistics indicate that the operating revenues from limestone and gypsum mining in 2016 were close to US$ 15 billion. So far as India, the second-largest producer of cement, is concerned, the mineral production statistics of the Indian Bureau of Mines (7), the apex body dealing with mines and minerals in the country, showed that the production of limestone from April, 2014 to March, 2015 was 293 million metric tons, although the total cement-grade limestone resources in the country were about 125 billion metric tons, spread across states such as Karnataka (28%), Andhra Pradesh (20%), Rajasthan (12%), Gujrat (11%), Meghalaya (9%), and Chattisgarh (5%). The actual production of cement-grade limestone in different states, of course, follows the demand pattern of the cement industry, which has been showing an increasing trend. It is tentatively presumed that about 1.5 tons of limestone are required to produce 1 ton of plain Portland cement, and about 0.8 tons of limestone for every ton of blended cement. Thus, for the present capacity of about 425 million metric tons per year—composed in India of approximately 25% plain Portland cement and 75% blended cements—the total annual limestone demand is estimated at about 580 million metric tons on hundred percent utilization of cement production capacity.

1.4 Assessment of limestone deposits

Occurrence of a mineral is an indication of mineralization that is worthy of further investigation, although it does not imply any measure of volume or grade. However, before a cement plant is conceived, an adequate quantity of limestone should be identified and explored, based on the geological reports of mineral occurrence. There is no shortcut to exploration or prospecting, which are the terms used in a more-or-less interchangeable manner, defined as a systematic process of searching for a mineral deposit by narrowing down the areas of promising mineral exploitation potential. Only after completing the necessary quantum of prospecting can a mineable deposit be defined with certainty and a high level of confidence. An exploration program is a carefully designed activity with a view to obtaining optimum information and data for decision making with minimum investment. Programs generally consist of the following steps—not necessarily in the same order, and, of course, not including statutory requirements, which are outside the purview of the present discussions:

- Reconnaissance
- Surveying
- Geological mapping
- Trenching and pit sinking
- Core drilling
- Preparation of plans and sections

- Exploratory mining
- Sampling and analysis
- Preparation of grade control maps, mine layout plans and sections

By and large, the exploration program of limestone deposits is subdivided into four sequential phases, namely, reconnaissance survey (phase 1), prospecting (phase 2), regional exploration (phase 3), and detailed exploration (phase 4).

Classification of deposits and exploration intensity

According to the "Norms for Proving Limestone Deposits for Cement Manufacture" published by the National Council of Cement and Building Material in India (8), limestone deposits can be classified as follows, depending on their geological features:

a. Simple deposits: large, continuous, bedded, horizontal to low-dipping and geologically undisturbed (Figure 1.3).
b. Complex deposits: medium to large consistent deposits, moderately to steeply dipping, gently folded (Figure 1.4).

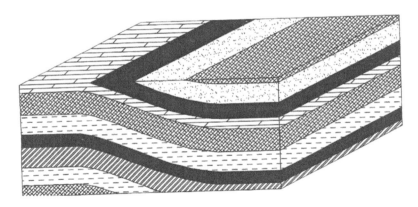

FIGURE 1.3 "Simple" gently dipping bedded deposit.

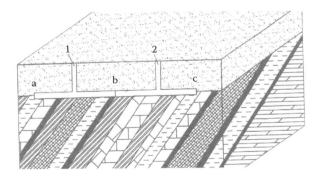

FIGURE 1.4 "Complex" steeply dipping bedded deposit under thick overburden. (1 & 2: pits, ab & bc: crosscuts.)

c. Intricate deposits: highly complicated, highly folded and faulted, dislocated bodies, highly lenticular, often interbedded with clay or shale (Figure 1.5).

The intensity of prospecting and exploration is decided depending on the geological complexity of a deposit—a tentative idea of which can be obtained from Table 1.4. The ultimate objective of any such exploration program is to ascertain the quality and reserves of the deposits with some level of precision.

Resource, reserve, and exploitability

During the 1990s, the Economic Commission for Europe (ECE) took the initiative of developing a simple, user-friendly, and uniform system for classifying and reporting reserves and resources of solid fuels and mineral commodities, primarily to harmonize the information and data originating from different countries and multiple agencies. The result of these efforts was the creation of the United Nations Framework Classification of Reserves and Resources and Minerals and Commodities (UNFC 1997),

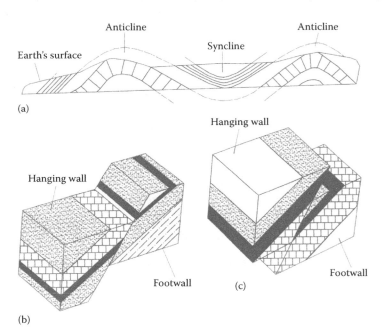

FIGURE 1.5 Geological disturbance in "intricate" deposits: (a) folding, (b) and (c) faulting of different types.

Table 1.4 Tentative quantum of exploration estimated for different types of deposits

Type of deposit	Quantum of drilling	No. of pits	Trenching interval
Simple	200–300-m interval; 12–15 m per million ton	2–4 per km^2	200–300 m
Complex	100–200-m interval; 20–80 m per million ton	3–6 per km^2	100–200 m
Intricate	50–100-m interval; 100–150 m per million ton	5–8 per km^2	50–100 m

which was endorsed by the United Nations Economic and Social Council (ECOSOC) in the same year. In 2004, the classification was extended to oil, natural gas, uranium, etc. Later, for worldwide adoption, a simpler version was prepared in 2009, known as UNFC 2009 (9). It is a universally acceptable and internationally applicable system and it is currently the only classification in the world to enjoy reasonable alignment with the templates and formats prevailing in a large number of international organizations and committees engaged in the extraction of mineral wealth. It is a generic principle-based system in which quantities are classified on the basis of three fundamental criteria—e.g., economic and social viability (axis E), field project status and feasibility (axis F), and geological knowledge (axis G)—and it uses a numerical and language independent coding scheme. Combination of these criteria creates a three-dimensional system, as depicted in Figure 1.6. The E-axis designates the degree of favorability of social and economic conditions in establishing consumer viability of the project, including the coordination of market prices and relevant legal, regulatory, environmental, and contractual conditions. The F-axis designates the maturity of studies and commitments necessary to implement mining plans or development projects. These considerations extend from the early exploration efforts to the stage of obtaining the project status, through all the intermediate phases. The third axis (G) designates the level of confidence in the geological knowledge and potential recoverability of the quantities. The categories and sub-categories are the building blocks of the system, as shown in Figure 1.7. Commercial projects in this context are the ones that are confirmed to be technically, economically, and socially feasible. The potentially commercial ones are the ones that can be developed in the foreseeable future. The non-commercial projects include those that are at an early stage of evaluation and also those that are unlikely to be viable within a reasonable time, unless the restricting factors change.

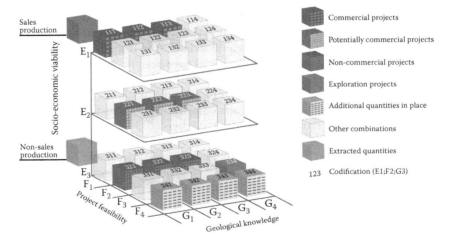

FIGURE 1.6 Schematic diagram of the UNFC system of classifying mineral resources. (Reproduced from Economic Commission for Europe, *United Nations Framework Classification for Fossil Energy and Mineral Reserves and Resources 2009*, ECE Energy Series No. 39, United Nations, New York and Geneva, 2010. With permission from the United Nations.)

	Extracted	Sales production			
		Non-sales production			
		Class	Categories		
			E	F	G
Total commodity initially in place	Future recovery by commercial development projects or mining operations	Commercial project	1	1	1, 2, 3
	Potential future recovery by contingent development projects or mining operations	Potentially commercial projects	2	2	1, 2, 3
		Non-commercial projects	3	2	1, 2, 3
	Additional quantities in place associated with known deposits		3	4	1, 2, 3
	Potential future recovery by successful exploration activities	Exploration projects	3	3	4
	Additional quantities in place associated with potential deposits		3	4	4

FIGURE 1.7 Classes and subclasses of UNFC system. (Reproduced from Economic Commission for Europe, *United Nations Framework Classification for Fossil Energy and Mineral Reserves and Resources 2009*, ECE Energy Series No. 39, United Nations, New York and Geneva, 2010. With permission from the United Nations.)

In UNFC 2009, terminologies such as "resource" or "reserve" have not been used, but in adopting the system in India the Indian Bureau of Mines, the apex organization engaged in discharging some of the important regulatory functions pertaining to the mineral and mining industry in the country, has merged the common terminologies with the numerical system (10). A "mineral resource" has been defined as a concentration of naturally non-renewable material of the Earth, which is economically amenable to industrial utilization now or in the foreseeable future. Mineral resources have been further divided into "inferred," "indicated," and "measured" categories, in order of increasing geological confidence. On the other hand, mineral reserves, defined as economically mineable parts of measured or indicated resources, have been categorized into two main subdivisions, namely, "proved" and "probable." The IBM guidelines also include the broad directions of improving the levels of geological knowledge, the status of mining feasibility, and conformity with the economic viability requirements of the project. In line with international practice the highest category of mineral resource has the code (111) and the lowest category (334). A few illustrations of interpretation of the digital designations are given in Table 1.5.

The UNFC system-based categorization of mineral resources has been in practice in India for some time and the total limestone resources of all categories and grades as per the UNFC system were estimated in 2010 at 185 billion metric tons, as shown in Table 1.6 (7). Of the total quantity, about 69% is taken as cement-grade.

Table 1.5 Illustrative explanations of UNFC digital codes for reserves and resources

Serial no.	Codes	Interpretation
1	(111)	Proved reserve being the economically mineable part of the measured mineral resource.
2	(121) (122)	Probable reserves being the economically mineable parts of an indicated or, in some cases, of a measured mineral resource.
3	(211)	Feasibility mineral resource being that part of the measured mineral resource, which after feasibility study has been found to be economically not mineable. It may turn viable if there are changes in technological, economic, environmental, or other relevant conditions.
4	(221) (222)	Prefeasibility mineral resource being that part of the indicated mineral resource that is not found economically mineable from the prefeasibility study. It may turn viable on further investigation or with change of relevant factors.
5	(331)	Measured resource that has undergone enough sampling such that a competent person (as defined in mining codes) could declare it acceptable in terms of quality and quantity.
6	(332)	Indicated resource signifying a part of the total resource for which the volume, density, shape, grade, and physical characteristics have been estimated with a high level of confidence based on sampling of outcrops, trenches, pits, and drill holes.
7	(333)	Inferred resource being that part of the mineral resource for which the volume and grade estimates are based on only geological evidence and assumptions, not verified by exploration.
8	(334)	Reconnaissance mineral resource for which the estimates of volume and quality are based on regional geological studies and extrapolation.

Table 1.6 Limestone reserves in India presented in accordance with UNFC system

Nomenclature	UNFC designations	Quantity (thousand metric tons)
Proved	111	8978583
Probable	121	3650576
Probable	122	2297234
Feasibility status	211	1827583
Prefeasibility status	221	3739470
Prefeasibility status	232	6309489
Measured	331	6858999
Indicated	332	22040640
Inferred	333	124835558
Reconnaissance status	334	4396981
Total	–	184935393

Reliability of different categories of reserves

According to the norms for proving limestone deposits (8), it has been specified that the cement-grade limestone reserves will be designated as possible, probable, and proved in the order of increasing reliability of estimates as indicated below:

- Possible reserve: $V_g \leq 4.0\%$ and $V_r \leq 50\%$
- Probable: $V_g \leq 3.0\%$ and $V_r \leq 30\%$
- Proved: $V_g \leq 1.0\%$ and $V_r \leq 10\%$

where

V_g = anticipated deviation in CaO content from the actual value as mined, and

V_r = anticipated difference between the estimated and mined reserves.

In addition to the above categories of reserves, which are estimated in the course of different phases of prospecting and exploration, another category of reserve, designated as "developed reserve," is important for some complex and intricate deposits. This category of reserve is estimated after exploratory mining and should have $V_g = 0.5\%$ and $V_r \leq 5.0\%$. It is advisable to ensure developed reserves for about ten years for any modern large-capacity plant. In a working mine, the reserves should periodically be revised, audited, and reviewed based on the feedback data of mining and advance proving.

Industrial implications of categorization of reserves

The basic information of limestone resources, which is based on regional geological studies or reconnaissance surveys, as they are often called, is generally used for the purposes of obtaining the prospecting license from the authorities empowered to give such permission in a country. The "proved" and "probable" categories of reserves obtained as a result of exploration are used for filing applications for mining leases from the concerned authorities. These categories of reserves are necessary for the preparation of project reports and for all financial decisions on project investments. The "possible" category of reserves is considered for future expansion purposes.

The requirement of the "proved" category of limestone reserves is arrived at after considering 15% mining losses, 30 years of plant life, 330 days of working of the kilns in a year, and specific consumption of 1.5 tons of limestone per ton of clinker. The "probable" reserve is expected to yield additional limestone of the "proved" category for an additional 15 years of plant life. This is computed by multiplying the estimated quantity of proven reserves for 15 years by a factor of 1.5 so that the required quantity is recovered from the probable reserves in the long run. No specific norms are as yet fixed for the "possible" category, but generally it is estimated at about 80% of the "proved" plus "probable" quantity. Following the above normative procedure, the required reserves for new large rotary kiln plants of various capacities are given in Table 1.7. In the case of small cement plants having capacities of up to 600 metric tons per day, mechanized mining is not a pre-requisite and the investment decisions may be taken on the basis of probable and

Table 1.7 Requirement of different categories of reserves for large plants

Category of reserves	Quantity (million metric tons)				
	12000 t/d	9000 t/d	6000 t/d	3000 t/d	2000 t/d
Proved	204	153	102	61	34
Probable	156	117	78	39	26
Possible	284	213	142	71	47

Sampling of limestone deposits for evaluation

possible reserves only. The tentative requirement of reserves for small plants is given in Table 1.8.

Sampling refers to the process of extracting small portions of limestone and associated rocks in the unprocessed state from a deposit such that the physical and chemical properties as determined for the small portion become representative of the whole. Depending on the phase of exploration, samples are drawn from outcrops of the rocks, natural sections of exposures, pits sunk, trenches excavated, boreholes drilled, and exploratory mining work carried out.

In the reconnaissance of a deposit the rock samples are drawn by grab sampling at a particular place, which is unlikely to be representative, or by chip or point sampling, which is a collection along a line or a grid at regular intervals, or (even more representatively) by digging strips (1–2 m long × 10–30 cm wide × 5–10 cm deep) or grooves (2–5 m long × 10–30 cm wide × 15–30 cm deep). In the subsequent prospecting or exploration stages, recourse is taken to channel sampling and borehole sampling of different magnitudes. The channel samples are drawn by cutting channels normal to the planar structures of a deposit and aligned in the direction of maximum variation. The depth and width of the channel are kept uniform to the extent possible. The amount of cuttings to be collected per unit width of the bed depends on the nature of the rock and degree of variation. The borehole sampling is done in the form of cores drilled into the rock, along with the sludge that comes out. High core recovery of the order of about 90% is considered necessary for the proper assessment of a deposit. Cores are usually split longitudinally and one half is preserved for future reference. The sludge sample is collected along with the cores in at least 10% of the boreholes drilled. The bulk samples are collected from pits or experimental excavations for pilot-scale technological testing. When the sectional

Table 1.8 Requirement of different categories of reserves (in million metric tons) for small plants

Category of reserves	600 tpd	300 tpd	150 tpd
Probable	10	5	3
Possible	16	8	4

or individual samples drawn at close intervals are required to be combined into a composite sample, they are mixed in such a manner that the composite sample typically represents a bench height. There are other modes of sampling that exist essentially for operational or quality control purposes, such as from the blast holes of a quarry, which are not discussed here.

The length of the channel samples should have a relation with the borehole samples of the same deposit, as shown in Figure 1.8 and expressed by the following equation:

$$l_c = l_b \cos \beta + [\sin \beta / \tan \alpha]$$

where
l_c = length of the channel
l_b = length of the borehole
β = angle of borehole
α = dip of the bed

In the initial stage of exploration, when the effective thickness of the limestone bed is not known, sampling should be quite extensive. The thickness of beds, bands, and intercalations showing variation in their lithological composition and structural disposition should be sampled individually at intervals ranging from 0.5 to 1.0 m. Surface sampling at this stage should be done for every 10.0 m of length. In the subsequent phase of exploration, the sampling interval should be related to the thickness of a bed, its angle of dipping, and the bench height requirements in actual mining. For example, if it is assumed that a deposit would be

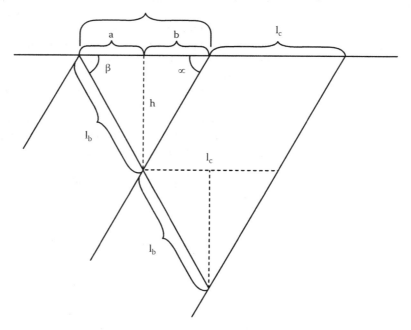

FIGURE 1.8 Relation between the sectional channel and borehole samples.

BASICS OF MINERAL RESOURCES FOR CEMENT PRODUCTION

mined in benches of 10.0 m height, then field practice is to consider the sampling interval of 1.5 to 2.0 m for a bed dipping at 10 degrees or 2.0 to 3.5 m for beds dipping from 10 to 45 degrees.

Similarly, there are also norms for borehole samples. With large thicknesses of lithologically homogeneous beds, the borehole samples are drawn at intervals of not more than half the planned or customary bench height, which might mean a limit of 5.0 m. The topmost bench should have a greater number of samples as it generally contains more intercalations of clay and other rocks. The sampling pattern is modified when a deposit shows thick intercalations or bands of clay or shale or if there are irregular intrusions of other rocks.

Dimension, quantity, and preparation of samples

The grab, chip, or channel samples primarily for chemical analysis may vary from $2 \times 2 \times 2$ cm to $10 \times 10 \times 10$ cm in size and 5–10 kg in sample size. For mineralogical, physical, and mechanical tests the sample requirements are larger. The dimensions range from $20 \times 20 \times 20$ cm to $30 \times 30 \times 30$ cm or cores of 0.5–1.0 m length and correspondingly the sample size may vary in the range of 20–50 kg.

The bulk samples that are collected for technological tests should obviously be representative of the deposit, particularly when it is uniform and homogeneous. For deposits showing considerable variation in quality, the bulk sample should represent the average material to be mined for feeding the plant. Hence, a bulk sample should be artificially prepared by mixing rocks from different parts of a deposit in such proportions that a sample would simulate the actual run of mine ore. The quantity of the bulk sample will depend on the type of test, the quality of the limestone, and the requirements of the testing organization. As a general guideline, an amount of 2.5 to 4.0 tons may be sufficient for running pilot tests. One should also take note of the fact that the minimum quantity of bulk samples to be collected will depend on the diameter of the largest particle and the degree of heterogeneity, as given in Table 1.9.

Table 1.9 Relation between the sample quantity and particle characteristics

Maximum diameter of particles, mm	Minimum quantity for homogeneous to slightly heterogeneous limestone, kg	Minimum quantity for heterogeneous limestone, kg
0.7	0.025	0.05
1.0	0.05	0.10
2.0	0.20	0.40
3.0	0.45	0.90
5.0	1.25	2.50
7.0	2.45	4.90
10.0	5.0	10.0
15.0	11.25	22.5
20.0	20.0	40.0
50.0	125.0	250.0

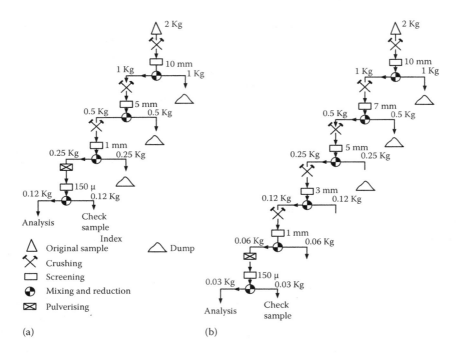

FIGURE 1.9 Reduction steps in sample preparation. (a) For homogeneous deposits. (b) For heterogeneous deposits.

The reduction of all samples, other than the bulk sample, is carried out by sequential crushing, grinding, screening, and cone-quartering, as depicted in Figure 1.9. In such preparation of samples, proper care should be taken to retain one-half of the samples at each stage of processing for future reference and to pass the final portion of material at each stage through the screens of relevant dimensions.

1.5 Mineral composition and quality of limestone

The composition and quality parameters of limestone rocks have been considered in detail in (4) and the salient points are discussed here. Limestone is by definition a rock that contains at least 40% calcium carbonate in the form of calcite mineral. All limestone occurrences contain small quantities of other minerals, such as quartz (SiO_2), feldspar (alkali or lime silicate), clay minerals (hydrous aluminum silicates), pyrite (FeS_2), siderite ($FeCO_3$), etc., in the form of small dispersed particles. Some of the minerals, such as quartz or chert (an amorphous silica) and pyrite, may also occur as large nodules or veins in the deposits. Because of the carbonate content, however, limestone is easily detected in the field by its effervescence in contact with a cold solution of 5% hydrochloric acid. The major calcium carbonate minerals and their important properties are furnished in Table 1.10.

Table 1.10 Calcium carbonate minerals

Mineral names with formulae	Crystal system	Common substitution	Specific gravity	Dissociation temperature (°C, at 760 mm pressure and 100% CO_2 removal)
Calcite ($CaCO_3$)	Hexagonal rhombohedral	Mn, Fe, Mg	2.72	898
Aragonite ($CaCO_3$)	Ortho-rhombic	Sr, Pb, Zn	2.94	425 (to calcite) 898
Dolomite ($CaMg(CO_3)_2$)	Hexagonal rhombohedral	Fe, Mn, Co, Zn	2.86	725 ($MgCO_3$) 890 ($CaCO_3$)

Based on the mineral composition, limestone is classified as:

- Dolomitic limestone, when the percentage of calcium carbonate lies in the range of 40–80%, along with the presence of magnesium carbonate.
- Siliceous, ferruginous, or magnesian limestone, when the percentage of calcium carbonate lies in the range of 80–95% with associated minerals of siliceous, ferruginous, or magnesian types, respectively.
- Calcitic limestone, when the calcium carbonate percentage is greater than 95%.

There are various other ways in which limestone is named and classified but for all practical purposes the above mineral-based classification is made use of in the field practices. In addition, on the basis of the mode of occurrence and texture, the limestone rocks are classified as follows (11):

- Chalk: a soft limestone with a very fine texture that is usually white or light-grey in color. It is formed mainly from the calcareous shell remains of microscopic marine organisms, such as foraminifera, or calcareous remains of numerous types of marine algae.
- Coquina: a poorly cemented limestone that is composed mainly of broken shell debris. It often forms on beaches where wave action segregates shell fragments of similar size.
- Fossiliferous limestone: a limestone that contains abundant fossils. These are normally shell and skeletal remains of organisms that produced the limestone.
- Oolitic limestone: a limestone composed mainly of calcium carbonate "oolites," small spheres formed by the concentric precipitation of calcium carbonate on a sand grain or shell fragment.
- Travertine: a limestone that forms by evaporative precipitation, often in a cave, to produce formations such as stalagtites, stalagmites, and flowstone.
- Tufa: a limestone produced by precipitation of calcium-laden water at a hot spring, lake shore, or similar location.

Some typical microstructures of limestone are shown in Figures 1.10 through 1.14.

Specification of cement-grade limestone

The National Council for Cement and Building Materials in India has drawn up the specification for limestone that is considered suitable for making Portland cement. The specification is given in Table 1.11 (8).

The reasons for specifying the limits of the constituents in Table 1.11 are as follows:

1. Sometimes there are opportunities of enriching the lime content in limestone by employing viable processes or to make available to a plant a sweetener-grade limestone at economic cost. Hence, the cut-off grade for the limestone is kept at 40%, although the desirable range of lime content is defined as 44–52%. If a limestone is near-calcitic in composition, it can still be used as a raw material

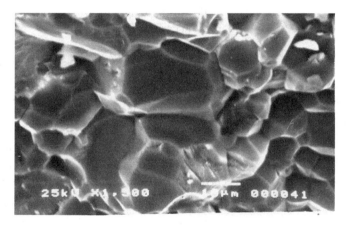

FIGURE 1.10 Coarse-grained compact limestone.

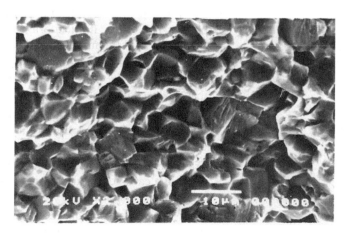

FIGURE 1.11 Fine-grained moderately compact limestone.

BASICS OF MINERAL RESOURCES FOR CEMENT PRODUCTION

FIGURE 1.12 Limestone with inter-granular pores.

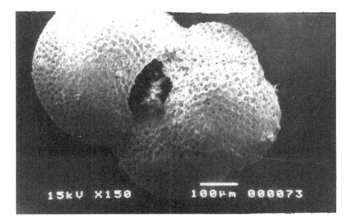

FIGURE 1.13 "Foraminifera" fossil in limestone.

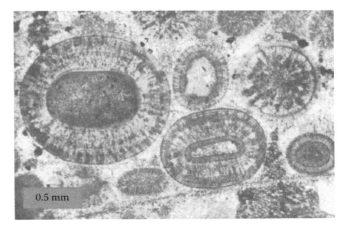

FIGURE 1.14 Limestone with "oolitic" microstructure.

Table 1.11 Specification of cement-grade limestone

Constituents	Desirable values for Portland cement (%)	Limiting values with scope of beneficiation, blending, etc. (%)
CaO	44.0–52.0	Min. 40.0
MgO (max)	3.5	5
$SiO_2 + Al_2O_3 + Fe_2O_3$	To satisfy LSF* and SM**	
TiO_2	0.5	1.0
Mn_2O_3 (Max)	0.5	1.0
R_2O (Max)	0.6	1.0
Total S as SO_3 (Max.)	0.6	0.8
P_2O_5 (Max)	0.6	1.0
Cl (Max)	0.015	0.05
Free SiO_2 (Max.)	8.0	10.0

*Lime saturation factor; **silica modulus

but the need for diluting the quality with corrective materials then increases substantially.

2. Since some of the standard global specifications for Portland cement permit composition of up to 6 per cent magnesia, the limiting value for this oxide in limestone is set at 3.5%. If, in some cases, it becomes unavoidable to admit higher contents of magnesia, for example, up to 5%, this can only be done by lowering the proportion of magnesia to 3.5% by beneficiating the limestone or by blending it with a high-purity limestone.
3. The limiting values of R_2O, SO_3, and Cl^- have been defined as they are volatile in nature and create "volatile cycles" in the kiln systems, leading to various operational and quality problems.
4. The limiting values of titanium dioxide, manganese oxide, and phosphorus pentoxide are dictated by the likely adverse impacts of these constituents on clinker quality.
5. The limit of free silica in the limestone is due to its adverse effects of abrasion on the crusher plates and in grinding mills, on one hand, and difficulties caused in the clinker burning process, on the other.

Typical physico-chemical properties of some limestone occurrences

It is important to note that a comprehensive characterization and evaluation of limestone includes the following:

- Hand specimen studies
- Chemical composition
- Mineral phase composition
- Microstructural characteristics
- Physical properties
- Thermal properties
- Mechanical characteristics

The overall assessment scheme for carbonate rocks is presented in Figure 1.15 and the scope of evaluation of the samples is given below:

- Chemical analysis for individual samples: $CaO + MgO + CO_2$ or $CaO + MgO +$ Insoluble residue.
- Chemical analysis for composite samples: $SiO_2 + Al_2O_3 + Fe_2O_3 + CaO + MgO + SO_3 +$ Loss on ignition; if the sum total is 98.5% or less, P_2O_5 and R_2O (alkali oxides) should be determined; if necessary, FeO, Mn_2O_3, Cl^- may be determined.
- Petrographic analysis of parent rock and inseparable intercalations: quantification of MgO, R_2O, SO_3, and P_2O_5-bearing mineral phases, grain size, cementing material, and special textural features, if any.
- Granulometric fractions obtained by mechanical analysis: complete characterization and estimation of free silica in all fractions and also details of mineral contents of inclusions, particularly with sizes greater than 0.09 mm.
- Physical properties: specific gravity, bulk weight, porosity, natural moisture content, and water absorption.
- Mechanical properties: hardness, compressive strength, crushing index, grinding index, and abrasion index.

In order to help understand the pattern of physical and chemical properties of limestone rocks belonging to geological basins of different ages, some relevant Indian data are furnished in Table 1.12.

From Table 1.12 it is quite apparent that the physical properties of limestone rocks cannot be easily correlated with their chemical composition. Hence, in most of the situations, limestone samples need to be characterized as comprehensively as possible. By and large, it is observed that the Mohs hardness of sedimentary limestone rocks falls in the range of 3–4, and the density in the range of 2.5–2.7 g/cm³. The porosity of compact limestone rocks may range from 1–20% and it may go up to 30% in porous limestone. Water absorption depends on the pattern and level of porosity. While it is generally negligible for compact limestone, water absorption may go up to 20% in porous limestone.

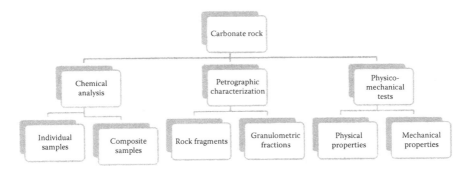

FIGURE 1.15 An assessment scheme for the carbonate rocks.

Table 1.12 Physical and chemical characteristics of some Indian limestone rocks of different geological ages

Parameters	\multicolumn{5}{c}{Limestone rocks of different geological ages}
	1	2	3	4	5
CaO (%)	43.0	44.5	46.6	47.7	48.7
SiO$_2$ (%)	13.0	18.0	11.7	12.6	10.5
LOI (%)	35.5	34.4	38.1	37.5	37.5
Quartz (%)	10.3	14.5	8.2	8.7	7.9
Dolomite (%)	7.9	3.3	5.0	–	–
Calcite (%)	72.5	77.7	80.5	85.2	87.0
Sp. Gravity	2.70	2.62	2.63	2.59	2.70
Bulk density (g/cm^3)	1.46	1.45	1.41	1.53	1.65
Porosity (%)	2.65	1.31	3.73	2.24	4.45
Abrasion (%)	22.96	18.45	30.64	23.74	65.38
Compressive strength (MPa)	110.2	113.6	140.0	94.9	48.0
Bond's grindability (kWh/short ton)	10.1	12.35	6.82	12.61	4.64

The compressive strength of cylindrical specimens prepared from the lumps of limestone samples vary from 10 to 200 MPa, and the Bond's grindability index from 4 to 14 units.

In general, the desirable characteristics of limestone from a cement manufacturing point of view can be summarized as follows:

a. Average calcite crystal size: less than 0.25 mm, which is considered characteristic of fine-grained rocks
b. Absence of coarse quartz and silica veins
c. Low moisture content: less than 5%
d. Low compressive strength: less than 100 MPa

1.6 Limestone mining

Limestone is predominantly mined from a quarry that is an open pit exposed to the surface. However, underground limestone mines can be found in the central and eastern USA, Brazil, and some other countries, especially in and near cities. Underground mining obviously has some advantages over surface quarrying but, since it is more expensive and capital-intensive, the cement industry depends mostly on opencast mining. The suitability of opencast mining is also due to the fact that the sedimentary limestone deposits are quite extensive, covering hundreds of square kilometers, and are reasonably uniform in thickness and quality.

The selection of mining method is dependent on several factors. The more important criteria are:

- Shape and configuration of the deposit
- Aerial extent and dimension
- Dip (inclination) and thickness of the bed
- Thickness of the overburden
- Physiographic and topographic conditions
- Climatic conditions including rainfall, snowing, wind pattern, etc.
- Daily and yearly raisings required for feeding the plant

Some variations in the opencast mining methods under different conditions are briefly described below.

Variations in opencast mining

For deposits in plain land the mine is opened at a point where the thickness of overburden is minimal. Inclined trenches and ramps are driven from one bench level to another in order to create and develop each bench. In hilly deposits, mining has to start from the top downwards, by slicing or decapping. The benches are developed at pre-determined locations. For thick limestone deposits that are horizontally bedded with a thick layer of overburden, the mining operation is mostly mechanized with large-capacity mining equipment, while for deposits with thin mineralization and thick overburden, or vice-versa, the operation is mechanized but with medium-capacity equipment. For deposits with a thin bed of mineralization and a thin layer of overburden, only a semi-mechanized scheme is adopted. The same principles hold good for deposits with dipping beds but a major consideration in this case is the economic depth of mining. The "stripping ratio," signifying the ratio of limestone to overburden or waste rock, is normally very important in mining but turns out to be critical in steeply dipping deposits. The economic or cut-off stripping ratio (COSR) is defined by:

$$COSR = \text{value of a unit quantity of limestone} - \text{average cost of production}/\text{average cost of overburden removal}$$

and the formula brings out the break-even point at which the mining operation is no longer profitable. An important alternative method of working for dipping deposits is to mine along the strike direction or perpendicular to the dip direction of the bed, instead of advancing along the dip. This method of mining is known as "terrace mining" and is often preferred for both economic and environmental reasons.

Most opencast mining operations are discontinuous. Nowadays, a continuous system is preferred in appropriate situations by using a special piece of mining equipment called a "surface miner." This equipment cuts the rock continuously and feeds the conveying system. One great advantage over conventional systems is that it does not require drilling and blasting as well as primary crushing of rock. At the same time, one must take into consideration its limitations. It can cut freely only softer limestone rocks with compressive strength of up to 50 MPa, or up to 80 MPa if the rock has cracks, crevices, and fissures, or if the seismic velocity of the rock is low.

Operational issues in mining

Normally, cement plants are located near the limestone deposits, while shale or clay is either substituted by the overburden, if it is chemically suitable and economical to use, or mined locally elsewhere and transported to the plant. Other corrective materials are usually brought from outside.

Mining plans of the limestone quarry are developed according to the nature of the deposit. If the limestone is not homogeneous, it may be necessary to blend rock from different benches and faces of the mine in order to maximize the recovery from the mine. In some mines, it may be necessary to undertake selective mining in order to avoid low-grade material or problems of associated harmful constituents like sulfates and alkalis.

The mining operation proceeds with initial close-spaced advance drilling and analysis of drill hole samples in order to ascertain the spatial distribution of quality. The areas to be blasted are decided based on such advance drilling. The conventional explosive used for limestone quarrying is ANFO (ammonium nitrate with 5% fuel oil). The specific consumption of explosives varies depending on the blasting behavior of limestone, but tentatively it is in the range of 200 g/t.

If, after primary blasting, the limestone boulder size is larger than the crusher feed size, there could be a need for secondary blasting or mechanical size reduction of the biggest boulders. The run-of-mine material, after blasting, is transported to the crusher. Mining and haulage operations are generally monitored by

a. Blasting factor: grams of explosive per metric ton of rock.

b. Overburden ratio: tons of waste rock removed per ton of useful rock.

c. Loading rate: tons of rock per hour of loading equipment availability.

d. Hauling factor: tons of rock transported per truck and truck availability.

It should also be borne in mind that all mine output and inventory records are kept on the basis of dry rock data but the moisture levels of mined, hauled, and crushed rocks are used for assessing the equipment efficiency.

Typical public concerns about limestone mining include dust, noise, blasting vibration, and vehicular traffic. Some limestone rocks are aquifers and there can be concerns about the contaminants from quarry operations escaping into the groundwater. In humid climates, large amounts of limestone dissolve and are carried away in the flowing water, creating sinkholes.

1.7 Quarry design and operational optimization

The design of a pit for opencast operation may look simple but it involves not only proper utilization of exploration data through build-up of a 3D model of the deposit, but also reasonable integration of various other factors such as geotechnical, economic, operational, and environmental considerations. The configuration of the quarry, along with its boundary limits, has to be defined through the development of a 3D model from the exploration data. Parameters like slope stability, the dimensions of the working benches and their compliance with the prevailing mining regulations, water seepage and drainage, layout of the road network, determination of blasting behavior of rocks, etc. fall primarily under the purview of geotechnical modeling. Environmental planning includes, apart from the minimization of dust and noise pollution at the operational stage, the design of a waste rock dump, the systematic closure of mines, and the rehabilitation of the quarry after closure. The economic model involves assessment of the economic stripping ratio, determination of the net present value of future mining operations, and the overall viability of the project. Thereafter, operational issues such as grade control, scheduling of mining sequence, opening-up of mine faces, planning of blasting pattern, advance drilling program, etc. are decided. Long-term integrated mine planning is, therefore, complex and has many components, as shown in Figure 1.16.

Regarding actual operations, in a quarry the planning system must pass information to drill operators. Drilling operations need to send samples from advance drilling operations to the laboratory, and the

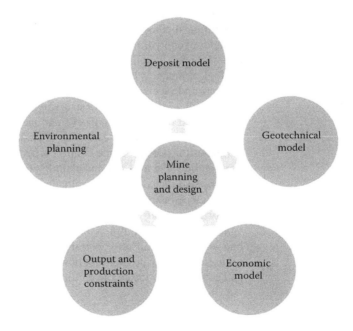

FIGURE 1.16 Components of integrated long-term mine planning.

grade control geologist needs to match up the quality of blast holes with their actual positions and depths. The blasting engineer needs to know which drill holes are in limestone and which are in waste rocks. The dispatch system needs to know which faces are to be mined, what the tonnages are, and where the crusher is. It is obvious, therefore, that the quality control of run-of-mine limestone and the optimization of quarry operation are quite involved and complex. Present-day mining software packages allow rapid and accurate iterative investigations of mining scenarios to determine the most appropriate course of action. Among the numerous software packages, those commonly used include GEOVIASurpac, Maptek Vulcan, Datamine, XPAC, iGantt, Micromine, and Promine.

The basic feature of these software packages includes a relational geological database that allows storage and query of all geochemical quality parameters, in addition to physical attributes such as minimum mining thickness. Complex 3D geological and quality models are developed to assist in deposit visualization and estimation of fairly accurate volumes. Statistical and multiple-element compositing tools can be used to investigate the quality distribution and undertake element combinations using established algorithms. Estimation using a polygonal, grid, or block model approach is available in such software. The effect of lateral quality variation can be minimized using the proper scheduling approach. A scheduler allows an engineer to examine multiple mining scenarios graphically and with ease. Such software can assist the plant manager in optimizing a blend of raw material from available quarry faces. These solutions use linear programming to achieve the required blend and are useful in ensuring that short-term goals are met. The long-range quarry scheduling problems require a rigorous solution and make use of dynamic programming methods. The plant quality targets for cement making can include several parameters such as directly measured minor constituents like MgO, SO_3, R_2O, etc. and other critical modulus values. While the 3D geological model is built to contain quality values, interpolated from drill cuttings and face samples, the schedule optimizer is able to utilize these basic quality attributes in constructing the target ratio of blending limestone mined from different blocks and faces. On the whole, therefore, the quarry production plan is essentially based on geological exploration data, 3D deposit modeling, grade control, scheduling of mining sequence, and ore blending (Figure 1.17), and planning functions include blast design and layout. Also included is the evaluation of blast operations and schedules are developed to plan production from the working benches. It should also be borne in mind that conventional economic analysis tells us that the value of an ore block mined today is worth more than an ore block mined one year from now. Hence, certain software packages have developed programs to solve the problems of how to discount the value of mining blocks and consequently locate the optimal limit of mining.

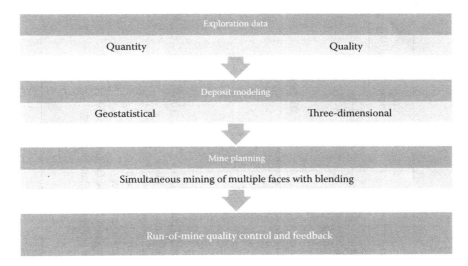

FIGURE 1.17 Sequential steps in quarry planning.

1.8 Argillaceous materials

The expression "argillaceous materials" refers to all fine-grained natural earthy substances that are alternatively known as "clay," including shale and argillite. It is well known that the clay minerals are part of a larger family of phyllosilicates and are characterized by interlinked tetrahedral and octahedral sheets. The chemical composition and structural configuration of clay minerals have been discussed in detail in (4,12). The structure and properties of the major clay mineral groups are presented in Table 1.13.

From the table it is evident that, chemically, these materials are hydrous aluminum silicates, with magnesium or iron substituting aluminum wholly or in part in certain minerals, and with alkalis or alkaline earth also present as essential constituents in a few others.

Some argillaceous materials are composed of a single clay mineral but in many there is a mixture; in addition to the clay minerals, many argillaceous materials contain non-clay minerals like quartz, calcite, feldspar, sulfides, and so on. Many others may contain organic substances as well as water-soluble salts, and some clay materials may contain phases that are x-ray amorphous. The multitude of variation in clay minerals is caused by substitution in the octahedral and tetrahedral layers, resulting in charge deficits. The manner in which the charge deficit is balanced leads to many of the useful and unique properties of clay minerals. As a result of such diversity of constitution of argillaceous materials, their technical assessment for specific applications is rather complex. Broadly

Table 1.13 Structure and properties of the major clay mineral groups

Structural grouping	Major mineral phase	Broad composition	Structural features	Cation exchange capacity meq/100g
2-layer kaolin group	Kaolinite	$(OH)_8Si_4Al_4O_{10}$ SiO_2 46.54% Al_2O_3 39.50% H_2O 13.96%	1:1 layer type (001) = 7.21 Å B = 8.99 Å Flaky habit	3–15
3-layer smectite group	Montmorillonite (expanding lattice)	$(OH)_2Si_8Al_4O_{20}\cdot nH_2O$ Composition without interlayer material: SiO_2 66.7% Al_2O_3 28.3% H_2O 5.0% Some substitution of Si by Al in tetrahedral layer and of Al in octahedral layer by Mg, Fe, Zn, Ni, Li, etc. Lattice always unbalanced	2:1 layer Equidimensional, extremely thin flakes	80–150
	Illite (non-expanding lattice)	Muscovite-like dioctahedral $(OH)_4K_2\,Si_6Al_2)$ Al_4O_{20}. Biotite type is trioctahedral with incorporation of Mg and Fe. Illites differ from mica in having less replacement of Al for Si, less K and less randomness of silicate layers	Structural characterization same as micas, 2: 1 layer type. Small, poorly defined flake commonly grouped in irregular aggregates	10–40
Chain-structure palygorskite group	Palygorskite–Attapulgite–Sepiolite	Composition of the balanced ideal cell of Attapulgite$(OH)_2\!_4(OH)_2\,Mg_5Si_8O.4H_2O$. Sepiolite and palygorskite show variations in the composition of Al, Mg, and Si	2:1 inverted ribbons. Generally seen as bundles of lath-shape units	20–50

BASICS OF MINERAL RESOURCES FOR CEMENT PRODUCTION 31

speaking, the property-controlling factors in argillaceous materials are the following:

i. Clay minerals
ii. Non-clay minerals
iii. Organic substances
iv. Exchangeable ions and soluble salts
v. Particle characteristics including shape, size, and orientation
vi. Structural assembly formed by 1:1 or 2:1 linkage of tetrahedral and octahedral sheets as well as neutralization of excess layer charge by various interlayer materials

Thermo-chemical reactivity of clay minerals

Argillaceous rocks generally show the presence of multiple clay minerals. Shale generally contains illite and chlorite. Montmorillonite is also a common constituent of many shale occurrences of Mesozoic age and younger. Kaolinite is a common mineral of certain types of shale but usually in minor proportions. Slates are also composed of illite and chlorites but with a higher degree of crystalline structure. The carbonate rocks show the association of a wide range of clay minerals, illite and chlorite being more predominant than kaolinite and montmorillonite.

Experience has shown that an aluminum silicate phase, when present in raw mixes from the argillaceous component, turns out to be significantly more reactive than fine-ground silica. It has been generally observed that the water vapor and hydroxyl ions released from the argillaceous materials show some catalytic effect on the dissociation of calcium carbonate and the subsequent solid-state oxidation reactions. The trend of decomposition of a few clay minerals is shown in Table 1.14. It is

Table 1.14 Decomposition behavior of some clays of varying mineral composition based on thermogravimetric analyses

Mineral composition	Temperature range (°C)	Weight loss (%)	Remarks
Kaolinite (90%), Anatase	Up to 350	1.20	Moisture loss
	350–650	13.34	Dehydroxylation of kaolinite
Kaolinite (70%), Quartz, Muscovite	Up to 250	0.42	Moisture loss
	250–750	10.05	Dehydroxylation of kaolinite
Kaolinite (40%), Quartz, Anatase, Goethite, Hematite, Anorthoclase	Up to 200	8.65	Moisture loss
	200–350	1.03	Breakdown of goethite
	350–750	4.36	Dehydroxylation of kaolinite
Quartz, kaolinite (50%), illite (14%), anatase	Up to 400	1.30	Moisture loss from illite
	400–550	5.70	Incomplete dehydroxylation of kaolinite
	550–1000	4.00	Prolonged breakdown of clay minerals
Montmorillonite (80%), kaolinite (6%), hematite, quartz, cristobalite, anatase	Up to 100	12.00	Moisture loss
	100–700	7.00	Dehydroxylation of kaolinite further weight loss for reactions not deciphered
	700–1000	1.06	

evident that the temperature ranges over which different clay minerals release water and hydroxyl ions to become amorphous and reactive are not the same. It is obvious, therefore, that their reactivity with lime and other oxides would not be identical in different situations. However, there is still no unanimity in the views of different investigators regarding the order of reactivity of different clay minerals. According to certain investigations (13), montmorillonite as a clay mineral is more reactive than kaolinite, which in turn shows higher reactivity than chlorite. Minerals like muscovite, biotite, vermiculite, pyrophyllite, etc. in general have low reactivity.

It may also be borne in mind that the argillaceous materials have high probability of containing titania, alkalies, sulfides, sulfates, and phosphates. The reactivity may get strongly influenced by the presence of the above constituents.

1.9 Corrective materials

When the primary components of a raw mix do not jointly permit the desired range of modulus values to be achieved, a third or even a fourth component is added, known as corrective materials. It has been a practice to recognize that a material with more than 70% silica, 40% iron oxide, or 30% alumina can be termed as siliceous, ferruginous, or aluminous corrective material (14).

Sand, sandstone, or quartzite acts as the source of siliceous corrective material. Other conditions being equal, the grain size and specific surface of silica in free form, and particularly of the least reactive forms like quartz and chalcedony, determine the rate of reaction in a kiln feed. The reactivity of different types of silica, free or combined, increases in the following order:

quartz < chalcedony < opal < α – cristobalite and α- tridymite
   < silica from feldspars < silica from mica and amphibole
   < silica from clay minerals < silica from glassy slags

On the whole, silica in amorphous state or derived from silicates or hydrosilicates is preferable to silica in other forms.

The correction of iron oxide in a raw mix is generally done with iron ore, which may either be magnetite or hematite. A hematite with colloidal texture or martitized magnetite is quite reactive with lime and alumina. Limonite ($FeO.OH.H_2O$) often associated with laterite is more reactive than the ferric oxide hydrate phases like goethite and lepidocrocite. It has been observed that the reactivity of raw material is often favorably influenced by the presence of iron oxide in the ferrous state, the appearance of which is obviously dependent on the parent mineral. For example, chlorite and glauconite may release FeO below 500°C, while goethite and lepidocrocite yields Fe_2O_3 at about 300°C. The iron-bearing minerals, thus, play some important role in shaping the reactivity of kiln feed.

The correction of alumina is done with the help of bauxite or aluminous laterite. These rocks contain such aluminous minerals as gibbsite ($Al_2O_3.3H_2O$), bohemite ($\alpha\ Al_2O_3.H_2O$), and diaspore ($\beta Al_2O_3.H_2O$). Generally, these phases show low crystallinity and high energy in the green state, dehydrate at 300–500°C, and give rise to different forms of alumina that ultimately define their reactivity at higher temperatures.

1.10 Natural gypsum

Gypsum ($CaSO_4.2H_2O$) is a common rock-forming mineral with thick extensive beds formed by the evaporation of extremely saline water. It is often associated with other minerals like halite or sulfur. It is deposited from lakes, seawater, hot springs, volcanic vapors, etc. and usually forms a solid nonporous rock near the Earth's surface, and consequently the mining of gypsum ore is generally carried out in quarries or shallow underground mines. It is a mineral belonging to the monoclinic system and mostly occurs as flattened crystals that are often twinned. It is a very soft mineral, having a Mohs hardness of 1.5–2.0 and density in the range of 2.31–2.33 g/cm³. Gypsum is white, grey, or reddish in color, depending on the presence of clay or iron oxide; the color may be black or nearly black if bitumen is present. Depending on the appearance and crystalline characteristics, gypsum is known by certain other names. A fine-grained white or slightly tinted variety is known as "alabaster"; a transparent colorless variety with pearly luster is called "selenite"; and a silky fibrous variety is named "satin spar." When the white variety is used for agricultural purposes, it is known as "terra alba" or "land plaster." An unusual and interesting occurrence of white gypsum is in New Mexico in the USA, where it covers an expanse of 710 km² with a dune-like structure. This deposit is not being exploited and is preserved as an uncommon occurrence (15).

Major gypsum producing countries

World production of gypsum in 2012 was 150 million metric tons, with China being the largest producer, followed by Iran, Spain, Thailand, and the USA. The production statistics of the first ten countries of the world in the same year are given in Table 1.15 (16).

According to IBM statistics, in 2010 gypsum production in India totaled only 2.5 million metric tons. Over the years, the resource position of the mineral gypsum has declined and the emphasis has been shifting towards the use of chemical gypsum primarily from the fertilizer industry. The state of Rajasthan alone accounts for over 81% of resources, and Jammu and Kashmir 14%. The remaining 5% of resources are distributed in Tamil Nadu and other states.

Gypsum is worked by opencast manual mining, except in a few semi-mechanized mines in Rajasthan. The deposits are found at shallow depths and scattered over large areas. Production is classified into four grades based on the calcium sulfate ($CaSO_4.2H_2O$) content: (i) above 90%; (ii) 85–90%; (iii) 80–85%; and (iv) less than 80%. High-grade gypsum is mined in the Bikaner and Jaisalmer districts of Rajasthan.

Table 1.15 Major gypsum producing countries and their production in 2012

Serial no.	Country	Quantity (thousand tons)
1	China	48000
2	Iran	14000
3	Spain	11500
4	Thailand	10000
5	USA	9900
6	Japan	5700
7	Italy	4100
8	Mexico	3840
9	Australia	3000
10	Canada	2200

Some gypsum mines in the Bikaner district also produce the selenite variety. Rajasthan gypsum is used in cement plants in northern and eastern states in India, while gypsum produced in Tamil Nadu is mainly of cement grade and is consumed in plants located in the southern parts of the country.

Chemical properties of gypsum

The $CaSO_4 - H_2O$ system is represented by three solid phases that coexist at room temperature in air containing water vapor, namely calcium sulfate dihydrate ($CaSO_4.2H_2O$), calcium sulfate hemihydrate ($CaSO_4.1/2H_2O$), and insoluble anhydrite ($CaSO_4$ – insol.). In addition to these phases, another phase is identified as soluble anhydrite ($CaSO_4$ – sol.), which has the same crystal structure as hemihydrate. Two transformations are very important for these substances. They are (1) dehydration from the dehydrated state and (2) rehydration to the dehydrated state. Dehydration refers to the stepwise loss of water of hydration that is accomplished by increasing the temperature of dehydration for the purpose of obtaining one of its dehydration products, i.e., hemihydrate, soluble anhydrite, and insoluble anhydrite. The temperature ranges applicable for the individual dehydration steps depend on whether the process is static or kinetic. In general, the static conditions appear to proceed at a lower temperature but they require more time. For example, the dihydrate phase begins to dehydrate at about 46°C but conversion takes months or even years to complete. Rehydration is the exposure of these dehydrated materials (hemihydrate, soluble anhydrite, and insoluble anhydrite) to liquid water for the purpose of obtaining the dihydrate state again at standard conditions. This process proceeds by the crystallization of the dihydrate phase in water. If the starting material is a hemihydrate, a supersaturated solution is formed prior to crystallization. In the case of an insoluble anhydrite as the starting material, the rate of conversion to gypsum is affected by the degree of its solubility and the temperature. The solubility of the dihydrate, hemihydrate, and insoluble anhydrite phases are an important consideration for a variety of reasons but primarily because they allow a distinct separation between the three phases on a theoretical basis. Their approximate solubility curves are shown in Figure 1.18 (17).

The intersection between two curves gives the equilibrium temperature, below which the less-soluble phase is stable. For example, the curves show that above 40°C (point A) insoluble anhydrite is the most stable phase. However, above this point, the dihydrate phase, if in contact with water, should not convert to insoluble anhydrite, but in practice It may do so but at a very slow rate. Below 100°C (point B) the dihydrate phase is more stable than the hemihydrate. When the saturated solutions of hemihydrate are supersaturated with respect to dihydrate, the dihydrate phase precipitates from the solution. This dehydration-rehydration process appears to be relatively simple but a number of critical conditions are to be met that finally ensure the transformation process. For example, water vapor plays a significant role in the dehydration and rehydration of soluble anhydrite. Also, the hemihydrate phase occurs in the α-form and ß-form depending on the method of dehydration. When the hemihydrate phase is formed under normal atmospheric conditions it is referred to as β-hemihydrate, while the α-form is obtained from the dihydrate phase by autoclaving. Although the normal solubility of the dihydrate phase is reckoned as 2.1 g/L at 20°C, the solubility of the β-hemihydrate is much higher at about 8.8 g/L at 20°C but it reduces to 6.5 g/L for the α-phase. For all practical purposes, the conversion temperature of dihydrate to hemihydrate lies in the range of 100–180°C, hemihydrate to soluble anhydrite in the range of 180–350°C under favorable water vapor pressure conditions and to insoluble anhydrite above 1180°C. Insoluble anhydrite also occurs in natural deposits.

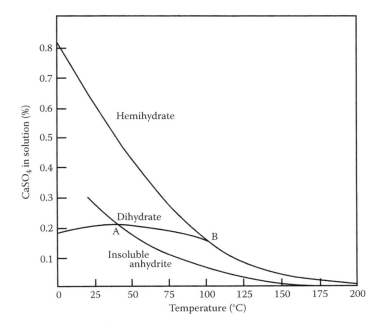

FIGURE 1.18 Solubility curves of calcium sulfate phases.

1.11 Influence of raw materials on unit operations

In the cement production process, since there are quite a few unit operations employing different equipment and machinery, it is important to examine the influence that the raw materials yield in this perspective. For example, all three principle unit operations such as crushing, grinding, and burning are strongly influenced by the characteristics of limestone and other raw materials. Some key requirements include:

- Crushing strength (kg/cm^2) or crushability index
- Grinding index
- Hardness
- Abrasion index
- Impurities present such as clay, quartz, etc.
- Degree of coarseness in terms of crystal size of the constituent minerals
- Particle size distribution of the crushed and ground materials
- Moisture content (minimum and maximum)

While the first three parameters are important for selection of the type and materials of construction of the crushing and grinding installations, the other parameters may have a strong influence on the process. It is observed that the handling of wet sticky materials may affect the throughput and increase the maintenance cost of the crusher. Similarly, most materials to be crushed contain inclusions or impurities, which behave quite differently from one another when introduced into a crusher or a screening system. The characteristics of these inclusions play an important part in the selection of the right equipment.

Coming to the grinding process, the objective is to achieve the targeted particle size distribution, average particle size, and specific surface with least consumption of energy and other operating costs. Ball mills are cost-effective only when a high degree of wear is expected due to, say, high quartz content. High pressure grinding rolls can profitably be used for relatively dry raw materials. Roller mills are generally preferred for raw grinding due to their low energy consumption and the option of simultaneous drying. So far as burning is concerned, it is important to note that different raw mixes with more or less the same chemical composition and similar fineness may have greatly differing burning behavior. The reason for the difference lies in the variation of the mineral composition of the raw mix constituents, their crystal size, and their particle size. Worldwide experience has shown that the poor burning is primarily caused by the presence of coarse grains of quartz (+ 44 µm), calcite (+ 125 µm), feldspar (+ 63 µm), and shale (+ 50 µm).

The volatile matters and, more particularly, alkalis and chlorides present in the raw materials have a very strong bearing on the process of clinker formation in the present-day preheater-precalciner rotary kilns. The circulation of these volatiles in the system without any bypass imposes certain upper input limits for these constituents in raw mixes.

The determination of alkali-, sulfur-, and chloride-bearing minerals in the raw materials becomes essential in this context. Some of these aspects will be discussed in more details in chapters dealing with size reduction and burning operations.

1.12 Summary

Cement production is dependent on a wide range of sedimentary and metamorphic rocks, limestone being the essential one. A limestone resource required for setting up a cement plant needs to be first examined from the point of view of its quality and quantity. The geological disposition defines the complexity of a deposit and an appropriate exploration program is necessary for proving the deposit and arriving at a reliable estimate of reserves. A general approach is to classify the deposits in terms of their geological complexity so that the quantum of exploration could be decided for a given deposit, which in turn will ensure that proper reserve estimates are made in "proved," "probable," and "possible" categories of reserves. The limestone reserves can be more rationally categorized by applying the United Nations Framework Classification system, using three parameters, viz., status of geological proving, extent of mining feasibility established, and compliance of economic and social viability of the project studied. This system provides a more dependable approach to transform a resource into a viable mine through various intermediate stages of investigation.

The assessment of limestone deposits involves sampling of different types at different stages of investigation. Standard operational procedures have been evolved in practice for such sampling exercises, which need to be observed in the studies. The dependability of technological assessment of limestone deposits is related to the representativeness of the samples drawn and evaluated.

Limestone mining is done mostly by the opencast method. The details of the mining method depend on the shape, size, and configuration of the deposit, terrain of its occurrence, disposition and thickness of the bed, overburden or waste rock to be removed and dumped, and several other factors. Parameters like the blasting factor, overburden ratio, loading rate, and so on are required to be monitored in the mining operation. Quarry design and operational optimization are essential steps in mining and several software packages are available to undertake these steps.

After proving a limestone resource, it is essential to undertake its thorough technological assessment. Limestone with a minimum of 44–45% CaO and maximum 3.0–3.5% MgO, 0.6% R_2O, 0.6–0.8% SO_3, and 0.015–0.05% Cl is regarded as a cement-grade limestone, provided its SiO_2, Al_2O_3, and Fe_2O_3 contents satisfy the desired modulus values of raw mixes. The specification of clay and other corrective materials cannot be very precisely defined as they have to have compatibility with the principle carbonate component. In general, an argillaceous component with more than 3% R_2O and 1% SO_3 may be considered, prima facie, unsuitable. For most of the minor constituents, 0.5% is regarded to be a safe limit. If some of the constituents exceed this limit, a special examination is called for.

The size reduction and thermal behavior of raw materials primarily depend on their mineral phases and microstructural features. The crystal size of calcite and the associated mineral assemblage define the dissociation characteristics of limestone in a very effective manner.

The mineralogy of argillaceous materials is quite complex due to their layered structure, frequent ionic substitutions in the layers, unbalanced charges, and the inclusion of interlayer materials. Because of these complications the choice of clay can be guided more by Si/(A,F) ratio, volatile contents, fusibility, particle size distribution, cation exchange capacity, etc. In the ultimate evaluation, it is important to note that the concurrence of carbonate dissociation and thermal decomposition of clay and other corrective materials is considered a basic necessity for high reactivity and proper burning.

The above concepts are equally applicable for all other siliceous, ferruginous, or aluminous corrective materials, when their use becomes unavoidable due to stoichiometric needs. Different forms of silica, aluminum hydrates, ferriferous minerals, etc. present in the above corrective materials influence the burning process quite significantly and hence demand careful evaluation in terms of overall impact on the burning behavior of the raw mix.

Mineral phases carrying volatile oxides like alkalis, sulfates, chlorides, etc. assume criticality in determining the likely nature of the volatility cycle in a rotary kiln system. Hence, the selection of raw materials depends on the quantitative contribution of such volatile components from the raw materials under consideration.

In cement production, size reduction is an important material preparation step. The amenability of limestone to size reduction processes is apparently controlled, inter alia, by the free and fixed silica content and the crystal size variations of calcite and quartz phases. The selection of hardware for grinding is also dependent on the above mineralogical and microstructural features.

Since the ultimate burning process is dependent on the prior size reduction steps, there has been a progressive evolution of the concept of limiting particle size for different mineral forms, supported by experimental findings. In the selection of raw materials, the feasibility of attaining such particle size distribution patterns requires specific attention.

All in all, it should be realized that the production of cement is solely dependent on natural raw materials and, more specifically, on the quality and quantity of limestone. The scale of operation is large and the location of a plant for its life is decided by the occurrence of a limestone deposit. The technological suitability and consistent supply of raw materials are, therefore, of paramount significance. Proper selection of raw materials is the first step towards setting up a cement plant.

References

1. http://en.wikipedia.org/wiki/limestone, retrieved on June 21, 2017.
2. E. S. DANA AND W. E. FORD, *A Textbook of Mineralogy*, John Wiley & Sons. Inc., New York (1955).
3. V. RZHEVSKY AND G. NOVIK, *The Physics of Rocks* (Translated from Russian by A. K. Chatterjee), Mir Publishers, Moscow (1971).
4. A. K. CHATTERJEE, Raw materials selection, in *Innovations in Portland Cement Manufacturing* (Eds J. I. Bhatty, E. M. Miller and S. Kosmatka). Portland Cement Association, USA (2004).
5. J. A. H. OATES, *Lime and Limestone*, John-Wiley-VCH Verlag GmbH, Weinheim (1998).
6. http://geology.com/usgs/limestone, retrieved on June 21, 2017.
7. Indian Bureau of Mines, *Indian Mineral Year Book 2013*, IBM, Nagpur, India (2015).
8. National Council of Cement & Building Materials, *Norms for Proving Limestone Deposits for Cement Manufacture*, NCB, New Delhi (2003).
9. Economic Commission for Europe, *United Nations Framework Classification for Fossil Energy and Mineral Reserves and Resources 2009*, ECE Energy Series No. 39, United Nations, New York & Geneva (2010).
10. Indian Bureau of Mines, *Guidelines under MCDR for United Nations FRAMEWORK Classification of Mineral Reserves/Resources*, Nagpur, India (2003).
11. http://geology.com/rocks/limestone.shtml, retrieved on June 21, 2017.
12. A. K. CHATTERJEE, Pozzolanicity of Calcined Clay, in *Calcined Clay for Sustainable Concrete*, Proceedings of the International Conference on Calcined Clay for Sustainable Concrete (Eds K. Scrivener and A. Favier), Springer (2015).
13. V. V. VOLKONSKII, S. D. MAKASHEV, AND N. P. SHTEIRT, *Technological, Physico-Mechanical and Physico-Chemical Studies of Cement Raw Materials* (in Russian), Izd-vo Literatury Po Stoitel'stvu, Leningrad (1972).
14. A. K. CHATTERJEE, Chemico-Mineralogical Characteristics of Raw Materials, in *Advances in Cement Technology* (Ed. S. N. Ghosh), Pergamon Press, Oxford (1984).
15. http://www.newworldencyclopedia.org/gypsum, retrieved January 24, 2014.
16. USGS Mineral Commodities Summary – Gypsum (2016).
17. H. F. W. TAYLOR, *Cement Chemistry*, Academic Press, London (1990).

CHAPTER TWO

Raw mix proportioning, processing, and burnability assessment

2.1 Preamble

It has already been stated in Chapter 1 that two generically different natural raw materials—calcium carbonate (limestone) and aluminum silicate (clay, shale, etc.)—are basically required to produce the Portland cement clinker. These two raw materials complement each other in stoichiometric proportions to give rise to the compounds or phases present in the clinker in the required quantities. At times, when such stoichiometric needs are not met by the above two primary components, certain corrective materials such as bauxite, laterite, iron ore or blue dust, sand or sandstone, etc. are used to compensate the specific chemical shortfalls in the composition of the raw mix. Additionally, for processing advantages, certain chemicals are used as grinding aids in the milling process and as mineralizers in the burning process. For resource conservation, a large number of industrial wastes or byproducts are used as raw materials, mostly for correction purposes. The most commonly used raw materials are summarized in Table 2.1.

2.2 Stoichiometric requirements in raw mix computation

A clinker, as will be explained later, comprises four major compounds or phases: 50–55% tricalcium silicate (C_3S), also known as "alite"; 25–30% dicalcium silicate (C_2S), also known as "belite"; 9–11% tricalcium aluminate (C_3A); and 12–15% tetracalcium aluminoferrite (C_4AF), also called "brownmillerite." In terms of the major oxides, this phase assemblage of the clinker approximately corresponds to CaO (expressed as C in cement chemistry) = 62–65%, SiO_2 (expressed as S) = 19–21%, Al_2O_3 (expressed as A) = 4–6% and Fe_2O_3 (expressed as F) = 3–5% on a loss-free basis.

Table 2.1 Raw materials for clinker making

Oxide components	Natural raw materials	Industrial wastes/ byproducts
CaO	Limestone, chalk, marble, seashells, marl	Carbonate sludge from paper, sugar, and fertilizer industries
$Al_2O_3 + SiO_2$	Clay, soil, shale, argillite, phyllite, slate, volcanic rocks	Fly ash from thermal power stations
$CaO + SiO_2$	Calcium silicate rocks or minerals, viz., wollastonite	Metallurgical slags
SiO_2	Sand and sandstone	Foundry sand
Al_2O_3	Bauxite	
$Al_2O_3 + Fe_2O_3$	Laterite	
Fe_2O_3	Iron ore, blue dust	Pyrite cinders, red mud, mill scale
Grinding aids	Surface-active chemicals like di- or tri-ethnolamine, sulfate lye, sodium polyphosphate, etc.	
Mineralizers	Fluorides, magnesia, calcium sulfate, etc.	

A carbonate rock containing all the four major oxides in the required proportions is available in nature, and is known as "cement rock," but is rarely encountered. Hence, for all practical purposes, raw mixes are prepared from one or two grades of limestone, argillaceous materials, and other corrective substances, as mentioned earlier. Generally, three to five component mixes are prepared, in which the proportion of limestone varies from about 80–95%, depending on its quality. In most countries, the stoichiometric requirements of the four major oxides are expressed by the following three ratios (1):

a. Alumina Modulus (AM) = A/F (2.1)

b. Silica Modulus (SM) = S/(A + F) (2.2)

c. Lime Saturation Factor = C/(2.8 S + 1.18 A + 0.7 F)
(when AM < 0.64) or = C/(2.8 S + 1.65 A + 0.35 F)
(when AM > 0.64) (2.3)

But in some countries by convention the ratio for lime saturation is expressed differently, as illustrated below:

I. $\Delta = 100 \times (2.8 S + 1.65 A + 0.35 F - C)/(S + A + F + C)$ (2.4)

II. KN = C − (1.65 A + 0.35 F + 0.7 Sulfate)/2.8 S (2.5)

$$\text{Kalkstandard} = 100(C + 0.75\,MgO)/(2.8\,S + 1.18\,A + 0.65\,F),$$
when MgO < 2% or the numerator changes to (CaO + 1.5),
if MgO > 2%. (2.6)

The above three variants of the lime saturation ratio owe their origin to the practices in France, Russia, and Germany, respectively, although the theoretical basis behind all the variants remains more or less the same. A more simplified ratio (C/(S + A + F)), called "hydraulic modulus," ranging from 1.9–2.4, was also in practice in some countries. But since, in this expression, all the oxides are considered without their reaction coefficients, the ratio eventually waned in its practical application due to its lack of effectiveness. The three main ratios of practical significance today are AM, SM, and LSF.

The Alumina Modulus governs the ratio of aluminate to ferrite phases in clinker and also determines the development of the melt phase (also known as clinker liquid) that is formed in the burning zone. For a given total of $Al_2O_3 + Fe_2O_3$ the quantity of liquid formed at 1338°C theoretically passes through a maximum at AM of 1.38. The effect of magnesia on the optimization of AM is discussed later in this chapter. The Silica Modulus essentially governs the proportion of silicate phases in clinker. The higher the ratio, the lower is the quantity of melt formed at the burning zone temperatures, making the clinker harder to burn. The Lime Saturation Factor is a measure of the degree of conversion of silica, alumina, and iron oxide into the corresponding calcium-bearing compounds. It also governs the alite:belite ratio in clinker and gives a reasonable idea of whether there could be a high proportion of unreacted free lime in the clinker. Although this ratio is used widely for clinker, it may be extended to cement, if it is corrected by subtracting 0.7 SO_3 from the total CaO.

The relation of clinker burning to LSF and SM is shown in Figure 2.1 (2), which demonstrates that the LSF is the most influential parameter to control the burning of clinker. The higher the LSF, the harder it is for the clinker to burn, becoming even harder with a simultaneous increase in SM, as shown by the light gray corner in the diagram. In the low LSF range, the effect of SM is shown by the blue segment. The influence of SM on a moderate range of LSF is fairly strong to make the mix progressively harder to burn, as seen in the middle medium gray portion. Although a high AM means lower burnability, it does not show a clear relation with free CaO. Ultimately, the three modulus values play an interactive role in the burning of clinker. In practice, SM and AM are maintained in the range of 2.0–2.8 and 1.0–1.8, respectively, while LSF is kept in the range of 0.92–0.98. LSF values in excess of unity are also encountered in the plants. In the Russian practice, KN is targeted in the range of 0.87–0.92 and in the French expression of lime saturation (Δ), the target is 5–7 for an approximate C_3S level of 60%.

Potential phase computation by the Bogue equations

In addition to the above ratios, or "modulus" values, as they are more commonly known, the recalculation of the oxide composition into the expected or potential clinker phase composition by the application of

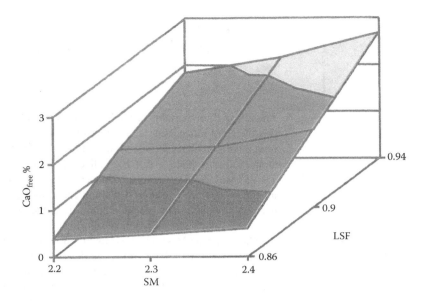

FIGURE 2.1 Interrelation of LSF and SM ratios with free lime in clinker. (From L. Opoczky, L. Sas, and F. D. Tamas, Preparation and quality assurance of raw meals, in *Modernization and Technology Upgradation in Cement Plants* (Eds S. N. Ghosh and Kamal Kumar), Akademia Press International, New Delhi, 1999.)

Bogue equations (first introduced by the outstanding cement chemist R. H. Bogue in 1929) is also important from the perspective of stoichiometric requirements of raw mixes. These calculations are carried out based on loss-free oxides and the following assumptions:

a. There are four major clinker phases.

b. Fe_2O_3 appears only in C_4AF.

c. Al_2O_3 occurs first in C_4AF and then in C_3A.

d. CaO that is considered to form C_3S and C_2S is the amount remaining after apportioning lime to C_4AF, C_3A, and the unreacted CaO in clinker.

The above assumptions lead to the following equations (1,3):

- $C_3S = 4.071\,C - 7.602\,S - 6.719\,A - 1.430\,F - 2.852\,SO_3$ (2.7)

- $C_2S = 2.867\,S - 0.754\,C_3S$ (2.8)

- $C_3A = 2.650\,A - 1.692\,F$ (2.9)

- $C_4AF = 3.043\,F$ (2.10)

The above expressions are valid when the A/F ratio is more than 0.64. In situations when the A/F ratio is less than 0.64, the entire quantity of alumina and iron oxide is expected to form a solid solution phase compositionally lying in the system $C_4AF - C_2F$ and no C_3A is likely to form in the absence of alumina to react with lime. The coefficient for Al_2O_3 in the above C_3S expression stands reduced. Thus, the expressions for the C_3S, C_2S, and the ferrite solid solution phase are given as follows:

- $C_3S = 4.071\,C - 7.602\,S - 4.479\,A - 2.859\,F - 2.852\,SO_3$ (2.11)

- $C_2S = 2.867\,S - 0.754\,C_3S$ (2.12)

- $(C_4AF - C_2F)\text{solid solution} = 2.1\,A + 1.702\,F$ (2.13)

In the above phase formation reactions, there is a further assumption that the entire MgO occurs only in the oxide state or as "periclase," which is the mineral name for magnesium oxide. Since the minor oxide components are ignored in this calculation, the total of the four phases plus free CaO will not add up to 100%.

The Bogue calculation is also applicable to cement, if CaO is further corrected by deducting 0.7 SO_3. A deduction from total CaO may also be made if the soluble alkalis are found to consume some lime. In some cases, a correction for the magnesia content may also be necessary. In other words, the amount of CaO that combines into the four clinker phases may be computed as follows:

$$CaO_{cor} = CaO_{total} - CaO_{free} - 56/44\,CO_2 \\ - 0.7(SO_3 - 0.85\,K_2O_{sol} - 1.29\,Na_2O_{sol})$$ (2.14)

With the corrected CaO, the potential amounts of the four phases may be calculated as shown below:

- $C_4AF = 3.043\,F$ (2.15)

- $C_3A = 2.650\,A - 1.692\,F$ (2.16)

- $C_2S = 8.60\,S + 1.08\,F + 5.07\,A - 3.07\,C - 3.790\,M + 2.58$ (2.17)

- $C_3S = 4.071\,C - 7.62\,S - 1.430\,F - 6.718\,A + 5.03\,M - 3.4$ (2.18)

The potential phase composition differs from the actual composition of the clinker, as determined by X-ray diffractometry or optical microscopy, due to the following reasons:

a. The clinkering reactions seldom reach equilibrium during cooling, which happens to be the basic theoretical assumption in developing the Bogue calculations.

b. The compositions of the clinker phases differ considerably from those of the pure compounds that are considered in the calculation due to the incorporation of impurities present in the raw materials.

The potential phase compositions are, however, quite useful in raw mix calculations. Though not quite accurate, they are an easy way of arriving at the likely phase composition of clinker of a given raw mix of known chemical composition.

Calculation of clinker liquid phase

The formation of clinker in a kiln requires the presence of a liquid or melt phase, which is a result of the chemical reactions that take place inside the kiln before sintering. The quantity of the melt phase likely to form at various temperatures has been estimated from the phase equilibrium relations of the chemical system $C_3S - C_2S - C_3A - C_4AF$ and is primarily governed by the A/F ratio, further contributed by MgO and alkalis. The formulas for estimating the quantity of liquid change, depending on whether the A/F ratio is less or more than 1.38 and whether the temperature considered is lower or higher than 1338°C. The fomulas adopted in practice are presented below:

$$1450°C: 3.00\,A + 2.25\,F + MgO + K_2O + Na_2O \qquad (2.19)$$

$$1400°C: 2.95\,A + 2.20\,F + MgO + K_2O + Na_2O \qquad (2.20)$$

$$1338°C: 8.20\,A - 5.22\,F + MgO + K_2O + Na_2O, \text{ when A/F} \leq 1.38 \qquad (2.21)$$

$$1338°C: 6.10\,F + MgO + K_2O + Na_2O, \text{ when A/F} \geq 1.38 \qquad (2.22)$$

In all the above expressions MgO denotes an upper limit of 2.0%. However, without considering the contributions of magnesia and also of alkalis, the effects of the two major oxides, e.g., alumina and ferric oxide, on the quantity of liquid are additive and are fairly similar weight-for-weight at temperatures greater than 1338°C. At 1338°C, which is an invariant point in the above-mentioned quaternary system, the effects are not additive and the maximum amount of liquid for a given total content of alumina plus ferric oxide is obtained at A/F = 1.38 (Figure 2.2) (4). There has also been an attempt to relate liquid content, temperature, and A/F ratio for compositions having LSF = 0.95 and SM = 2.6 (Figure 2.3) (5). So far as the minor constituents are concerned, MgO has been reported to depress the liquidus temperature by about 50°C, shifting the eutectic composition in terms of the A/F ratio to 1.63 (6). Alkalis and sulfate may depress the liquidus temperature further but the effects have not been thoroughly studied. It is known that alkali sulfates melt at low temperatures but these melts are not miscible with clinker liquid. Alkalis that do not form alkali sulfates can enter the clinker liquid and retard the formation of alite (7).

RAW MIX PROPORTIONING, PROCESSING, AND BURNABILITY ASSESSMENT   47

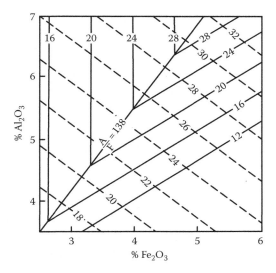

FIGURE 2.2 Melt formation due to alumina and ferric oxide in forming liquid at 1338°C (solid contours) and 1400°C (dashed contours).

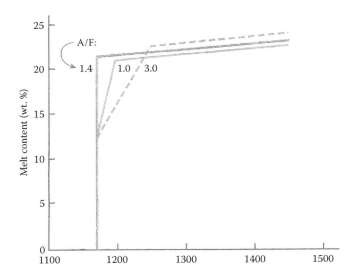

FIGURE 2.3 Relation between melt formation and temperatures for raw mixes having the same LSF and SM but varying AM.

Calculation of coal ash absorption

If coal or any high ash-bearing solid fuel is used in burning clinker, the ash generated inside the kiln is mostly absorbed in the clinker composition. A change in the clinker composition due to ash absorption may sometimes be significant, more particularly when the specific heat consumption is high.

Therefore, it is important to know the likely effect of ash absorption. This is computed on the basis of the four major oxides in the coal ash, raw mix, and clinker, as given below:

Coal ash: C_a, A_a, F_a, S_a
Raw mix: C_{rm}, A_{rm}, F_{rm}, S_{rm}
Clinker: C_{cl}, A_{cl}, F_{cl}, S_{cl}

The absorption of the individual oxides is calculated as illustrated below with the help of CaO:

$$x_1 = (C_{cl} - C_{rm})/(C_a - C_{rm}) \times 100 \tag{2.23}$$

Similarly, x_2, x_3, and x_4 can be calculated for Al_2O_3, Fe_2O_3, and SiO_2. Finally, one may obtain the ash absorption in percentage as an average of four individual oxide values:

$$x = x_1 + x_2 + x_3 + x_4 / 4 \tag{2.24}$$

Finally, the ash effect on clinker is calculated from the average ash absorption as shown above, the quantity of coal consumed in clinker making, and the ash content in the coal used.

2.3 Raw mix computation

Based on the oxide compositions of limestone and other raw materials, raw mixes are computed using the modulus values and Bogue equations mentioned earlier. This process is known as the "raw mix design." The primary goal is to prepare a kiln feed that permits the production of Portland cement clinker of the desired quality with the lowest possible energy consumption and overall cost. The raw mix design is done by pre-fixing certain target parameters for the resulting clinker, which include the following aspects:

- composition
- burning behavior, or "burnability"
- energy consumption
- cost

A two-component raw mix can easily be calculated on the basis of the target LSF of the clinker, if the oxide composition of both the components is known. For raw mixes with three or more components the calculations become more complicated, as the proportions of raw materials are computed by setting up and solving simultaneous equations to fix n parameters from n + 1 raw materials of appropriate composition. These calculations can be done either by trial and error or by step-wise matrix method, as explained below.

Table 2.2 Illustration of raw mix design by the trial-and-error method[a]

Parameters	Target Clinker	Loss-free and Ash-free Raw Mix Target	Raw Materials (loss free) Limestone 1	Limestone 2	Slag	Finally Designed Raw Mix
LOI	0.90	–	–	–	–	–
SiO_2	20.93	20.68	23.01	17.56	22.68	20.54
Al_2O_3	5.62	5.57	5.91	4.51	11.56	5.44
Fe_2O_3	3.74	3.63	2.39	1.88	34.26	3.12
CaO	64.55	64.51	61.80	69.88	14.82	63.89
MgO	2.88	2.85	3.21	2.63	0.44	2.86
C_3S	60.57	62.80	33.55	116.78	(-) 238.94	62.96
C_2S	14.36	11.86	40.74	(-) 37.69	(-) 14.96	22.44
C_3A	8.68	8.73	11.74	8.86	0.00	9.24
C_4AF	11.38	11.05	7.27	5.72	104.25	9.49
LSF	93.31	94.37	82.35	121.50	15.67	94.55
SM	2.24	2.23	2.77	2.75	0.49	2.40
AM	1.50	1.53	2.47	1.40	0.34	1.74
Percentage of raw materials			52.0	45.0	3.0	

[a] All values are in mass %, except the modulus values, which are ratios.

Trial-and-error Method

In the trial-and-error method, a definite clinker composition is targeted (see Table 2.2). From this the desired raw mix composition is arrived at on a loss-free basis after deducting the absorbed fuel ash oxides. Various percentages of the individual raw materials are tried, multiplying the percentage by the amount of the major oxides and summing up the values. The percentages of individual raw material are progressively adjusted to reach the target raw mix composition as closely as possible. In preparing Table 2.2, the fuel ash composition was taken as follows: SiO_2 31.54%, Al_2O_3 7.24%, Fe_2O_3 11.89%, CaO 23.50%, and MgO 3.50% (and its absorption in clinker was assumed to be 1.0%). It may also be noted that the finally designed raw mix deviated, to some extent, from the target composition due to certain limitations of the composition of the raw material. However, for all practical purposes the designed raw mix appeared to be a workable one.

Step-wise matrix method

The initial step in the matrix method of raw mix design is to fix up the target composition of the clinker by the number of parameters, which are 1 less than the number of raw mix components. This means that if one has three raw materials, one should use 2 target parameters for clinker. For 4 materials, one may design on the basis of 3 target parameters, such as, C_3S, C_3A, and clinker liquid percentage.

Let us take an example of designing a 4-component raw mix consisting of limestone, shale, sand, and iron ore. The oxide compositions of these raw materials are first converted to loss-free or ignited basis and then their phase compositions are calculated with the help of Bogue equations. The details are furnished in Table 2.3, which also shows the 3 target parameters pre-fixed for the design of the raw mix.

Table 2.3 Composition of the raw materials used to illustrate the stepwise matrix method

Parameters	Raw materials				Selected targets of the clinker composition
	Limestone	Shale	Sand	Iron ore	
LOI %	39.52	5.15	1.00	10.55	–
Loss-free composition (%)					
SiO_2	10.63	66.25	96.20	11.35	–
Al_2O_3	3.50	20.32	1.80	2.10	–
Fe_2O_3	1.73	6.40	1.60	86.45	–
CaO	82.96	5.34	–	–	–
MgO	1.85	2.12	–	–	–
Total	100.67	100.43	99.60	99.90	–
Modulus values					
LSF	229.49	2.41	–	–	–
SM	2.03	1.73	28.25	0.113	–
AM	2.02	1.20	1.20	0.02	–
Bogue composition (%)					
C_3S	231.30	(-) 627.55	(-) 745.55	(-) 224.01	55.00
C_2S	(-) 146.55	660.05	(-) 291.45	(-) 137.02	–
C_3A	6.35	43.03	2.30	0.00	8.00
C_4AF	5.26	16,43	4.56	26.31	–
Liquid phase	15.99	76.14	8.61	196.38	30.00

Now we choose pairs of raw materials to achieve the first target, i.e., $C_3S = 55.00\%$.

Step 1: Limestone and Shale

$$C_3S_{lst} - C_3S_{tgt} = 231.30 - 55.00 = 176.30 \tag{2.25}$$

$$C_3S_{sh} - C_3S_{tgt} = -627.55 - 55.00 = -682.55 \tag{2.26}$$

From the above equations, and ignoring the negative sign at this stage, the proportion of shale is 176.30/(176.30 + 682.55) = 20.52% and correspondingly that of limestone is 79.48%.

Step 2: Limestone and Sand

Following the same type of calculation, the proportions of limestone and sand work out to be 81.95% and 18.05%, respectively.

Step 3: Limestone and Iron Ore

Following the same mode of calculation, the proportions of limestone and iron ore are found to be 61.28% and 38.72%, respectively.

In all the above 3 steps, C_3S works out to be 55.00%, considering the positive C_3S in limestone and negative C_3S in shale, sand, and iron ore.

RAW MIX PROPORTIONING, PROCESSING, AND BURNABILITY ASSESSMENT 51

Now from the data in Table 2.3, one may calculate the other two clinker targets for steps 1, 2, and 3 as follows:

Step 1: 79.48% Limestone and 20.52% Shale

C_3A = 13.88%

Clinker liquid = 28.33%

Step 2: Limestone 81.95% + sand 18.05%

C_3A = 5.60%

Clinker liquid = 28.64%

Step 3: Limestone 61.28% + Iron ore 38.72%

C_3A = 3.89%

Clinker liquid = 85.84%

Since both the C_3A and clinker liquid targets are substantially deviating from the pre-set values, some more steps of combinations need to be explored, first, to arrive at the target C_3A value, and then at the target liquid phase.

Step 4: Achieving the Target C_3A

From Steps 1 and 2, one may observe that while the C_3S target was achieved in both the cases, the C_3A value in Step 1 was higher than the target and in Step 2 the C_3A value was lower than the target. Hence, there is a possibility that if both Steps 1 and 2 are combined in the correct proportions, the C_3A target is likely to be achieved, along with the C_3S target. Thus, in Step 4 the proportions of raw materials of Steps 1 and 2 can be calculated as follows:

$$C_3A \text{ in Step } 1 - \text{Target } C_3A = 13.88 - 8.00 = 5.88 \qquad (2.27)$$

$$C_3A \text{ in Step } 2 - \text{Target } C_3A = 5.60 - 8.00 = -2.40 \qquad (2.28)$$

Hence, the raw materials of Step 1 and Step 2 are to be combined by calculating the ratios in the same manner as explained earlier:

$$\text{Proportion of Step } 1 = 2.40/(240 + 5.88) = 28.99\% \qquad (2.29)$$

$$\text{Proportion of Step } 2 = 5.88/(5.88 + 2.40) = 71.01\% \qquad (2.30)$$

Thus, the raw materials proportions obtained at the end of Step 4 are:

$$\text{Limestone: } 230.5\% + 58.19\% = 81.23\% \qquad (2.31)$$

$$\text{Shale: } 5.95\% + 0 = 5.95\% \qquad (2.32)$$

$$\text{Sand: } 0 + 12.82 = 12.82\% \qquad (2.33)$$

With these proportions of limestone, shale, and sand the target clinker parameters would work out as follows:

$C_3S = 54.67\%$; $C_3A = 8.00\%$; liquid = 18.11%

From the above values it is obvious that, at the end of Step 4, the C_3S and C_3A targets could be met, but not the target of the liquid phase. Hence, another step is necessary.

Step 5: Achieving the Target of the Liquid Phase

From the abovementioned values, it is evident that in Step 4 the liquid phase (18.11%) obtained was lower than the target, while in Step 3 the obtained value was higher (85.84%) than the target. The liquid phase values in Step 1 and 2 were also lower than the target, viz., 28.33% in Step 1 and 14.65% in Step 2. Therefore, it is obvious that Step 4 and Step 3 need to be combined to achieve the target value in the same manner as described earlier:

Step 4: 30.00% − 18.11% = 11.89% (2.34)

Step 3: 85.84% − 30.00% = 55.84% (2.35)

Hence, the ratios in which the raw materials of Step 3 are to be combined with Step 4 are 17.55% and 82.45%. Using these ratios, the final raw mix composition at the end of Step 5 is:

Limestone: 77.72%

Shale: 4.91%

Sand: 10.57%

Iron ore: 6.80%

The raw mix of the above composition will yield on calculation 54.93% C_3S, 7.29% C_3A, and 30.42% liquid phases, which for all practical purposes come very close to the target values.

As explained earlier, the above proportions are obtained on a loss-free basis. If these proportions were to be converted into as-received values, the loss-free values would have to be multiplied by the factor 1/(100 − loss on ignition). In other words, the above raw mix components will be used in the following proportions:

$$\begin{aligned}
\text{Limestone} &= 77.72 \times 1/(100 - 39.52) = 1.285 \\
\text{Shale} &= 4.91 \times 1/(100 - 5.150) = 0.052 \\
\text{Sand} &= 10.57 \times 1/(100 - 1.00) = 0.107 \\
\text{Iron ore} &= 6.80 \times 1/(100 - 10.55) = \underline{0.076} \\
\text{Total} &= 1.520
\end{aligned}$$

The above decimal expressions of the raw mix components may be normalized to as-received raw materials percentages as follows:

$$\begin{aligned}
\text{Limestone} &= 84.54\% \\
\text{Shale} &= 3.42\% \\
\text{Sand} &= 7.04\% \\
\text{Iron ore} &= \underline{5.00\%} \\
\text{Total} &= 100.00\%
\end{aligned}$$

Adoption of computer programming

In order to eliminate the manual labor, most plants convert the computation logic of their own into a program that can be run on their computer facilities in the laboratory or the central control room. In addition to the chemical parameters, many of the programs take into account operational and economic factors such as specific heat consumption, landed costs of raw materials and fuels, optimum conversion cost, etc. Almost-ready-to-use software programs can be accessed from the market and then tailored to the specific inputs and outputs required in a given situation. The program inputs include the oxide composition of all the raw material components and their modulus ratios, coal ash composition, targeted specific heat consumption, landed costs of raw mix components and fuel, and potential clinker phase composition. The program calculates all possible variations, compares the acceptable solutions, and offers the most optimum solution as the output. A typical spread-sheet of such computer software is shown in Figure 2.4. The major advantage of such programming is to obtain multiple options in a short time by providing the correct inputs into the system.

2.4 Preparation process for raw mix

Once the raw mix has been designed, the preparation process starts with the crushing of lumpy raw materials, followed by pre-blending, grinding, and homogenizing. The homogenized raw mix as it enters into the kiln is known as "raw meal" or "kiln feed." In designing and setting up a plant, a buffer stock of raw meal is provided between the raw mill and the kiln as the raw mix leaving the grinding mill will rarely be homogeneous enough to be fed directly to the kiln. How large the buffer stock needs to be is a matter of process design but it is generally agreed that the standard deviation of the kiln feed evaluated by grab samples should not exceed 1.0 LSF unit. A detailed account of the unit operations for preparing the raw meal has been given by the author in (7) and the more salient features are described below.

Crushing operation

The crushing operation is the first stage of comminution to reduce limestone pieces of 1–2-m size obtained from the mines to the roughly 25-mm size necessary to form the feed to the grinding mills. There are four basic mechanisms for size reduction of materials – impact, attrition, shear, or compression, and most crushers employ a combination of

A typical spreadsheet of a raw mix design calculation, using a computer software

Raw materials

Parameter	Mat-1 Limestone	Mat-2 Low-grade	Mat-3 Sweetener	Mat-4 Iron ore	Mat-5 Bauxite	Raw mix (Expected)	Clinker (Expected)	Ash analysis	
SiO$_2$	13.30	21.12	6.05	9.95	18.59	13.17	20.50	43.00	
Al$_2$O$_3$	2.00	3.42	1.74	6.47	43.26	2.76	4.58	25.80	
Fe$_2$O$_3$	0.80	2.89	1.20	78.47	8.56	1.99	3.07	4.25	
CaO	43.54	39.05	50.09	0.57	2.91	42.42	64.06	23.09	
MgO	1.10	1.04	0.47	1.01	0.48	1.07	1.63	1.36	
P$_2$O$_5$	0.11	0.12	0.08	0.01	0.03	0.11	0.17	0.20	
K$_2$O	0.17	0.23	0.18	0.16	0.13	0.17	0.26	0.44	
Na$_2$O	0.07	0.00	0.09	0.00	0.00	0.07	0.10	0.12	
Cl–	0.005	0.006	0.005	0.00	0.005	0.005	0.007	0.00	
SO$_3$	0.000	0.000	0.000	0.00	0.000	0.000	0.22	2.00	
TiO$_2$	0.25	0.340	0.130	0.01	4.200	0.31	0.50	1.80	
LOI	35.42	31.83	39.88	3.18	21.60	34.51	0.30	0.00	
Sum	96.77	100.04	99.92	99.83	99.77	96.59	95.41	100.06	
Addition %	94.52	0.00	2.41	1.35	1.71			1.71	Ash abs.
LSF	108.52	60.03	253.32	0.66	2.68	102.39	98.61	685.0	KCal
S/R	4.75	3.35	2.06	0.12	0.36	2.77	2.68	62.36	CV
A/F	2.50	1.18	1.45	0.08	5.05	1.38	1.50	15.57	Ash

Bogues	
F-Cao	1.50
C3S	58.61
C2S	14.57
C3A	8.72
C4AF	9.32
Liq %	24.57
C3S WoF	64.71
2C3A+C4AF	26.76

Burnability factor	
Raw mix	126.33
Clinker	119.68
Heat of reaction	412.55
Diff RM to clinker	
	3.78
	0.09
	–0.11

Material	83.00	320.00	1565.00	421.00			
Raw mix	78.46	7.71	21.20	7.20		114.57	Clinker cost with fuel
Clinker	119.17	11.72	32.20	10.94		174.03	525.89

Details of fuel input

Source	% Addition	Ash (%)	SO$_3$ (%)	NCV (KCL/kg)	Moist (%)	Landed cost (Rs/MT)
Coal-1	0.00	16.00	1.00	5830	14.59	2760
Coal-2	100.00	15.57	2.00	6236	7.00	2979
Coal-3	0.00	8.66	2.00	6236	11.00	2979
Petcoke	0.00	1.00	10.00	8170	7.00	3690
Lignite	0.00	14.00	15.00	4000	37.00	1255
Calc...	100.00	15.57	2.00	6236	7.00	2979

Coal cons. %
Dry basis   10.98
As recd. basis   11.81

Fuel cost
Rs/ton of clinker   351.86

FIGURE 2.4 A typical computer-aided spreadsheet of raw mix design.

all these crushing methods. There are two variations of impact: gravity impact and dynamic impact. When crushed by gravity impact, the free-falling material is momentarily stopped by the stationary object. But when crushed by dynamic impact, the material is unsupported and the force of impact accelerates movement of the reduced particles toward breaker blocks or hammers. Dynamic impact has definite advantages for the reduction of many materials. Attrition is the mechanism of reducing materials by scrubbing between two hard surfaces. Shear consists of a trimming or cleaving action and is usually combined with other mechanisms. Crushing by compression is done between two surfaces.

Generally, the following types of crushers are used in the cement production process:

- Jaw crusher (single or double toggle), based on compression mechanism, for primary crushing.
- Gyratory crusher, based on compression mechanism, for primary crushing; and cone crusher, a modification of the gyratory crusher, for secondary or even tertiary crushing.
- Hammer crusher (single or double rotor), operating primarily on dynamic impact, and having many design variants such as fixed hammer type, swing hammer reversible impactor type, horizontal or vertical impact type, and others.
- Double roll crusher operating with combined mechanism of shear, impact, and compression.

Hammer or impact crushers, which are normally used in cement plants, are suitable for types of limestone that are not hard and abrasive. Jaw and gyratory or cone crushers are preferred for hard and abrasive types of limestone mostly in multistage modes. Double roll crushers are considered the most versatile for rocks that are either dry and dusty or wet and sticky, or hard with compressive strength up to about 140 MPa.

For the commonly used hammer crushers the specific power consumption is in the range of 0.5–1.0 kWh per metric ton of material crushed. The wear rate of hammers is known to be 1–10 g/t. In general, the crushing stage is not energy-intensive but the wear rate of metallic components is often a factor to monitor.

In cement plants the crushing operation is required mainly for the run-of-mine limestone, as all other supplementary and corrective materials are purchased in crushed lumps. It may be relevant to mention here that controlling the top size of the crushed limestone is very important in a plant, which is often accomplished by closed-circuit crushing, i.e., by screening the product and returning oversize material to the feed end of the crusher for another pass through the machine (Figure 2.5) (8). In single-stage crushing there could be overgrinding of a portion of the product with corresponding increase in fines and loss of efficiency. Hence, it is necessary in many situations to take recourse to two stages of crushing, where the second stage is generally a closed circuit, with the primary crusher producing a satisfactory feed size for the secondary one. In some circumstances, a three-stage crushing plant may be

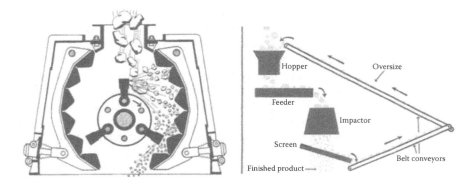

FIGURE 2.5 A schematic layout of a closed-circuit crushing plant.

necessary. In case the transportation of as-mined large limestone blocks to the stationary crusher plant turns out to be uneconomical, an option of installing small and mobile crushers near the mine faces is considered, which facilitates the crushed limestone to be conveyed downstream with the help of belt conveyors.

Pre-homogenizing systems

In cement production, it is often necessary to make, prior to milling, the crushed limestone mined from different mine faces to be reasonably uniform in composition, particularly when the face-to-face quality variation is high or when different grades of limestone are used. For this purpose, the crushed limestone—with or without other materials—is subjected to a process called "pre-homogenizing." This is achieved through "pre-blending beds" or "stacking-and-reclaiming systems." The blending effect $H = S_{in}/S_{out}$ of a pre-homogenizing stacker/reclaimer system is generally determined as the ratio between the standard deviation S_{in} of any one significant chemical parameter of the stacked input material and the standard deviation of the same chemical parameter S_{out} of the reclaimed output material. In principle, the standard deviation is reduced by stacking the material in a large number of layers (typically 100–400) and subsequently reclaiming these layers in cross-section across the height of the pile (end reclaim) or along the side of the pile from one end to the other (side reclaim) (see Figure 2.6). Without taking into account the particulate nature of the material, the blending effect is proportional to square root of the number of layers stacked and reclaimed. However, all reclaiming systems operate with some kind of scraper or bucket chain arrangement and the actual blending effect decreases from the theoretical value as a pair of scraper blades or a bucket would have limitations of holding coarse particles. The blending effect in these systems is generally in the range of 1:2 to 1:5 as the number of layers reclaimed at a time is limited.

The most commonly used longitudinal stacking methods are Chevron and Windrow. Basically, these methods consist of stacking a large number of layers on top of each other in the longitudinal direction of the stock pile. According to the Chevron method the crushed limestone is deposited by the stacker moving to and fro over the center line

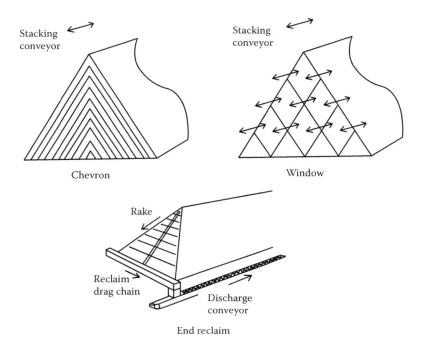

FIGURE 2.6 Schematic layout of pre-blending systems.

of the pile. The Chevron method causes segregation of the material. The fine particles remain in the central part of the pile and the coarse particles move to the sides and the bottom of the pile. To ensure proper blending, a Chevron pile should be reclaimed from the end of the pile, working across the entire cross-section. According to the Windrow method, the crushed limestone is deposited from a number of positions across the full width of the pile. The Windrow method prevents segregation by ensuring more even distribution of fine and coarse particles across the pile. A schematic diagram of the Chevron and Windrow methods is given in Figure 2.6. It may be relevant to mention here that a third method of stockpiling, known as the cone shell method, is sometimes used when blending is not the main objective. The pile is formed by depositing material in a single cone from a fixed position. When this conical pile is full, the stacker discharge point is moved to a new position and a shell is formed against the surface of the first one. The process continues in the longitudinal direction until the stockpile is complete.

Continuous Chevron stacking is the most commonly used "circular" stockpiling method, which is different from the longitudinal stockpiling practices described above. A circular stockpile has a round base with one pile being continuously stacked at one end and reclaimed at the other. Stacking is done in a fan-shaped arc of 120 degrees. With each sweeping movement corresponding to two layers of material the whole sector moves approximately half-a-degree ahead. The circular stockpile is considered primarily from space consideration but it has its limitation of not being amenable to change of composition without completely emptying the pile. The storage capacity cannot also be expanded. Hence, the longitudinal stockpile is more popular.

The reclaiming arrangement generally has constant-speed motors and the reclaimed material is carried by belt conveyors to a feed bin of large volume. Reclaiming capacity is designed to be higher than the mill requirement and the reclaimer therefore operates in an on/off mode controlled by maximum/minimum level indicators in the feed bin. If, in a particular situation, there is no provision for the feed bin between the stockpile and the mill, the reclaimer capacity has to match the mill requirement and the reclaimer will have to operate in a continuous mode with speed-regulated motors and weigh feeding arrangements.

It is evident that the aim of pre-homogenization is to store the raw materials in such a way that the pile at any position will have a chemical composition as close to what is required at the plant. This objective can be achieved only if the dosing of the material as well as the real chemical composition of the pile are monitored and controlled. An important step is sampling during the formation of the stockpile. Ideally each layer of the heap should be analyzed, which requires a complicated sampling station, where typically about 2% of the material flow is diverted as a spot sample. The accumulated sample over a given period of time is subjected to stepwise size reduction and infrared drying; the test samples are then subjected to on-line or off-line x-ray fluorescence spectrometry. A more convenient system has subsequently emerged with the development of the on-line cross-belt gamma ray analyzer for bulk materials, which has been dealt with in a different chapter.

Raw milling operation

Raw milling is the next step of the process. Usually, each raw component is stored separately in storage bins equipped with weighing and dosing systems for the mix control to the grinding mill. There are three main types of raw milling system in vogue:

- Ball mills
- Vertical roller mills (VRM)
- Hydraulic roll presses, often in combination with ball mills

New raw grinding installations are primarily VRMs. Roll presses are used particularly in upgrading existing ball mill circuits for increased production or decreased specific power consumption. All raw grinding systems are closed-circuited with separators (also known as classifiers) for efficient grinding. More details of these systems have been presented in the chapter dealing with clinker grinding. The schematic diagrams of ball mill, VRM, and hybrid milling systems for raw materials are given in Figures 2.7 through 2.9.

Performance of the raw milling systems depends on a few important parameters:

- Feed size to the mill
- Grinding behavior of the feed material

RAW MIX PROPORTIONING, PROCESSING, AND BURNABILITY ASSESSMENT 59

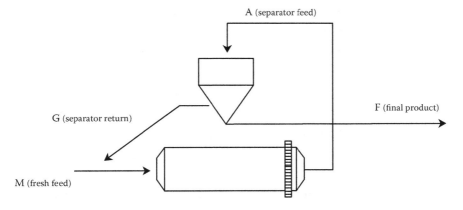

FIGURE 2.7 A ball mill circuit.

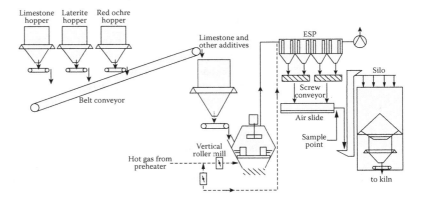

FIGURE 2.8 A vertical roller mill system.

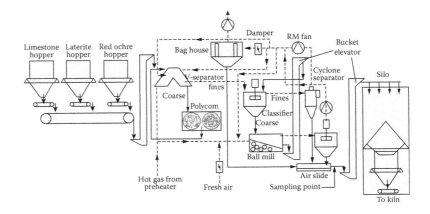

FIGURE 2.9 A roll press and ball mill combined installation.

- Drying capacity of the system
- Product fineness required
- Grinding power available

The optimum feed size requirements to the mill systems are:

Ball mills: 5% residue on 15–30 mm

Vertical mills: 5% residue on 50–100 mm

Roll presses: crushed and pre-homogenized raw mix

So far as the drying capability is concerned, the milling systems behave as follows:

a. Ball mills having a drying compartment and being air-swept with hot gas having 2.5–3 m/s velocity above the grinding media can handle up to 8% moisture.
b. Double rotator central discharge fully air-swept mills (5–6 m/s air velocity) can dry up to 12% moisture.
c. Roller mills can handle up to 25% moisture.

Since ball mills of more than 4 m diameter encounter serious operational and design difficulties, VRMs are being increasingly preferred in large capacity plants. Mills with capacities in the range of 400 to 600 metric tons per hour with drive ratings between 4000 to 5000 kW are in use in the industry. The other alternative is the adoption of a hybrid system combining both a ball mill and roll press for energy conservation. A typical illustration of a ball mill upgraded with a roll press is shown in Table 2.4.

Homogenization process

The post-milling homogenization process involves intimate blending of fine powdery material with a view to bringing down the standard deviation of LSF, SM, and AM or other chemical parameters, if used for controlling the quality of the ground raw mix (often called "raw meal"), prior to feeding the kiln. There are four methods of homogenization: the first and most traditional method is the slurry mixing in the wet process plants; other methods adopted in dry process plants are mechanical, pneumatic, and gravity systems. Since we are focused on dry process plants, we highlight the salient points of the latter three homogenization systems.

Table 2.4 A ball mill system modified with a roll press

Mill systems	Before	After
	Capacity, tph	
	198	214
	Specific energy, kWh/t	
Ball mill	13.2	8.3
Roller press	–	2.5
Total	13.2	10.8

The mechanical system consists of multiple storage silos, each of which is provided with regulated withdrawal facilities. Blending is achieved by orderly withdrawal rates from all silos. While this type of mixing consumes lower power, the system requires a great deal of material handling that increases power consumption. In addition, the required number of silos is obviously more than in other systems. As a result of these shortcomings, this kind of system is not in wide use.

The most common homogenization system is the pneumatic one, which is based on an air-fluidization method. The fluidization of the dry raw meal is a very innovative approach towards making it behave like raw-meal slurry in a mixer basin of traditional wet plants. Air introduced through a permeable medium in the silo bottom causes the raw meal to behave almost like a fluid. This method is known to be efficient but relatively more power-consuming.

The pneumatic systems can either be discontinuous (or batch type) or continuous. The batch-type blending system was once a standard feature of the dry-process plants of earlier design. Today, the most common practice is to have pre-blending beds followed by the continuous homogenization system.

A typical batch-type blending system is made up of a two-storied silo, with the homogenizing silo installed on top of the storage silo (Figure 2.10). The permeable silo bottom is divided into multiple segments such as quadrants or octants, which are earmarked alternately to introduce air flow with the help of compressors at high velocity and pressure for homogenization and at lower velocity and pressure for aeration and extraction of homogenized raw meal. The air flow through the homogenizing sectors creates active upward-moving, low-density columns of raw meal that spill into denser downward-moving materials located over the aeration sectors. In this process, the blending factor of batch-type homogenizing silos may reach a level of 1:20. The method is

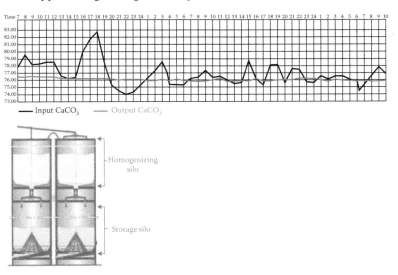

FIGURE 2.10 Typical batch blending silo system.

very effective but the energy consumption for the compressed air is quite high. As a result, this mode of blending is preferred for special products needing high blending factors for quality purposes.

In continuous homogenizing systems, the same silo performs both the storing and blending functions. Different designs are available (7). The incoming material is deposited layer by layer in the silo. While discharging, the layers are disrupted by downward funneling of the material and get intermixed. The silos are emptied via central chambers, which may be located inside or outside the silos, depending on the equipment suppliers and their design features. Multiple continuous homogenization silos can be connected in parallel, into which the material flow can be divided, while feeding, and the discharge flows can be recombined before further downstream movement. The operational experience indicates that continuous homogenization with one silo may have the blending factor in the range of 5–7, and this improves to 7–12 with two silos.

Despite the wide adoption of pneumatic blending systems, high energy consumption (1.0–1.5 kWh/t of material), maintenance problems of the porous media, and the need to install oil-free compressors have been the basic concerns for this technology. Solutions to these problems have been found in the innovation of gravity for blending, which is possible even in one single silo. Energy consumption reduces to 0.2 kWh/t of material or even lower. Homogenization is achieved by keeping the entire raw meal in the silo in a constant sinking movement but at controlled and different flow velocities over the cross-section of the silo. The raw meal leaving the raw mill during the course of one hour is retained inside the silo for different lengths of time with prolonged residence time, mixed with raw meal produced during the subsequent hours

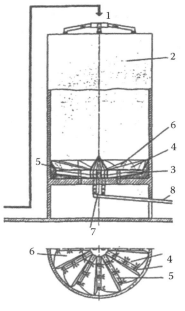

FIGURE 2.11 A continuous blending silo.

(Figure 2.11). In the overall system design the raw meal leaving multiple outlets is mixed in a mixing vessel below the silo and mounted on load cells, and it forms a part of the kiln feed. The silo is typically designed for a capacity corresponding to consumption of raw meal for three days.

For effective homogenization, therefore, various options are available and the specific process requirements can be easily met.

2.5 Burnability features of raw meal

Product fineness

There are no rigid standards for raw meal fineness. It is determined empirically and should be as coarse as a given kiln system can tolerate. Typically, a raw meal is ground to about 15% residue on an 88-μm sieve, and correspondingly to 1.5–2.5% residue on a 212-μm sieve. But with improved burning systems and techniques, the residue for raw meal can be raised to about 25–30% on an 88-μm sieve and 6% on a 212-μm sieve in some plants operating with large preheater-precalciner kilns. It should be borne in mind that a narrow particle size distribution is optimally required as fines tend to increase dust loss by entrainment in exhaust gases, while coarse particles are harder to react in the kiln, resulting in high free lime or high fuel consumption.

From the considerations of easy burning, it is highly beneficial to limit the top sizes of the following mineral phases in the raw meal as indicated below:

- Silica minerals (e.g., quartz, chert, acid insoluble residue, etc.): 44 μm
- Shale particles: 50 μm
- Silicate minerals (e.g., feldspar): 63 μm
- Carbonate minerals (e.g., calcite, dolomite): 125 μm

It has often been experienced in practice that every 1% increase in coarser quartz particles and every 1% increase in carbonate particles results in 0.93% and 0.51% increase in free lime, respectively. Therefore, it may be advantageous to characterize the particulate fineness on the 125-μm sieve and then acidify the mix and examine the coarse fractions (+ 44 μm) for their content of grains of concern. This may help to find out the ways and means of tackling hard burning characteristics of raw meals.

Another aspect in this context is worth mentioning. For all practical purposes, it may be desirable to check if the + 88-μm fractions of raw meals are lime-rich or silica-rich, relative to the total sample. For uniform and stable burning the raw meals should be ensured to –

a. be consistently either lime-rich or silica-rich
b. have coarse mineral phases in the total raw meal distributed as follows:
   i. Silica phases (> 200 μm): not more than 0.5%
   ii. Silica phases (90–200 μm): not more than 1.0%
   iii. Silica phases (> 45 μm): not more than 2.0%
   iv. Carbonate phases (> 125 μm): not more than 5%

Chemical and mineral characteristics

It has already been mentioned that the three modulus values, e.g., lime saturation factor (LSF), silica modulus (SM), or alumina modulus (AM), are fundamentally considered to ascertain the expected burning behavior of a kiln feed. From the extensive practices in the cement plants the following ranges are considered optimal:

LSF: 1.0 ± 0.05

SM: 2.4 ± 0.2

AM: 1.4 ± 0.2

Values of individual modulus beyond the above desirable ranges are not uncommon, but are relatively rare. Since the lowering of LSF directly reflects on the reduction of the potential C_3S in clinker, in many situations, particularly while dealing with large-diameter rotary kilns, attempts are made to make the kiln feeds soft-burning by bringing down SM to 2.0 and AM to 1.0.

In addition to the above chemical characteristics, the mineral phases present in kiln feeds also influence their burning behavior in some way. The more influencing features are:

- Presence of ankerite (Ca(Mg, Fe) $(CO_3)_2$) or dolomite (Ca, Mg $(CO_3)_2$) causes an early release of lime in a reactive state, which may require reactive clays to enter into combination reactions.
- Presence of magnesia in the silicate form ensures the formation of small periclase crystals in clinker. Bigger crystals of periclase are likely to be present, when magnesia comes from the carbonate phase.
- Different forms of silica show different levels of reactivity, as shown below:

    β - quartz < chalcedony < opal

    < α - cristobalite & α - tridymite

    < silica from silicates < amorphous silica

- Ferruginous minerals like chlorite, glauconite, etc., release iron oxide in the ferrous state at early stages to give mineralizing effects.
- Alkali- and sulfate-bearing minerals result in the formation of volatile cycles during the burning process.

Effect of minor constituents

The more commonly encountered minor constituents have been specified in the cement-grade limestone (see Table 1.10 in Chapter 1) but the raw materials are known to contain many other minor constituents. Despite many investigations in various laboratories on the effects of minor constituents on clinker making processes and clinker properties, the results are often conflicting. Nevertheless, one might have some idea from the author's chapter in (3). As a precautionary measure, it is always advisable to conduct special investigations, if a raw mix is found to contain any unspecified minor constituent in excess of 0.5%. Some of the minor constituents in small amounts may be beneficial in clinker making but the same elements may turn harmful beyond a ceiling. Elements like barium and zinc are examples in this respect as they behave differently when present at different percentages.

Raw meal homogeneity

One of the parameters that bear a strong effect on the stability of the clinker burning operation is the homogeneity of the kiln feed.

In actual plant practice the homogeneity is determined on the basis of n numbers of hourly spot samples (normally 24 samples) and the targets of standard deviations (S) for the measured parameters of this set of samples are fixed as follows:

$S_{LSF} \leq 1\%$

$S\ CaCO_3 \leq 0.2\%$

$S_{SM} \leq 0.1$

In some plants, the measure of homogeneity is carried out on the basis of four or eight hourly samples. These samples are analyzed for major oxides and these data are converted to potential C_3S or LSF. A kiln feed should typically have an estimated standard deviation of less than 3% C_3S or 1.2% LSF.

Although the homogeneity of kiln feed is expressed in terms of standard deviations of certain parameters, it should be borne in mind that it has a limitation in the sense that the standard deviation of any parameter does not distinguish between a steady trend and constant fluctuation. Hence, the interpretation of data is crucial.

Reactivity and burnability

The reactivity of a raw mix is defined by the overall rate of chemical reactions of the constituents on burning at a certain temperature for a certain time. Generally, it is influenced by the chemical, mineral, and particulate characteristics of raw mix components. But in practice this parameter has not been of much significance as many other parameters come into play in the burning operation. Hence, another parameter known as "burnability" has evolved, which, to some extent, reflects the burning behavior of a raw meal. Although "burnability" itself is a qualitative parameter that signifies the ease or difficulty of burning of a kiln feed, attempts have been made to quantify this parameter by some indirect means—both theoretical and experimental.

Theoretical approaches One of the simplest and most traditional ways of determining the burnability of a kiln feed is to apply the ratio of its potential phase composition, known as the Kuel's Index, introduced back in 1929: $C_3S/(C_4AF + C_3A)$. The higher the value, the more difficult is the kiln feed to burn. Since this ratio did not take into account some of the fluxing oxides normally present in kiln feed, a modified version was presented in (9):

$$C_3S/(C_4AF + C_3A + MgO + K_2O + Na_2O). \tag{2.36}$$

Since both the above indices did not reflect the burning behavior of kiln feed in significantly differing practical situations, in the early 1970s a new index, called the Burnability Factor, was introduced (10):

$$BF = X(LSF) + Y(SM) - Z(MgO + K_2O + Na_2O) \qquad (2.37)$$

The coefficients in the above formula were determined by multiple regression analysis of results obtained from tests conducted in the laboratory. The originally published formula had the following coefficients: $X = 100$, $Y = 10$, and $Z = 3$. The concept of the Burnablity Factor was more widely accepted, but changes were introduced in the coefficients in some situations. An example of a modified formula is given below (9), although it may not be universally applicable:

$$LSF + 6(SM - 2) - (MgO + K_2O + Na_2O) \qquad (2.38)$$

The Burnability Factor may still be applicable in operating plants if the coefficients are properly worked out from the plant data in a given situation. The Burnability Factor, however, does not recognize the role of the liquid phase or the problems that might be encountered in burning due to the presence of coarse materials in the kiln feed. Another theoretical formula for burnability was developed to take care of these parameters (11):

$$B_{th} = 55.5 + 1.56(KS - 90)^2 - 0.43 SA^2 + 11.9 H_{g50} \qquad (2.39)$$

Where B_{th} is the time at 1350°C required to bring down free lime to 2% by weight, KS is the Kalkstandard or its LSF equivalent, SA is the weight % of liquid, and H_{g50} is the weight % residue coarser than 90 μm. The above equation shows that increasing melt content reduces the time required to obtain fixed free lime content in clinker.

A similar attempt was made to predict the weight percentage of free lime that would be expected to remain in clinker after the corresponding raw mix nodules had been fired at 1400°C for 30 min as a measure of its burnability without carrying out the actual experiment (12). The formula developed is given below:

$$CaO = 0.33 LSF - 0.349 + 1.18 SR + 0.56 K_{125} + 0.93 Q_{45} \qquad (2.40)$$

where SR is the same as SM, K_{125} is the weight percentage of calcite grains of + 125-μm size and Q_{45} is the weight percentage of quartz grains of + 45-μm size in the kiln feed. The above equation is justified for moderate variations from the reference sample but for larger variations a non-linear expression will be needed to fit data.

Experimental approaches All experimental approaches for the determination of burnability involve calcination of raw meal or making of clinker in the laboratory under specified laboratory conditions and measuring the degree of reaction achieved. The free lime content in clinker is the most common criterion followed in all experiments, although the methods and procedures of conducting the tests differ. Some alternative methods have also been tried. All the test methods are essentially relative to reference samples and the basic assumption is that they take care of all the known and unknown factors influencing the process of clinker making. Some of these methods are presented below.

a. Test method based on lime combinability: the raw meal, which may include the theoretically computed proportion of coal ash, is first wetted and then converted to 8–10-mm nodules by hand. The nodules are dried and then calcined at 950°C. The calcined nodules are isothermally treated: one part at 1350°C, another at 1400°C, and a third at 1450°C—all for 20 min. After cooling the fired nodules to room temperature, free lime is determined for each part. The trend of free lime reduction with temperature is graphically plotted. The temperature at which 1% free lime is extrapolated is called the "lime combinability temperature." The higher this temperature, the harder the burnability of the raw meal. A typical temperature-free lime curve is shown in Figure 2.12. A few other variants of this method have been reported (13,14) but they are not in practice as they are either too simple or too complex to adopt.

b. Test method based on firing shrinkage: another approach for assessing burnability is by determining the volume change of raw

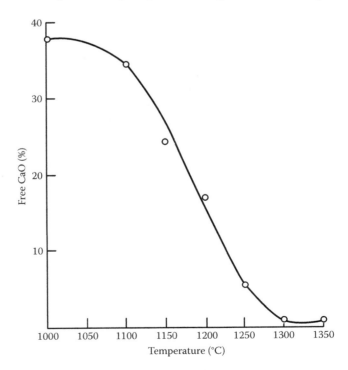

FIGURE 2.12 Free lime temperature relation for determining burnability.

mix pellets fired at different temperatures, developed in Russia (15). The steps involved are:

- Preparation of homogenized raw mix or kiln feed, including the likely incorporation of coal ash, maintaining the fineness as obtained in the plants or as desired.
- Preparation of pellets of 3.3Ø × 1.5 cm size with pressure of 500 kg/cm².
- Firing of pellets at temperatures ranging from 1000°C to 1450°C.
- Removal of pellets at intervals of 100°C in the above temperature range, after a soaking of 30 min and immediate measurement of their firing shrinkage/expansion.
- Plotting of pellet volume change against temperature, as shown in Figure 2.13.
- Interpretation of the plots.

The plots are characteristic of the raw mixes/kiln feeds, as no two patterns are exactly alike. The initial expansion of the fired pellets indicates the solid-state reactions forming the intermediate phases; the subsequent trend of shrinkage shows the appearance and quantity of the liquid phase; and finally the compressive strength of the pellets at 1400°C throws light on the expected granulometry of the clinker.

2.6 Use of mineralizers

In clinker burning the use of Al_2O_3 and Fe_2O_3 as fluxes is universally practiced. The fluxes are generally considered as materials that lower the temperature of liquid formation, while mineralizers accelerate the kinetics of reactions by modifying the solid and liquid state sintering. Fluorides, alkali or calcium fluorosilicates, alkali sulfates, magnesia, etc. often perform as mineralizers in clinker burning. The proportion of

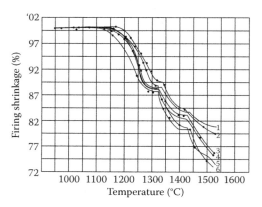

FIGURE 2.13 Volume shrinkage temperature relation for determining burnability.

mineralizers in raw mixes is generally very small and their compatibility and quantity for addition are determined through laboratory experiments before undertaking plant trials. The use of mineralizers results in energy saving due to the following factors:

- Lowering of burning zone temperature;
- Increase in reaction rates including calcination;
- Lowering of exhaust gas temperature;
- Reduction in dust losses;
- Reduction in exit gas volume;
- Lowering of kiln-shell losses due to both lower burning zone temperature, improved formation of coating, and increase in kiln throughput;
- Improvement in clinker grinding due to betterment of clinker granulometry as well as phase composition;
- Stability of kiln operation leading to minimum fluctuation in kiln output levels;
- Increase in mill output due to the coarser grinding option, taking advantage of improved reactivity of raw mixes.

However, the above gains may, to some extent, get diluted by lower secondary air temperature in kiln systems. It is interesting to note that in static firing the saving of heat input on account of bringing down the material temperature by 100°C from about 1450°C does not exceed 35 kcal/kg of fired product. In pulverized fuel-fired rotary kilns, where the mode of heat transfer is through combustion gases, the saving of heat varies from system to system.

2.7 Summary

Limestone and aluminum silicate materials are the primary constituents of a raw mix. Sometimes other corrective materials like bauxite, laterite, and iron ore are used in small proportions. The stoichiometric requirements of the four major oxides, viz., CaO, Al_2O_3, Fe_2O_3, and SiO_2, are determined with the help of three modulus values known as the lime saturation factor (LSF), silica modulus (SM), and alumina modulus (AM). The four major clinker phases as well as the melt phase that is likely to form in the kiln system can be computed with the help of Bogue equations, which are essentially empirical in nature.

Based on the oxide compositions of limestone and other raw materials, raw mixes are calculated using the modulus values and Bogue equations. This process is known as the raw mix design. The raw mix design becomes more and more complicated with increases in the number of components required for the target composition. The proportions of raw materials are calculated by setting up and solving simultaneous equations; to fix n parameters n + 1 raw materials of appropriate composition are required. The calculations can be manually carried out by trail and error or by step-wise computation. The approach can be faster and more convenient by adopting simple computer programming.

Several unit operations, such as crushing, pre-blending, grinding, and homogenizing are involved in the preparation of raw mixes and their burning behavior is defined by such features as product fineness, top particle size of the siliceous and carbonate phases, chemical composition including minor constituents, modulus values, degree of homogeneity, etc.

The practical ease of burning a raw mix is called its "burnability," which, in scientific connotation is different from "reactivity." Burnability of a given raw mix can be evaluated both theoretically and experimentally, as elaborated in this chapter. There are two different experimental procedures—one, to determine the unreacted free lime in a raw mix after it has been fired in a laboratory furnace under given conditions; and, two, by determining the firing shrinkage of the fired raw mix pellets.

Burning of a raw mix can be improved by using a mineralizer, which is different from a fluxing oxide in having the capacity to accelerate the phase-forming reactions at lower temperatures. The efficacy of mineralizers can be evaluated experimentally prior to their selection and use.

References

1. H. F. W. TAYLOR, *Cement Chemistry*, Academic Press, London (1990).
2. L. OPOCZKY, L. SAS, and F. D. TAMAS, Preparation and quality assurance of raw meals, in *Modernization and Technology Upgradation in Cement Plants* (Eds S. N. Ghosh and Kamal Kumar), Akademia Press International, New Delhi (1999).
3. A. K. CHATTERJEE, Chemico-physico-mineralogical characteristics of raw materials of Portland cement, in *Advances in Cement Technology* (Ed. S. N. Ghosh), Pergamon Press Ltd., Oxford (1983).
4. R. M. HERATH BANDA AND F. P. GLASSER, Role of iron and aluminum oxides as fluxes during the burning of Portland cement, *Cement and Concrete Research*, Vol. 8 (1978).
5. N. H. CHRISTENSEN AND V. JOHANSEN, Role of liquid phase and mineralisers, in *Cement Production and Use* (Ed. J. Skalny), Engineering Foundation, New York, Vol. 8, No. 6 (1979).
6. M. A. SWAYZE, System $CaO - 5\,CaO.3\,Al_2O_3 - 2\,CaO.SiO_2$ modified by 5% MgO, *American Journal of Science*, 244, 70 (1946).
7. A. K. CHATTERJEE, Materials preparation and raw milling, in *Innovations in Portland Cement Manufacturing*, Portland Cement Association, Skokie, Illinois, USA (2004).
8. Pennsylvania Crusher Corporation, *The Handbook of Crushing*, Bulletin 4050, USA (2000).
9. H. N. BANERJEA, *Technology of Portland Cement and Blended Cements*, ACC, Mumbai (1980).
10. K. E. PERAY AND J. WADDEL, *The Rotary Cement Kiln*, Chemical Publishing Co. Inc., New York (1972).
11. U. LUDWIG AND G. RUCKENSTEINER, *Einflusse auf die Brennbarkeit von Zementrohmehlen*, Westdeutscher Verlag, Opladen (1973).
12. E. FUNDAL, The burnability of cement raw mixes, *World Cement Technology*, Vol. 42, July/August (1979).

13. G. R. GOUDA, The cement raw materials, in *Proceedings of All India Seminar on Cement Manufacture,* Vol. IV, New Delhi (1981).
14. R. BLAISE, N. MUSIKAS, AND H. TIEDREZ, *Rev. Mater. Contr.* No. 674–675, 287 (1970).
15. V. I. SHUBIN, *Refractory Lining of Cement Rotary Kilns* (in Russian), Stroiizdat, Moscow (1975).

CHAPTER THREE

Fuels commonly in use for clinker production

3.1 Preamble

A fuel is defined as any combustible substance containing carbon as the main constituent that, on proper burning, produces heat that can be used economically for domestic and industrial purposes and in generation of power. Fuels are broadly classified into two types depending upon their genesis and occurrence:

- Primary fuels: these are the naturally occurring fuels found freely on the Earth's crust. They are further classified as solid fuels such as wood, peat, lignite, and coal; liquid fuels, viz., crude oil and petroleum; and gaseous fuels such as natural gas.
- Secondary fuels: these are derived from the primary fuels through industrial production. They can also be classified as solid, liquid, and gaseous. Examples of solid secondary fuels include coke, charcoal, petroleum coke, and so on. Gasoline, diesel, kerosene, LPG, and similar products come under the category of secondary liquid fuels, while coal gas, water gas, biogas, and similar products are known as secondary gaseous fuels.

Coal, oil, and natural gas are common primary fuels used globally for clinker production. Of the secondary fuels, petroleum coke (abbreviated as "petcoke") is also widely used in the cement industry. Price and availability are determining factors in the choice of fuel. Due to economic considerations, coal, lignite, and petcoke have become the most widely used fuels in the industry. This chapter will essentially focus on these fuels.

3.2 Characteristics of fuels

Before discussing the properties of specific fuels, it would be pertinent to recall the broad characteristics that make a fuel good for use. The following properties are important from the perspective of proper combustion:

- high calorific value
- low moisture content
- low content of noncombustible matter including ash
- moderate ignition temperature without causing problems of self-ignition in storage on one side and easy combustion in air on the other
- minimum pollution on combustion

Calorific value of a fuel

From the above points, it is evident that calorific value is the prime requirement of a fuel. It is defined as the amount of heat obtained by the complete combustion of a unit mass of the fuel. The calorific value is presented in two ways:

a. Gross calorific value (GCV): the total amount of heat generated when a unit mass of fuel is completely burnt and the products of combustion are condensed to 15°C or 288°K. When a hydrogen-bearing fuel is burnt, the hydrogen present produces steam. On condensation, its latent heat also gets included in the measured heat.

b. Net calorific value (NCV): the net heat produced when one unit mass of fuel is completely burnt and the products are allowed to escape.

Thus, NCV is obtained by deducting the latent heat of condensation of the water vapor produced from GCV. Since one part of hydrogen gives nine parts of water and since the latent heat of steam is 587 cal/g of water vapor produced,

$$NCV = GCV - 9 \times (H/100) \times 587 = GCV - 0.09 \times H \times 587 \qquad (3.1)$$

Calorific values are expressed in kilocalories per kilogram (kcal/kg) for solid and liquid fuels and in kilocalories per cubic meter (kcal/m^3) for gaseous fuels. The calorific value of solid and liquid fuels is determined with the help of a bomb calorimeter, following the standard procedure. The calorific value of gaseous fuels is mostly calculated from their volumetric combustible constituents, based on their known individual calorific values.

Ranks and properties of coal

Coal, as we all know, is the most important solid fuel, formed by the combined action of high temperature and high pressure on vegetable matter over geologic periods of time. In the process, the vegetable matter, like wood, passes through various stages of coalification and the solid fuels in different stages of formation are known as peat, lignite, sub-bituminous coal, bituminous coal, and anthracite. During this prolonged transformation process the proportion of hydrogen and oxygen decreases while that of carbon increases. Since peat is the first stage of coalification, it is mostly immature with high moisture and low carbon content and is not considered an economic fuel. The other categories are in industrial use in varying quantities in different regions, depending on their availability. The tentative composition and characteristics of different ranks of coal are compared in Table 3.1.

Lignite is not considered a good fuel, although it has better properties than peat. Sub-bituminous and bituminous coal are easy to handle and have good heating properties. They are the most widely used solid fuels in the world and are used both for domestic and industrial purposes. Anthracite is the highest-ranking coal and contains a maximum percentage of carbon and the lowest amount of volatile matter. It is regarded as a special-purpose solid fuel suitable only for certain applications.

For evaluation purposes, coal is analyzed in two steps:

a. proximate analysis (H_2O + volatiles + fixed C* + ash = 100%)

b. ultimate analysis (C + H + N + S + O* + ash = 100%)

(*Not normally determined, obtained by difference.)

There are standard analytical procedures for carrying out the above analyses. The proximate analysis gives quick and valuable information regarding commercial classification and suitability for a particular industrial use, while the ultimate analysis gives an accurate elemental composition of a coal sample. The calorific values are generally computed from the composition as follows:

$$GCV = 80.8C + 22.45S + 339.4H - 35.90 \tag{3.2}$$

$$NCV = 80.8C + 22.45S + 287(H-O/8) - 6W \tag{3.3}$$

where W is water content in percentage

$$gross - net = 5150H \tag{3.4}$$

where H* is total H_2 including water

Table 3.1 Analysis of different ranks of coal

Type	As received and free of mineral matter (%)			Gross calorific value (kcal/kg)
	H_2O	Volatiles	Fixed C	
Lignite	45	25	30	4,000
Sub-bituminous	25	35	40	5,400
Bituminous				
High Volatiles	5	45	50	7,200
Low Volatiles	5	20	75	8,500
Anthracite	3	5	92	8,000

As already explained, the calorific value depends upon the carbon content and, therefore, the greater the percentage of carbon, the higher the quality of coal. Nitrogen has no calorific value as it is inert in nature. The presence of sulfur is unwanted as it produces acid, which corrodes the equipment and also causes atmospheric pollution. Hydrogen is present in combination with oxygen as water and lowers the calorific value of the fuel. Ash, being a noncombustible matter, reduces the calorific value of coal. It also causes hindrance in the flow of heat. It forms cinders in boilers and may cause problems of disposal. The presence of excess volatile matter results in incomplete combustion. Coal with high volatile matter gives rise to long flames and relatively low heating values. Finally, the presence of moisture in coal samples reduces its calorific value and a considerable amount of heat is wasted in evaporating the moisture during combustion. Hence, a high percentage of moisture is undesirable.

The ash content in bituminous coal of good quality is normally in the range of 5–15% of air-dried weight, but the ash composition may vary widely according to the mineral contamination in the coal bed. However, in some parts of the world, such as India, the ash content in coal may be as high as 45–50%. The sulfur content in coal varies from 0.2–7% and comprises mainly organic sulfur and pyrites with traces of sulfate. Pyritic sulfur can to some extent be removed by standard washing techniques, while other forms of sulfur may have to be dealt with in the process of application.

In selecting coal there has to be specific attention to properties like calorific value, surface moisture, inherent moisture, and ash content. In addition, it is important to know the composition of ash and its fusion temperature. In clinker production, the composition and fusion temperature of ash are relevant as the ash available forms a part of the total reactions taking place inside the kiln, modifying the clinker composition to some extent.

It may be relevant to mention here that there is another parameter of coal combustion, called "calorific intensity," which may sometimes be important for evaluating the coal combustion process. It is defined as the maximum temperature reached when coal is completely burnt in the theoretical amount of air. It depends upon: (i) quantity, (ii) nature, and (iii) specific heat of the gaseous products of coal combustion. In most cases, the heat liberated due to combustion preheats the air and thus affects the calorific intensity. Calorific intensity, reflecting the flame temperature, can be calculated theoretically as follows

$$\text{flame temperature} = (\text{heat of combustion} + \text{sensible heat of air}) / \Sigma(\text{combustion products} \times \text{specific heat}) \quad (3.5)$$

The composition of coal varies widely and hence it is important to analyze it and interpret the results for classification, economic justification, and proper industrial application.

Another property of coal and other solid fuels is their grinding behavior, which is determined by an apparatus called a Hardgrove grindability tester, which is a small ball bearing mill used to grind 50 g of material. The feed size of the material is kept in the range of 590–119 μm and the product size after 60 revolutions of the mill is measured by the residue on a 74-μm sieve. The Hardgrove grindability index (HGI) is calculated as (13 + 0.93 × quantity of material (g) passing through the sieve). The lower the index, the harder it is to grind the material. For coal, the index may vary widely, from 40 to 70.

Types and properties of liquid fuels

Although less in use today in the firing of cement kilns, liquid fuels cover a wide range and are important in many industrial sectors. Examples of liquid fuels include petroleum, petroleum products, tar, alcohols, and colloidal fuels. Since petroleum and its derivatives are used in some countries for burning clinker, the salient features of these fuels are discussed below.

Petroleum, as we all now, occurs at various depths below the surface of the Earth and is often associated with large quantities of natural gas and some water. When water and natural gas are removed, the remaining liquid is called crude oil. The approximate composition of petroleum is: carbon 79–87%, hydrogen 11–15%, oxygen 0.1–0.9%, nitrogen 0.4–0.9%, and sulfur 0.1–3.5%. Crude oil, after the removal of dirt, water, sulfur, and other impurities, is subjected to a process of fractional distillation. This process breaks the crude oil into various fractions with different boiling points. Each part of the process attempts to impart specific properties in them. The whole process is better known as the refining of petroleum. The three most important liquid fuels obtained from crude oil are:

- petrol, consisting of a mixture of hydrocarbons ranging from C_5 to C_9, having calorific value of about 11,250 kcal/kg. The approximate overall elemental composition is 84% C and 15% H, with N + S + O making up less than 1%
- kerosene oil, consisting of hydrocarbons ranging from C_{11} to C_{16}, having calorific value of about 11,100 kcal/kg. The overall elemental composition is close to that of petrol
- diesel oil, consisting of hydrocarbons ranging from C_{10} to C_{18}, having calorific value of about 11,000 kcal/kg

Of the above three derivatives, diesel oil, also known as fuel oil, is used for combustion purposes. As per ASTM D 396, fuel oil is classified into six types on the basis of its properties, as shown in Table 3.2.

"Flash point" is an important property of a fuel oil parameter as it defines ignition conditions and storage requirements. It refers to the lowest temperature at which a fuel oil can start a fire, if it is exposed to a source of ignition, due to its volatility. Another parameter of significance is viscosity, which is a measure of internal resistance to flow and is dependent on temperature. Viscosity decreases with an increase in temperature. For liquid fuels, the minimum and maximum values of viscosity at

Table 3.2 Fuel oil characteristics

	Properties			Composition				
Type	Flash point (°C)	Specific gravity	API gravity*	C	H	N	S	O
#1	38	0.850	41.5	86.4	13.6	0.003	0.09	0.01
#2	38	0.876	33.0	87.3	12.6	0.006	0.22	0.04
#4	55		23.2	86.5	11.6	0.24	1.35	0.27
#5 light	55							
#5 heavy	55							
#6 low-S	60	0.986	12.6	87.3	10.5	0.28	0.84	0.64
#6 high-S	60		15.5	84.7	11.0	0.18	3.97	0.38

*API gravity = (141.5/specific gravity) − 131.5.

Table 3.3 Viscosity of fuel oils

	Viscosity (mm^2/s, at 40°C)		Viscosity (mm^2/s, at 100°C)	
Type	Min	Max	Min	Max
#1	1.3	1.4	–	–
#2	1.9	3.4	–	–
#4	5.5	24.0	–	–
#6	–	–	15	50

40°C are generally specified. For heavier and high-sulfur oil, the viscosity values at 100°C become important. Some illustrative values are given in Table 3.3 (1).

The third important parameter of any fuel oil is its pour point, defined as the lowest temperature at which it will pour or flow when cooled under prescribed conditions. It is an indication of the lowest temperature at which a fuel oil can be readily pumped. A few other parameters, such as ash, sediment, water, and sulfur content, are also specified for the supplies, as shown in Table 3.4 (2).

Usually, the heavier the fuel oil, the lower the price when compared on the basis of a unit of energy. Inclusive of preheating and other preparatory costs, fuel oil #6 appears more economical and, hence, has become the preferred fuel for the cement industry (1). However, sulfur content in this fuel oil may be as high as 5%. Hence, the refining process must address the emission of SO_2. Fuel oil #6 also has higher ash content—up to 0.08%—but this is lower than ash generated in coal-fired kilns and, hence, may not be a factor of concern. Nevertheless, economics-permitting, operating plants would always prefer a cleaner, easily pourable, and readily pumpable fuel oil.

Table 3.4 Comparison of the typical specifications of fuel oils in India

Properties	Furnace oil	Low sulfur heavy stock (LSHS)	Light diesel oil (LDO)
Density (g/cm³ at 15°C)	0.89–0.93	0.88–0.98	0.85–0.87
Flash point (°C)	66	93	66
Pour point (°C)	20	72	18
GCV (kcal/kg)	10,500	10,600	10,700
Sediment (wt %), max	0.25	0.25	0.10
Sulfur (wt %) max	4.0	0.5	1.8
Water (vol %), max	1.0	1.0	0.15
Ash (wt %), max	0.1	0.1	0.02

Properties of natural gas and synthetically produced gaseous fuels

Natural gas is the primary gaseous fuel obtained from underground strata either as such or associated with crude oil. When it occurs along with petroleum in oil wells, it is called "wet gas" and contains appreciable amounts of propane, butane, and other liquid hydrocarbons like pentanes, hexanes, etc. The wet gas is suitably treated to remove these hydrocarbons, which are used as LPG. When the gas is associated with crude oil, it is called "dry gas" and consists entirely of methane and ethane along with small amounts of impurities such as CO_2, CO, H_2S, N_2, and inert gases. The approximate composition of natural gas is methane (CH_4) (70–90%), ethane (C_2H_6) (5–10%), and hydrogen (H_2) (3%). Besides these, CO and CO_2 occur in small amounts. The calorific value varies from 12,000 to 14,000 kcal/m³. Natural gas is used both as domestic and industrial fuel, and its calorific value may be theoretically calculated from its composition as follows:

$$\text{gross calorific value (kcal/m}^3) = 90.3CH_4 + 159.2C_2H_6 + 229C_3H_8$$
$$+ 301.9C_4H_{10} + 373.8C_5H_{12} + 57.6H_2S$$

(3.6)

Synthetic gaseous fuels include producer gas, water gas, coal gas, and oil gas. Producer gas is prepared by passing a mixture of air and steam through red-hot coal or coke maintained at about 1100°C in a special reactor called a "gas producer." A somewhat different manufacturing process is adopted for water gas. Air is passed through the coke bed at about 1000°C; the outgoing gases CO_2 and N_2 are used to preheat the incoming air and steam, which is blown through the red-hot coke to produce CO and H_2. Coal gas is obtained by heating good-quality coal in the absence of air to a high temperature at a pressure somewhat higher than the atmospheric pressure; this is done by feeding coal into a large vertical silica retort heated externally to about 1350°C. Oil gas

Table 3.5 Comparative features of synthetic gases

Type of gas	Approximate composition	Calorific value (kcal/m³)	Remarks on application
Producer gas	CO 30% + H_2 10% + N_2 55% + CO_2 5%	900–1300	Cheap, clean, and insoluble in water, poisonous in nature, used in various process furnaces
Water gas	CO 43% + hydrogen 42% + CO_2 4% + N_2 4%	2580–2670	Burns with blue short flame and used for heating purposes in industrial operations
Coal gas	CH_4 32% + H_2 40% + ethylene 3% + other hydrocarbons 4% + CO 7% + acetylene 2% + CO_2 1% + N_2 4%	~ 4900	Colorless gas with characteristic odor, lighter than air, burns with a non-smoky flame; used for heating and providing reducing atmosphere
Oil gas	CH_4 25–30% + CO 10–15% + H_2 50-55% + CO_2 3%	4500–5400	Burns with a smoky flame and characteristic odor; used mainly as a laboratory gas

is obtained by cracking kerosene oil in an iron retort enclosed in a coal-fired furnace. The composition and properties of these synthetic gases are compared in Table 3.5 (3).

Synthetic gases have not been in much use in the cement plants, although they hold some promise for providing cleaner technology, as discussed in a later chapter.

After the oil crisis of the 1970s, the worldwide cement industry predominantly switched over to coal as fuel. However, natural gas has in recent years gained some popularity due to its economic viability in certain situations.

Comparative behavior of the three fuels

All three fuels require processing before use but the process and effects differ significantly. For example, oil may require preheating to reduce viscosity and is fired with a nozzle pressure of about 20 kg/cm². Gas is usually received in a pressurized state at 10–70 kg/cm². It is injected as an axial or as a combination of axial and swirl flow at a velocity of 3–10 kg/cm² with very high tip velocity (300–400 m/s). Gas requires turbulent diffusion and its heat flux tends to be released more slowly than from coal or oil. Peak heat release usually takes place at about 20m inside the kiln, while it occurs at 5–10 m for oil. It also requires higher ignition temperature than low-ash coal or oil. Coal is more prone to composition variations than oil or gas. Coal is usually dried and fine-ground before firing. Coal is injected with carrying air at a pressure of 120–150 g/cm² and tip velocity of 60/80 m/s. It should also be noted that the gas flame has the lowest emissivity. The kiln productivity typically increases by 2–3% when gas is replaced by coal. A broad comparison of the heating characteristics of the three fuels is given in Table 3.6 (4).

Table 3.6 Comparison of the typical thermal characteristics of coal, oil, and gas

Sl. No.	Fuels	Gross CV (kcal/kg)	Net CV (kcal/kg)	*Flame temperature (°C)	*Combustion gas (Nm³/ million cal)	*Total exhaust gas (Nm³/t clinker)
1.	Low-ash coal	5500–7100	5400–7000	2250	1.23	1360
2.	Oil	10200	9700	2350	1.31	1420
3.	Gas	6200	5600	2400	1.45	1550

* Assumptions: specific heat consumption of 850 kcal/kg clinker and excess air of 2%.

3.3 Coal resources of the world

Coal resources are available in almost every country in the world with recoverable reserves known from around 70 countries. The total coal reserves of the world at the end of 2016 stood at 1139 billion metric tons, as reported in (5). World reserves considered in the aforesaid report are those quantities that can be recovered in future from the known reservoirs under existing economic and operating conditions. The data presented here do not meet the rigid definitions of proved reserves as explained in a previous chapter. It is estimated that known coal reserves are currently sufficient to meet 153 years of global production, roughly three times the reserve-to-production (R/P) ratio for oil and gas. The R/P ratio is expressed as the number of years that the reserves would last at the current rate of production, which is arrived at by dividing the reserves remaining at the end of any year by the production in that year. The regional distribution of world reserves, as extracted from (5), is shown in Table 3.7.

From the table it is evident that the Asia-Pacific region has the major share in the world reserves of coal. Within the region, China, Australia, and India have larger proportions of reserves, as shown in Table 3.8. On the whole, the USA and China remain the first and second holders of coal reserves.

Table 3.7 Regional distribution of coal reserves (billion metric tons)

Regions	Anthracite + bituminous coal	Sub-bituminous coal + lignite	Total	Percentage of world reserves	R/P Ratio (years)
Asia-Pacific	413	117	530	46.5	102
Middle East & Africa	14	0.066	14	1.3	54
Europe & Eurasia	153	169	322	28.3	284
South & Central America	9	5	14	1.2	138
North America	227	32	259	22.8	366
World	816	323	1139	100	153

Table 3.8 Countries with the largest coal reserves in the Asia-Pacific Region (billion metric tons)

Country	Anthracite + bituminous coal	Sub-bituminous coal + lignite	Total	Percentage of world reserves	R/P Ratio (years)
China	230	14	244	21.4	72
Australia	68	76	144	12.7	294
India	89	5	94	8.3	137

Status of coal production

Total global production of coal in the year 2016 was 3651.4 mtoe (million tons of oil equivalent), which in effect registered the largest decline of as much as 6.2%, compared to production in the previous year. A higher decline in production, to the order of 7.9%, was recorded in China in the same year. The average annual growth of world coal production during the decade from 2005 to 2015, however, had been 2.5%, while that of China was 4% (5).

Coal inventory in India

According to authoritative sources in India (6), a total of 308.8 billion metric tons of geological reserves of coal have been recorded until the end of March 2016 up to a maximum depth of 1200 m, which is much larger than what has been considered in Table 3.8. The total reserves have been categorized as follows:

- measured: 138,687.20 million metric tons
- indicated: 139,150.87 million metric tons
- inferred: 31,563.77 million metric tons

The coal reserves in India are primarily of the non-coking type. In fact, of the total known reserves, 88.38% is of the non-coking variety, which is suitable for cement making and other non-metallurgical industry. The coal reserves occur predominantly in the eastern states, and four states, viz., Jharkhand, Odisha, Chattisgarh, and West Bengal, account for 79% of the reserves in the country. As reported by the Ministry of Coal, production was 638.05 million metric tons during the period from April 2015 to March 2016. Unlike the global trend, the growth of coal production in India recorded a growth of 2.5% during this period and the average annual growth of coal production in the last decade from 2005 to 2015 had been 4%. Most of India's non-coking coal is characterized by high ash content but it has other quality parameters, such as low sulfur content (generally below 0.5%), low iron content in ash, low refractory nature of ash, low chloride content, and low trace element concentration. Based on GCV, coal is graded as follows (2):

Grade	GCV (kcal/kg)
A	> 6454
B	6049–6454

(*Continued*)

Grade	GCV (kcal/kg)
C	5597–6049
D	5089–5597
E	4324–5089
F	3865–4324
G	3113–3865

Table 3.9 Comparison of Indian, Indonesian, and South African coal

Parameters	Indian coal	Indonesian coal	South African coal
GCV (kcal/kg)	4000	5500	6000
Moisture (%)	5.98	9.43	8.50
Ash (%)	38.63	13.99	17.10
Volatile matter (%)	20.70	29.79	23.28
Fixed carbon (%)	34.89	46.79	51,93
Carbon by ultimate analysis (%)	41.11	58.96	–
Hydrogen (%)	2.76	4.16	–
Nitrogen (%)	1.22	1.02	–
Sulfur (%)	0.41	0.56	–
Oxygen (%)	9.89	11.88	–

The manufacturing industries including the electricity generation sector are generally supplied with D, E, and F grades. A broad comparison of Indian coal with some foreign coal being used in the country is given in Table 3.9 (3).

Depending on the economics, some of the coastal cement plants in India occasionally make use of imported coal from other countries.

3.4 Basic chemistry and physics of combustion

Combustion engineering is a vast subject by itself and is beyond the scope of the present chapter. What is intended here is to recapitulate certain basics of the combustion process, particularly in the cement rotary kilns, as it involves a group of chemical reactions, facilitated by a set of physical actions. Combustion and heat transfer in rotary kilns have been discussed at length in (7–9), which may be referred to for details.

The combustion chemistry involves the oxidation reaction of the two elements of the hydrocarbon fuel, i.e., carbon and hydrogen. The complete oxidation of carbon is:

$$C + 2O_2 \rightarrow CO_2 + 394 \text{ kJ/mol (94 kcal/mol)} \tag{3.7}$$

In the event of incomplete combustion some carbon will be partially oxidized to carbon monoxide:

$$2C + O_2 \rightarrow 2CO + 221 \text{ kJ/mol (53 kcal/mol)} \tag{3.8}$$

In the production of carbon monoxide, the heat released is only half of the amount released in complete combustion of carbon. It should therefore be borne in mind that carbon monoxide retains considerable combustion energy in inefficient burning systems, apart from being an air pollutant and a poisonous gas when in high concentrations. Carbon monoxide may further be oxidized by the following reversible reaction:

$$2CO + O_2 \leftrightarrow 2CO_2 + 173 \text{ kJ/mol (41 kcal/mol)} \tag{3.9}$$

The combustion rate of dry carbon monoxide is very slow but it accelerates in the presence of H-containing radicals. The complete combustion of hydrogen is shown by the reaction:

$$\begin{aligned} 2H_2 + O_2 \rightarrow\ & 2H_2O + (572 \text{ kJ/mol of condensed water) or} \\ & +(484 \text{ kJ/mol of steam)} \end{aligned} \tag{3.10}$$

The differences in the physical states of water produced as a result of the oxidation of hydrogen are the reason for distinguishing the net and gross calorific values of hydrocarbon fuels, as explained earlier.

None of the above reactions can take place until oxygen in the air is in contact with fuel. When both fuel and air are available, combustion takes place in four stages:

- mixing
- ignition
- chemical reaction
- dispersal of products

In most combustion systems, ignition and chemical reactions are very fast, while the mixing and dispersal of products are slow. Hence, time, temperature, and turbulence are three basic requirements to achieve complete combustion, which are well provided in the rotary kilns. The rate and completeness of combustion is, however, controlled by the rate and achievement of homogeneous fuel/air mixing.

Fuel and air mixing

The ground coal in the kiln burner system is conveyed with the help of primary air, which first mixes very rapidly with the fuel at the nozzle. Further air required for combustion is mixed with the fuel through the process of jet entrainment. This is depicted in Figure 3.1, which shows an unconfined jet (9). A free jet exiting from a nozzle in an unrestricted air medium causes the surrounding fluid to be locally accelerated to the jet velocity, pulled into it, and expanded. The process continues until the

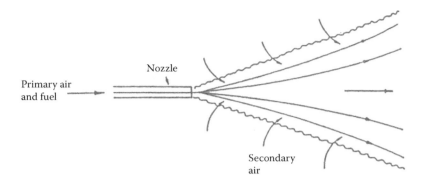

FIGURE 3.1 Entrapment pattern of secondary air into an unconfined jet.

velocity of the jet and that of the surrounding fluid equalize. The greater the momentum of the jet, the higher the entrainment of the surrounding fluid into it. In an unrestricted system, the free jet can entrain as much of its surrounding medium as is necessary to satisfy its entrainment capacity. But in the case of a confined jet, such as in a rotary kiln flame, there are two constraints—one, the quantity of secondary air fed into the kiln, and two, the kiln refractory surface limiting the flame. If the momentum of the jet is more than what is required for complete entrainment of the secondary air stream, the combustion gases inside the kiln will be pulled back to overcome the excess momentum. This phenomenon is known as external recirculation (Figure 3.2) (10). The presence or absence of recirculation has a significant effect on the flame characteristics. A moderate degree of recirculation is a positive indication that the fuel/air mixing is complete. On the contrary, its absence is a clear indication that not all of the secondary air has been entrained into the jet. In such a situation, there is a strong possibility of generation of carbon monoxide. Further, there could be a tendency for the flame to expand and impinge on the refractory brickwork, causing some damage.

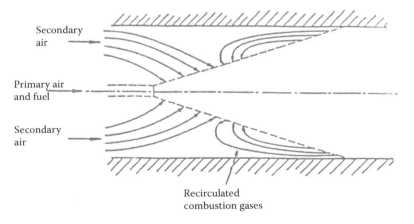

FIGURE 3.2 Recirculation of combustion gases in a confined jet.

Heat transfer

Since the purpose of burning a fuel is to either heat materials or to produce steam, it is imperative that heat is transferred from the flame to the process. If there is a temperature difference between two parts of a system, heat, as we all know, will be transferred by any one or more of the three means, viz., conduction, convection, and radiation. The general equation for heat transfer is:

$$Q = \phi(\Delta t^n) \qquad (3.11)$$

where
  Q = the heat transferred;
  ϕ = kA/x for conduction; hA for convection; and $\sigma\varepsilon A$ for radiation;
  n = 1 for conduction, 1 for convection, and 4 for radiation;
  k = thermal conductivity;
  A = surface area;
  h = convective heat transfer coefficient;
  σ = Stefan-Boltzmann constant;
  ε = emissivity;
  Δt = difference in temperature, and;
  x = separation distance of t.

In the context of rotary kilns, radiation becomes the dominant mechanism of heat transfer and over 95% of heat transfer in the burning zone takes place through this mechanism. The process is more complex and the basic equation for heat transfer gets modified to:

$$Q = A\,\sigma\varepsilon\left(T_F^4 - T_p^4\right) \qquad (3.12)$$

where
  T_F = flame temperature, °K, and;
  T_p = product temperature, °K.

The rate of heat transfer is governed by the temperature, emissivity, and relative geometry of the flame and surroundings.

Flame emissivity

The emissive properties of a flame are a function of the concentrations of the spectrally emissive and absorptive gases such as CO_2, CO, H_2O, etc. from the combustion process and the particulate burden in the flame. The emissivity of particulate matter is a function of the particle size, the radiation from the fuel drops, and the soot formed during combustion. On the whole, the physics and chemistry of emissivity and heat transfer in kilns are quite complex. Simple computation or even physical modeling may not always provide answers to the problems. Hence, the application of mathematical modeling in deciphering heat transfer in a kiln system is becoming increasingly common.

3.5 Coal preparation and firing

The firing of fine coal into the cement kilns is generally done in two ways: direct and indirect firing. In the direct firing system, coal is finely ground and fed directly to the burner, with both the drying and coal-conveying air as primary air (say, 15–30% of the total combustion air) (Figure 3.3). In the indirect firing system, there is an intermediate storage of fine coal as well as separate paths for cleaning and venting of the drying air and the conveying air (Figure 3.4). Within the two broad systems, different variations are possible (see Figure 3.5). It is obvious that the direct firing system requires one mill per burner, whereas an indirect firing system with a single mill can supply fine coal to multiple burners. It is generally believed that the indirect firing system yields higher thermal efficiency as it may operate with lower primary air and also by eliminating the water vapor given out by coal drying from entering the firing system. There is a difference of opinion on these issues and more particularly on the latter point, as technically the water vapor may have a beneficial effect on the flame.

Coal is generally ground in roller mills, ball mills, or Babcock E-mills. Generally, roller mills are designed with integral static classifiers. The feed size for the roller mills is generally maintained at less than 25 mm, with one-third of the crushed coal being below 10 mm. Roller mills can dry coal up to 10% moisture and for achieving the designed capacity coal should not be excessively hard to grind. A Hardgrove grinding index (HGI) of about 55 is often suggested for easy grinding.

So far as the mill operating parameters are concerned, the mill inlet temperature of hot air should not exceed 350°C and the outlet temperature is restricted to 65°C for indirect systems and 80°C for direct systems. Hot air for coal drying can come from the cooler exhaust or preheater exhaust.

Coal handling demands a lot of safety measures. The explosion hazards increase with volatile content and fineness, and decrease with water content and with inert dust diluents like limestone. The presence of pyrite in coal in amounts in excess of 2% enhances the tendency for spontaneous combustion of coal due to oxidation. Stockpiling of such coals requires particular vigilance. Smoldering coal should be dug out, the site spread with limestone dust, and the coal layer compressed. If long-term storage is required, the pile should be compacted and sealed with coal tar emulsion.

The design of a coal firing system including burners is critically important and should take into account

  i. the ignition temperature of air/fuel mix, which may range from 200 to 700°C generally;
  ii. a minimum explosive concentration of fuel in air, which is generally taken as 40 g/m^3;

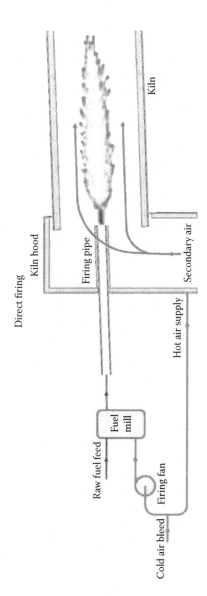

FIGURE 3.3 Direct system of coal firing.

FUELS COMMONLY IN USE FOR CLINKER PRODUCTION

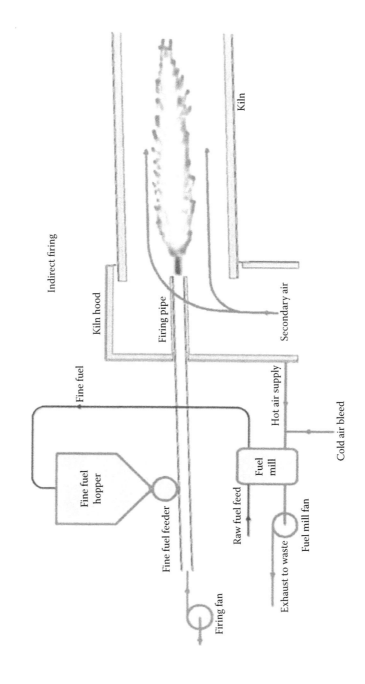

FIGURE 3.4 Indirect system of coal firing.

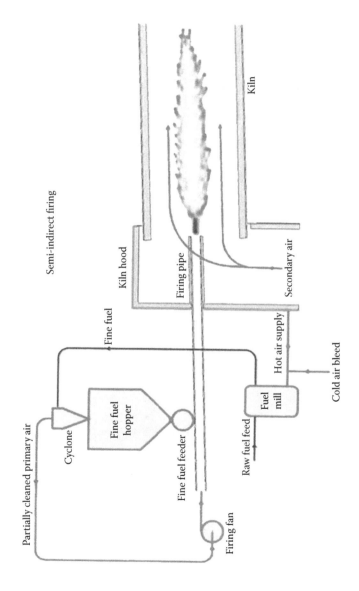

FIGURE 3.5 Semi-direct system of coal firing.

iii. oxygen concentration to prevent ignition (commonly considered as 12%);
iv. conveying air required for pulverized fuel (approximately 2–4%);
v. total estimated primary air (6–8%);
vi. axial velocity required (140–450 m/s);
vii. swirl velocity required (110–200 m/s).

The last four parameters obviously refer to burners for indirect firing systems of different designs.

Effects of coal characteristics on combustion

The characteristics of fine-ground coal as fed to the kiln system play an important role in the efficient combustion of coal. The effects of some of these characteristics are briefly described below.

Moisture content: as mentioned above, the normally crushed coal is pulverized and dried in air-swept or E-type mills. In this preparatory process, moisture content should be reduced to the optimum levels for given quality of coal. Coal exhibits a spontaneous combustion tendency when dried to very low moisture (< 1.0%). A higher moisture content causes a delay in the start of the combustion process and tends to make the flame longer. Furthermore, the water in the coal increases the amount of waste gas to be carried through the process and may thus impair the performance of the kiln system.

Volatile matter: the reactivity of coal increases with the increase in volatile matter. The volatile matter is an indirect indicator of the rate at which the solid carbon reacts to form CO and CO_2 streams. At higher volatile contents, the coal has higher porosity, thus offering a higher surface area for combustion, as a result of which it requires a lower ignition temperature. A coal with volatile matter above 30% decomposes with high speed and promotes fast combustion, generating a short flame; on the other hand, low volatile coal (less than 20%) generates a longer flame in a rotary kiln. In practice, a reactive coal need not be ground as fine as a low volatile coal.

Particle size: an increase in the fineness of coal increases the total surface area and reduces the ignition temperature and ultimately helps in achieving a higher rate of combustion. Thus, for finer coal particles, the flame gets shortened and has a higher temperature. However, for high volatile coal, if the fineness is reduced beyond a certain limit there is possibility of explosion. As a general practice the percentage residue on 90 microns is maintained at approximately 0.5 to 0.7 times that of volatile content.

Ash content: internationally, coal is classified as low-ash (3–8%), medium-ash (8–15%), and high-ash (> 15%). The high ash content can affect the process in two ways. It being an inert substance, the concentration of energy will vary with the content of this substance, and so an increase in ash will make it difficult to maintain the desired constant rate of thermal energy into the process. The second problem is that at the same moisture level for every percent increase in ash the heat value will decrease by about 130 kilocalories. The higher the ash content of coal, the longer the burning time of the char, causing increase in flame length. As the fuel value decreases, there will be a reduction in flame temperature.

Calorific value: as already discussed, the calorific value is a function of carbon, sulfur, moisture, and ash content. Coal with low calorific value increases the specific heat consumption for clinker burning while decreasing the specific kiln throughout.

Effect of ash absorption on clinker

It is evident from the above that the major effect of coal properties is on the preparation of fine coal for firing. If done incorrectly, the clinker burning process is affected. In addition, if the coal ash absorption is high in the system due to firing of high-ash coal or due to high specific heat consumption in the kiln or both, the clinker quality may get modified. Further, there could be certain other effects due to the presence of minor constituents like sulfur, alkalis, chlorides, and phosphates in coal and coal ash.

Depending on the quality of the available limestone and correctives, every plant has a tolerance limit for ash content in coal, above which the use of high-ash coal is not feasible. However, in plants using high-grade limestone, the use of coal of higher ash content is feasible with necessary modification of a raw mix design and choice of suitable correctives, to achieve the optimum clinker parameters. The use of high-ash coals in such cases would require a complete optimization of kiln feed and process parameters. The presence of minor constituents like sulfur, alkalis, and chlorides above tolerable limits in coal would necessitate installations of bypass systems. In some situations, it may become necessary to search for suitable correctives/additives to be used in raw mixes for controlling and balancing the molar ratio of sulfates to alkalis, in order to prevent the formation of undesired buildups at the kiln inlet and preheater cyclones.

3.6 Relation of process parameters with combustion

In addition to the material characteristics of coal, its combustion is heavily dependent on the process parameters and to a large extent on the system design features. The cooler throat, kiln hood, tertiary air off-take, secondary air temperature and velocity, air leakage points, etc. are some of the salient process and design features of the kiln system that influence the combustion of fuels in general and coal in particular. The important aspects of these relations are discussed below.

Primary air

Primary air transports the fuel and mixes with it at the nozzle. Depending on the coal quality and heat consumption in a kiln, the primary air quantity may vary in the range of 8 to 15% of the total air needed for combustion—the balance quantity coming from the kiln system in the form of secondary and tertiary air flows. The proportion of primary air for coal transport is related to the volatiles content of the coal, a low volatile coal requiring relatively more primary air. In the context of primary air, it should also be noted that in large-capacity kilns with multi-channel burners it is important to maintain specific flame momentum above a certain threshold value. The burner momentum is calculated as follows:

$$\begin{aligned}&\text{specific axial momentum (N/MW)} = \\ &\quad \text{mass flow of primary air at the burner (kg/s)} \\ &\quad \times \text{ velocity of primary air at burner tip (m/s)} \\ &\quad /\text{fuel heat input at the burner (MW)}\end{aligned} \quad (3.13)$$

A threshold value of 6 N/MW is generally recommended, although much higher values in the range of 9–11 N/MW are encountered in actual practice. The higher the momentum, the more rapid the fuel and air mixing, and the greater the development of an intense and short flame. However, it should be borne in mind that a greater momentum may lead to an increase in the proportion of primary air, which is not desirable from a thermal efficiency angle. Hence, the aspect of kiln burner momentum requires a carefully balanced operational approach.

Secondary air

Since the secondary air has to be entrained into the primary air and fuel jet, the aerodynamics of the secondary air can have a huge effect on the fuel/air mixing. In the case of cement rotary kilns, the secondary air flow is determined by the design of the cooler uptake and the kiln hood system, or, in the case of the planetary coolers, by the cooler elbows. The differences in the aerodynamics of a low-momentum and a high-momentum burner in a kiln with a grate cooler are shown in Figures 3.6 and 3.7. It is evident from the figures that the dimension of the recirculating zones decreases with a decrease in jet momentum. This, in turn, affects the flame shape and heat transfer.

Excess air

The kiln systems almost always operate with some excess air, including the false air that enters due to improper sealing or leakage. The excess air is mostly admitted in order to ensure complete combustion of fuels. The effect of excess air on the thermal efficiency of the system can be

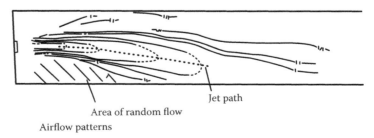

FIGURE 3.6 Aerodynamics of a low-momentum burner.

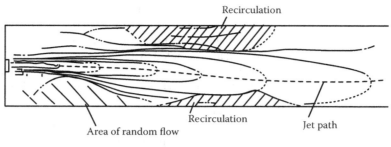

FIGURE 3.7 Aerodynamics of a high-momentum burner.

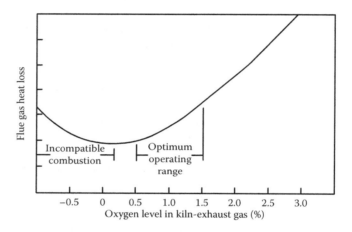

FIGURE 3.8 Relation between excess air and flue gas heat loss.

easily understood from the fact that additional fuel would be required to heat up the excess oxygen and nitrogen passing through the system, first to the burning zone and then to the kiln back-end. Figure 3.8 shows a relationship between excess air and fuel efficiency.

Flame temperature The clinker forming reactions occurring in the burning zone primarily depend on the temperature of the flame. The volatiles liberated from coal in the burning zone will ignite only when there is sufficiently high temperature, along with high concentration of oxygen. The flame temperature actually occurring in a kiln depends on:

  a. temperature of combustion air
  b. quantity of excess air
  c. rate of mixing of fuel and combustion air
  d. coal particle size
  e. moisture content in fine coal

The temperature of the combustion air depends on the ratio of secondary to primary air flow rates and temperature of secondary air. Contribution of heat by combustion air could be up to 35% of the heat of fuel combustion. Thus, a significant amount of heat is imparted to the flame by combustion air. An increase of 100°C in secondary air temperature can result in an increase of 50–70°C in flame temperature. A reduction in primary air quantity from 20 to 10% can give a similar benefit. The quantity of excess air includes leakage air from the kiln seal at the burner end. A 10% rise in excess air quantity results in a 50–100°C fall in flame temperature.

Poor rate of mixing of air and fuel, coarse coal particles, or high moisture content in coal increases the reaction time of combustion. Thus, sometimes complete combustion is not possible within the burning zone. In such cases combustibles, such as carbon monoxide, are found in flue gases. This incomplete combustion also results in a major loss of heat and a reduction in flame temperature.

An unstable flame may appear if the fuel has delayed and varying ignition characteristics and if, on firing, the fuel leaves a flame-out space inside the kiln between the burner tip and the tail of the flame (Figure 3.9). A substantial amount of unburnt fuel remains in this space and gives rise to a situation of high explosion risk. Thus, it is important to ensure the early ignition of fuel entering the combustion zone; this is often achieved by ensuring the recirculation of hot gases, which can be developed by the use of swirls in either the fuel or the primary air jet stream (Figure 3.10).

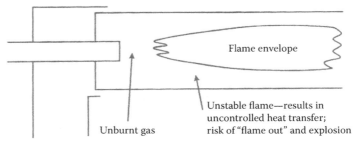

FIGURE 3.9 Unstable flame in a kiln. (Reprinted from Con G. Manias, Kiln Burning Systems, Chapter 3.1, in *Innovations in Portland Cement Manufacturing* (Eds J. I. Bhatty, E. M. Miller, and S. H. Kosmatka), Portland Cement Association, USA, 2004. With permission.)

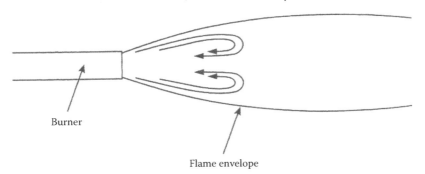

FIGURE 3.10 Stabilization of flame by swirling in the burner. (Reprinted from Con G. Manias, Kiln Burning Systems, Chapter 3.1, in *Innovations in Portland Cement Manufacturing* (Eds J. I. Bhatty, E. M. Miller, and S. H. Kosmatka), Portland Cement Association, USA, 2004. With permission.)

3.7 Petcoke as a substitute fuel

Petroleum coke, or petcoke, is obtained from the oil refining industry. The refining process generates gasoline, diesel oils, lubricating oils, waxes, etc., leaving some residual crude that undergoes additional catalytic treatment along with some other distillates at high temperature and pressure in a special piece of equipment called a coker. This process ultimately generates petcoke as a residue, after driving off gases and volatiles and separating the light and heavy oils. Petcoke, thus produced, is a carbon-rich material with very low ash content and varying sulfur content. The petcoke, which is high in sulfur and metals, is categorized as "fuel-grade," while the low-sulfur variety without metals is called "anode-grade." The fuel-grade petcoke is used in cement kilns and also for electricity generation, while the anode-grade petcoke is used for non-fuel application in the steel and aluminum industry. Already, many cement plants in the world are successfully using petcoke as a secondary fuel. As long as petcoke will retain its cost advantage over coal, the application as a fuel substitute in the cement industry will keep on increasing.

World status of petcoke production

Currently, petcoke is produced at more than 140 refineries in the world and the top ten petcoke-producing countries are the USA, China, India, Venezuela, Brazil, Spain, Canada, Mexico, Russia, and Germany. The USA continues to be the lead producer and exporter of petcoke. The total petcoke market is estimated at 127 million tons per year. From 2009 to 2014, the petcoke market grew by 4%; but the growth rate is seen to be declining and is estimated to be at about 3% between 2015 and 2020 (11).

Classification of petcoke

Depending on the production process, fuel-grade petcoke is classified as "delayed coke," "fluid coke," and "flexi-coke" (1). The delayed petcoke is a relatively high-volatile variety that comes from a semi-continuous production process developed originally to minimize the generation of heavy liquid waste fuels through thermal cracking. The fluid coke is the petcoke resulting from a continuous fluidized-bed coking oven specially developed to treat a high-sulfur feed. The flexi-coke is a product also from a continuous fluidized-bed coking furnace, in which most of the coke is turned into a gaseous fuel of low calorific value for use in the refinery itself; hence, the residual coke has the lowest volatiles content. Upon leaving the coking process, the above varieties of coke are also known as "green coke," but when coke is further heat-treated in rotary kilns in a reducing atmosphere and at temperatures above 1200°C, it becomes virtually free of volatiles and is known as "calcined coke," suitable for anode making. Nearly 75% of the world's petcoke production finds use as a source of energy in the following industrial segments (1):

- catalytic cracking (40%)
- cement manufacture (16%)
- power generation (14%)
- heating processes (5%)

The remaining 25% is used mainly for making anodes and in some miscellaneous industrial processes.

Properties of petcoke

Compared to coal, petcoke is characterized as a low-volatile fuel with high carbon, high sulfur, and very low ash content. Depending on the type of crude oil, the sulfur content may vary widely. But, as already mentioned, the fuel-grade petcoke is mostly high in sulfur content, typically in the range of 2 to 8%. The range of composition of the delayed-type petcoke, along with its thermal and physical characteristics, is given below (12):

Moisture (%): 6 to 12

Ash content (%): 0.2 to 2

Volatiles (%): 5 to 14

Fixed carbon (%): 88 to 90

Sulfur (%): 2 to 8

Calorific value (kcal/kg): 7500 to 8000

Hardgrove grindability index: 35 to 70

Bulk density (kg/m^3): 0.77 to 0.88

Lump size (mm): 0 to 75

Although a part of the ash originates from the crude oil, it is mainly derived from the catalyst residue from the refining process. As a result, the ash is relatively high in calcium oxide, iron oxide, sulfate, and titanium dioxide. A typical composition of an ash sample is: SiO_2 (32.49%), Al_2O_3 (8.04%), Fe_2O_3 (11.78%), CaO (21.0%), MgO (3.61%), SO_3 (15.05%), Na_2O (0.90%), K_2O (0.28%), TiO_2 (5.78%), P_2O_5 (0.15%), and Mn_2O_3 (0.16%). In addition, the ash often contains metals like vanadium, silicon, nickel, platinum, etc. The nature of ash is abrasive and it makes the material very difficult to handle and pulverize. Petcoke is available in different size range varying from large lumps to fine dust, as shown above. However, its major availability is usually in intermediate sizes of 2–5 mm. Although the grindability index and sulfur content could lie within a wide range, as shown above, the HGI and sulfur content determine the price of petcoke in the market. The lower the HGI, the harder it is to grind and, consequently, the cheaper its price. For the same HGI, the higher the sulfur content, the cheaper is the petcoke. In the prevailing market the HGI is generally observed to be in the range of 40 to 45, and the sulfur content varying from 3.0–6.5%.

Grinding of petcoke

Petcoke is always used in pulverized form, both in kiln and precalciner burners. Due to lower volatile content in petcoke, it is difficult to ignite, burn, and maintain a proper shape of flame. The ignition temperature of petcoke is 590–620°C, compared to a coal ignition temperature of 400–500°C. An important factor influencing the rate of combustion is the specific surface area of the fired petcoke. So, the fineness of petcoke is increased to improve its burning behavior. Petcoke fired in kilns and calciners is generally ground to a residue level of 4–6% on a 90-micron sieve, and sometimes even lower.

Petcoke, being highly abrasive and hard, is difficult to grind. The grinding system, therefore, has to be designed to minimize the wear of mill components, the transport system, and ducting. Petcoke can be ground either in a ball mill or a vertical mill. While a ball mill system is simple and rugged, it cannot handle higher moisture contents and its specific energy consumption is high; the vertical roller mill system can handle higher moisture content in petcoke and is energy efficient but the expected wear of its components would be high. Depending on the hardness of petcoke and the fineness required for firing (4–6% residue on a 90-µm sieve), the expected capacity of the basic mill after conversion from coal firing may reduce by 50–80% of the basic output of the coal mill.

Firing of petcoke

The level of substitution for petcoke in an existing kiln (having no bypass arrangement) is decided on the basis of sulfur content in petcoke and clinker. This is because of the fact that higher concentrations of sulfur cause coating formations in cyclones and kiln inlet areas, resulting in operational disturbances. In extreme cases, the kiln has to be stopped to enable cleaning of these build-ups. Consequently, the availability and hence productivity of the kiln is adversely affected. The acceptable limit of sulfur content in petcoke depends on the type of kiln and raw mix design. It should be borne in mind that with the use of petcoke it may not always be possible to maintain the molar ratio of alkali oxides to sulfur trioxide within the desired range of 0.8 to 1.2, unless special efforts are made to use alkali-rich additives.

Petcoke can be used by two methods. One is to mix it with the raw meal. In this method, the reducing atmosphere is created. This may result in a change in temperature profile, an increase in the preheater outlet temperature, and an increase in uncontrolled CO formation. In preheater cyclones, the reducing atmosphere may cause clogging. Another method, which is being widely adopted, is to blend petcoke with coal that is normally used as the fuel. In this strategy, adequate care has to be taken to ensure that the petcoke is burnt properly and completely. It is also possible to combine both the methods and to introduce petcoke along with the raw meal on one side and simultaneously fire it along with the normal fuel in precalciners and kilns.

If the petcoke is introduced with raw meal, it is first taken to the homogenizing silo and then, along with the raw meal, it is fed to the preheater. Hence, the feeding system in this case must be suitable to handle petcoke. It is desired that the burning of petcoke starts in the kiln inlet and volatiles are not driven off earlier. Hence, this method is preferred for the petcoke of very-low-volatile contents.

For the use of petcoke with fuel firing, two separate feeding systems are installed for firing in both kiln and precalciner strings to prevent fluctuations in the rate of firing. Each system must have separate feed bins (located close to the point of use), a transport system, and dosing equipment like weigh feeders. The system design should take care of the abrasive nature of fuel.

If the plants have single-channel burners for the kiln then these will have to be replaced by specially designed multi-fuel burners. Such a burner will ensure the complete mixing of petcoke and air, the burn-out time of individual particles, and create the required flame momentum. The burner must be suitable to handle highly abrasive material. A little quantity of support fuel like coal with high-volatile matter or oil or gas can be used to ensure ignition. Several plants have successfully used petcoke up to 100% in main firing.

The desired conditions for the proper combustion of petcoke in precalciners need to be created by appropriate measures, such as the following (12):

- The design of a "hot spot" in the "in-line calciner" by shifting upstream the raw meal entry by about 1.5–2 m. A better solution may be the provision of a pre-combustion chamber.
- The precalciner is designed for a gas residence time of minimum 3.5 seconds.
- The creation of adequate turbulence in the calciner.

Merits and demerits of using petcoke

As compared to coal, which has higher volatile contents than petcoke, the requirement of combustion air for petcoke is less. Similarly, the generation of the flue gas is lower, which results in a decrease in the energy consumption of the fans handling it. Alternatively, the reduction in gas volume can be used to increase the production from the kiln, provided there are margins in other equipment in the system. Further, as the petcoke is a high-heat-value fuel, the temperature of the flame increases and hence the heat transfer in the kiln improves. But at the same time, the petcoke flame tends to be longer than the oil or coal flame, which may result in finer granulometry of the clinker and may make the clinker more difficult to grind. In certain specific situations, where the alkali content in the raw meal is high, the use of petcoke, because of its high sulfur content, may be of great help in the process of adjusting the alkali/sulfur molar ratio. On the other hand, under normal circumstances, with low alkali in the raw meal, the sulfur content in the petcoke creates a volatile cycle in the kiln and the quantity of fuel substitution by petcoke is limited by the molar ratio of alkalis to sulfur in the hot meal entering the kiln. Under such circumstances, the clinker entraps sulfur as $CaSO_4$. The sulfate content in petcoke-fired clinker may increase to about 1.4%, which may often be hydraulically reactive, requiring less addition of external gypsum during cement milling. Experience shows that, all other conditions remaining the same, the replacement of coal by petcoke does not alter the NO_x emission. With incomplete combustion of petcoke, however, there are chances of increased CO formation. Further, toxic elements like vanadium, nickel, etc. present in petcoke are carried into the dust. Hence, before disposing of the dust care has to be taken to ensure that the concentration of these elements does not exceed the limits defined by environmental laws.

Finally, exposure to green petcoke may call for health and safety precautions, as it may cause irritation to the eyes, skin, and respiratory system. This happens because of the abrasive nature of the fuel and also due to the presence of polycyclic aromatic hydrocarbons (PAHs). The precautionary measures include, apart from personal safety gears, the installation of local exhaust systems and filters.

3.8 Summary

All combustible substances that contain carbon as the main constituent are called fuels. The fuels are broadly classified into two types: primary and secondary. Primary fuels occur in nature and can be solid, liquid, or gaseous. Coal and lignite, petroleum and crude oil, and natural gas are common examples of primary fuels. Secondary fuels are the ones that are artificially manufactured or are derived from the primary fuels. Secondary fuels can also be solid, liquid, and gaseous. Coke, diesel, and producer gas are examples of secondary fuels. So far as the cement industry is concerned, coal, fuel oils, and natural gas are the main primary fuels in use. In addition, petroleum coke is the most widely used secondary fuel in cement production.

Depending on the stage of maturity in the long process of geologic development, coal is ranked as peat, lignite, sub-bituminous, bituminous, and anthracite. Anthracite is the highest-ranking coal but is of limited occurrence. The dependence of the industry, therefore, is mainly on bituminous and sub-bituminous coal. Lignite is also used in the industry by adopting special technology but the availability is limited. Peat is not used as an industrial fuel. Among the liquid fuels, the diesel oil fraction, obtained from the refining process, is used as a fuel. Of all gaseous fuels, natural gas has wide industrial application both as a fuel and as a feed stock. The composition and properties of the above-mentioned fuels have been presented in the chapter. Emphasis, however, is on coal, as it is the most extensively used fuel in the cement industry.

Fuels in general and coal in particular are characterized by proximate and ultimate analysis. Proximate analysis refers to determination of moisture, volatile matter, ash and fixed carbon contents in weight percentage by standard methods. The ultimate analysis provides the information about the elemental composition of coal, determined by the prescribed methods. The selection and use of coal are done by taking into account the gross and net calorific value, moisture content, ash composition, and characteristics and ignition temperature. For liquid fuels, in addition, properties like flash point, pour point, viscosity, etc. become important.

Coal resources are found in almost all countries and world reserves are tentatively estimated at over 1100 billion tons, which may be sufficient to meet the global production of coal for 153 years at the current rate. The Asia-Pacific region holds more than 46% of world reserves and, within the region, China, Australia, and India are the countries with the three highest coal reserves. India has a total geological coal reserve of over 300 billion tons.

Coal is dried and ground in vertical roller mills. Fineness is controlled in terms of residues on 90-µm and 200-µm sieves. The transport of coal requires primary air, which forms a small percentage of the total combustion air. The bulk air requirement is met by the kiln system in the form of secondary and tertiary hot gases. Combustion refers to the rapid oxidation of fuel accompanied by the production of heat. Complete combustion is possible only in the presence of adequate oxygen. Incomplete combustion results in higher energy consumption and lower productivity. Apart from primary and secondary air, the combustion process is also dependent on the quantity of excess air and the resultant formation of a flame with proper shape, length, and stability. Radiation is the dominant process of heat transfer inside the kiln. The salient combustion process aspects have been discussed in this chapter.

The use of petroleum coke (abbreviated as petcoke) plays an important role as a secondary fuel in cement production. There is a world market of about 127 million tons per year of petcoke, 75% of which is fuel-grade. Fuel-grade petcoke is a high-carbon, low-volatile, low-ash substance available from the oil refining industry but it has the disadvantage of containing a high percentage of sulfur. Notwithstanding certain demerits, the burning technology of petcoke has been so improved that it is considered a viable secondary fuel in cement production.

References

1. C. GRECO, G. PICCIOTTI, R. B. GRECO, and G. M. FERREIRA, Fuel Selection and Use, in *Innovations in Portland Cement Manufacturing* (Eds J. I. Bhatty, E. M. Miller, and S. H. Kosmatka), Portland Cement Association, USA (2004).
2. *Guide Books – National Certification Examination for Energy Managers & Energy Auditors, Fuels and Combustion* (Book 2, Chapter 1), Bureau of Energy Efficiency, New Delhi (2015).
3. A. PAHARI and G. CHAUHAN, *Engineering Chemistry*, Laxmi Publications Private Limited, New Delhi & Infinity Science Press LLC, Massachusetts, USA (2007).
4. P. A. ALSOP, HUNG CHEN, and HERMAN TSENG, *The Cement Plant Operations Handbook*, Fifth Edition, Tradeship Publications Limited, UK (2007).
5. BP Statistical Review of World Energy, June (2017). www.bp.com/bp-statistical-review-of-world-energy-2017-full-report.pdf.
6. www.cmpdi.co.in/coalinventory.php.
7. G. MARTIN, *Chemical Engineering and Thermodynamics applied to Cement Rotary Kiln*, The Technical Press, London (1932).
8. P. WEBER, *Heat Transfer in Rotary Kilns*, Bauverlag GmbH, Wiesbaden-Berlin (1963).

9. P. J. MULLINGER, B. G. JENKINS, AND J. P. SMART, Combustion and Heat Transfer in Rotary Kilns, in *Modernisation and Technology Upgradation in Cement Plants* (Eds S. N. Ghosh and Kamal Kumar), Akademia Books International, New Delhi (1999).
10. Con G. MANIAS, Kiln Burning Systems (Chapter 3.1), in *Innovations in Portland Cement Manufacturing* (Eds J. I. Bhatty, E. M. Miller, and S. H. Kosmatka), Portland Cement Association, USA (2004).
11. http://roskill.com/product/petroleum-coke-global-industry-markets-outlook-7th-edition-2015.
12. V. K. BATRA, Kamal KUMAR, AND P. N. CHHANGANI, *Use of petcoke as an alternate fuel in cement production*, Indian Cement Review, April 2005.

CHAPTER FOUR

Alternative fuels and raw materials

4.1 Preamble

In cement plants, the use of agricultural and industrial wastes is becoming increasingly important, on account of both economic considerations and environmental compulsions. The economic considerations arise from the fact that the costs of raw materials and fuels comprise as much as 50% of the total operating costs in most of the plants and, therefore, alternative fuels and raw materials with cheaper cost implications find an important role to play. The environmental compulsions occur due to enormous increases in the generation of wastes that require safe and gainful modes of recycling.

Wastes are generated from diverse sources, which include our mode of living, agricultural practices, and industrial operations encompassing mining, processing, and manufacture. Plastic wastes, municipal solid wastes, organic solvents, scrap motor tires, electronic garbage, and animal carcass wastes are just a few examples. The cement industry has a proven track record of being well adapted to absorbing and utilizing these waste products in the production of cement. The problems of recycling the waste have essentially been threefold: environmental concerns associated with the transport of such materials to the facilities, consequences of their use in the cement process, and contending with various operational and quality issues online.

Looking at the abundance and variety of wastes around us, one must accept that alternative fuels and raw materials are changing both quantitatively and qualitatively. In such a dynamic situation of waste management, the legislation and technologies for processing and treatment are undergoing rapid development, which is clearly influencing the material flows. Therefore, the material data presented in this chapter cannot be considered universal and have to be verified for each situation. Further, it is important to mention that the availability of wastes depends on the cultural, social, and economic activities of a country or a region. The net effect of all these factors is the emergence of a new operational strategy in the cement industry, designated as "AFR" or "Alternative Fuels and Raw Materials."

4.2 Broad classification

The agricultural and industrial wastes may either be incombustible or combustible. The incombustible wastes include coal ash—more commonly identified as fly ash, bottom ash, and mixed pond ash—from the thermal power generating plants; slags from the metallurgical industry, of which granulated blast furnace slag is more widely used, and steel slag from the iron and steel industry and copper slag and lead-zinc slag from the nonferrous industry is less widely used; lime-containing sludge from paper and chemical production; used pots from the aluminum industry, etc. This chapter focuses on combustible waste and does not deal with the above material in detail.

Combustible wastes may come from both the agricultural and industrial sectors. Agricultural waste, often called biomass residues, mainly consists of rice husk, bamboo dust, bagasse, straw, coconut husk, peanut husk, wood waste, etc. Industrial wastes cover both non-hazardous and hazardous varieties and include carbon slurry from fertilizer plants (that use coal for the production of CO_2), petrochemical waste, spent pot-liners from the aluminum industry, waste oils from gearboxes and machines, hazardous wastes from pesticides and pharmaceutical industries, scrap motor tires, rubber wastes, plastic waste, cardboard, graphite dust, old battery boxes, etc. A third variety of combustible waste is broadly identified as household waste, animal waste, municipal refuse, sewage sludge, landfill gas, etc. The composition and properties of a few of these alternative fuels are given in Table 4.1 (1).

In terms of the physical state in which the combustible wastes are generated, they are classified as follows:

- Gaseous: coke oven gas, refinery waste gas, pyrolysis gas, landfill gas, etc.
- Liquid: low-chlorine spent solvents, lubricating oils, vegetable oils, fats, distillation residues, hydraulic oils, insulating oils, etc.
- Pulverized, granulated, or fine-crushed solids: waste wood, sawdust, wood shavings, dried sewage sludge, finely shredded used tires, residues from food production, etc.
- Coarse crushed solids: plastic waste, discarded wood strips, re-agglomerated organic matter, etc.
- Lumpy solids: whole scrap tires, plastic bales, plastic bags, and plastic drums, etc.
- Refuse-derived fuel of specified dimensions.

Definition of hazardous wastes

A hazardous waste is defined as any waste that by virtue of its physical and chemical properties causes danger or is likely to cause danger to

Table 4.1 Physical, calorific, and chemical characteristics of typical alternative fuels used in the cement industry

Properties and constituents	Unit	Petroleum coke	Tires	Fluffy refuse-derived fuel	Sewage sludge	Meat and bone meal
Moisture	%	0.11	1.00	17.80	6.60	3.4
Volatiles	%	10.90	61.00	65.00	45.30	68.70
Hardgrove index	–	40-50	–	–	–	–
Calorie value	kJ/kg	34,830	29,480	14,650	9849	19,990
Ash	%	1.3	7.5	17.7	46.2	20.5
Carbon	%	86.4	81.0	53.1	26.6	43.8
Hydrogen	%	3.5	6.7	7.3	4.9	5.3
Oxygen	%	1.9	3.0	19.5	16.0	16.9
Nitrogen	%	1.6	0.3	0.5	5.7	8.9
Chloride	%	0.01	0.1	1.2	0.05	0.6
Sulphur	%	5.4	1.7	0.8	0.46	0.5
P_2O_5	%	–	–	–	< 10	< 5
Hg	ppm	< 0.01	< 0.01	< 1	< 5	< 0.1
Cd + Tl	ppm	< 10	< 20	< 20	< 10	< 10
Trace elements total*	ppm	< 3000	< 2000	< 2000	< 5000	< 1000

*Total of Sb, As, Pb, Cr, Co, Cu, Mn, Ni, V.

health and environment, alone or in contact with any other waste material or substance. The basic characteristics of hazardous waste are:

- Corrosivity: the ability to destroy tissue by chemical reaction. Known examples are rust removers, waste acid, alkaline cleaning fluids, and waste battery fluids. Corrosive substances have pH value either below 2.0 or more than 12.5.
- Ignitability: ready oxidization by burning; spontaneously combusting at 54.3°C in air or at any temperature in water; strong oxidization leading to combustion; any non-liquid material capable of causing fire under standard temperature and pressure due to friction, moisture absorption, or spontaneous chemical change. Common examples are coating wastes, paint, certain degreasers and solvents.
- Reactivity: the ability to react, detonate, or decompose explosively at environmental temperature and pressure. This property can often be found in wastes from cyanide-based plating operations, bleaches, waste oxidizers, and waste explosives.
- Toxicity: the capability of causing harm to organisms. This is characterized by the toxicity characteristics of 37 elements (such as arsenic, cadmium, chromium, mercury, lead, etc.) and organic compounds (such as nitrobenzene, hexachlorobenzene, etc.) exceeding

the limiting values as determined by the Toxicity Characteristics Leaching Procedure (TCLP US-EPA SW 46). Toxicity is graded as acute, high, or chronic depending on the behavior of the toxic substances. Pesticides, heavy metals, and mobile or volatile compounds display toxicity. Wastes from halogenated phenols and pesticide production and even the containers used for handling hazardous or toxic substance show toxicity.

There are recommended standard tests to determine the above properties in various national and international standards and guides. The Central Pollution Control Board in India, in its guide of February 2010, specified certain parameters for hazardous wastes, as given in Table 4.2.

Biological parameters in classifying hazardous wastes

Hazardous wastes can have organism-mediated chemical activity leading to infectious properties. In order to classify such materials biologically, some terminologies are used, which are briefly explained below (2):

- Bioconcentration: a process by which living organisms concentrate a chemical to levels exceeding the surrounding environmental media.
- Lethal dose (LD): a dose of a chemical calculated to expect a certain percentage of mortality in a population of an organism exposed through a route other than respiration.

Table 4.2 Specification for hazardous AFR

Serial no.	Parameters	Fuel	Raw materials
1	CV as received (kCal/kg)	> 2500	–
2	Ash (%)	< 5 for Liquid < 20 for solids	–
3	Cl⁻ (%)	< 1.5	< 1.5
4	Halogens (F +Br +I) (%)	< 1.0	–
5	S (%)	< 1.5	< 1.5
6	PCB/PCT (ppm)	< 5.0	< 5.0
7	Hg (ppm)	< 10.0	< 10.0
8	Cd + Tl + Hg	< 100.0	< 100.0
9	pH	4–12	–
10	Viscosity for liquid CSE	< 100	–
11	Flash point for liquid (°C)	< 60	–
12	Volatile organic hydrocarbon (ppm)	–	< 5000
13	Total organic carbon (ppm)	–	< 1000
14	$CaO + SiO_2 + Al_2O_3 + Fe_2O_3 + SO_3$ (%) in ash	–	< 80
15	As + Co + Ni + Se + Te + Sb + Cr + Sn + Pb + V (ppm)	–	< 10,000

- Lethal concentration (LC): a calculated concentration of a chemical in air, a 4-hour exposure to which through respiration by a population of an organism will kill a certain percentage.
- Phytotoxicity: a chemical's ability to elicit biochemical reactions that harm flora.

Based on the chemical and biological criteria mentioned above, hazardous wastes have been listed in three schedules in the guide of the Central Pollution Control Board in India.

4.3 Feasibility of an AFR project

There are certain basic social, economic, and technical factors to be considered for developing an AFR project for a cement production unit. Based on the experience in different countries, it appears that the importance and relevance of the factors vary from country to country. While some factors become irrelevant in one situation, certain others turn out to be more pertinent in another. Keeping this in view, one may consider the following factors as generally important in the assessment of an AFR project for the cement manufacturing process:

1. availability and cost
2. source and transportation
3. characteristics, e.g., moisture, ash content, volatile matter, and calorific value
4. present mode of disposal
5. drying and grinding characteristics
6. storage, handling, feeding, and firing systems
7. ignition and combustion aspects
8. effects on kiln operation and refractory lining
9. impact on clinker quality
10. environmental and social considerations
11. economic viability

The above list is only illustrative and not exhaustive. For liquid and hazardous wastes, one will have to evaluate certain special properties as mentioned earlier. Although lowering the unit cost in cement production is the key driver when developing an AFR project at a cement plant, the feasibility study is never complete with the development of a business case with an acceptable level of capital investment, attractive return, and manageable technical risks. It has to be effectively supported by the strategies to be adopted for the health, safety, and environmental issues (3):

- odor
- noise pollution
- emissions to air

- emissions to water and land
- visual impact
- carbon footprint
- stakeholder engagement

Odor

Odor issues become important particularly with the use of municipal waste, perhaps due to residual organics that continue to break down. What constitutes an offensive odor is obviously subjective as it depends on the intensity, character, frequency, and duration of the odor. A standard has been developed to attempt to deal with the issue objectively: "EN 13725:2003 Air quality – determination of odor concentration by dynamic olfactometry." When odor risks become an issue, the approach is to undertake a baseline study at sensitive receptor sites, followed by measurements when the fuel is in use with the help of the standard mentioned above. Various mitigation measures need to be adopted to enclose and vent the storage vessel and building, along with the treatment of the vented air. A useful alternative may also be to vent the odorous air to the inlet compartments of the clinker cooler.

Noise pollution

The noise pollution arises from increased traffic on the approach roads to the plant and also inside the plant to the storage area. The noise issue can normally be dealt with by restricting the delivery time during a day but this may have implications for the storage capacity and needs to be fully assessed during the design stage. The necessity of minimizing the interaction between the pedestrians and delivery trucks inside the plant by appropriate provision of internal roads is another important consideration in the design phase.

Emissions to air

Kiln emissions that are of prime concern are particulates and nitrous oxide. It is well established that particulate emissions are not dependent on the fuel being burnt. The data available also show that with firing of AF the emission of nitrous oxide may not change in most cases; in some situations, it may be reduced. However, heavy metal content in most alternative fuels is expected to be higher than in bituminous coal but it tends to be trapped during the clinkering process. Consequently, no significant impact occurs due to emission of heavy metals. Nevertheless, a comparative desk calculation is essential for emissions due to change in fuels.

Emissions to water and land

A major risk to water and land in AF storage and handling is fire and the prevalent regulations for fire hazard management in a region will apply also to the facility. Further, care must also be exercised in containing the residues, should such an incident occur. Provision should be made in the design of the storage for the containment of water and fire retardants used to quench fire.

Visual impacts

For the benefit of employees and the community it is always desirable to assess the AF storage area for its visual impact, although it may not be an imperative requirement. Key considerations are the visibility of the proposed facility, the choice of cladding materials, and the height of the facility. An important housekeeping issue is the escape of loose materials from the receiving station and spillage from the feed delivery system. A well designed and engineered storage and handling system will properly address these issues and minimize the impact.

Carbon footprint Calculation of the greenhouse gas saving should be undertaken in accordance with the national technical guidelines and aggregated with the direct savings to arrive at a true greenhouse measure of the project. One such document that one may like to refer to for estimating CO_2 savings is "Technical guidelines for the estimation of greenhouse gas emissions by facilities in Australia," which provides some standard factors and methodologies. It may be pertinent to mention here that the presence of biomass in the AF will result in a reduction in carbon emissions as the CO_2 emissions from materials like wood and paper are considered part of the natural carbon cycle (biogenic), unlike the emissions from coal or natural gas, which are classified as anthropogenic. This is a simple example of factors that need to be considered in estimating the greenhouse benefits.

Stakeholders engagement In most situations, there are no universal processes to engage with the stakeholders, it may be useful to make a reference to the community engagement document entitled "Communications and stakeholder involvement in the cement industry," prepared by the Cement Sustainability Initiative of the World Business Council for Sustainable Development. Feedback from the surrounding community and involved stakeholders is quite important in developing an AF project. Formation of a community consultative committee may be a facilitating step in the AFR strategy (3).

4.4 Inventory and material characteristics

Although the inventory and quality of alternative fuels vary from country to country and even region to region, the past use pattern globally indicates that the following types make up the major trend of application:

- municipal solid waste
- biomass
- scrap tires
- hazardous wastes
- sewage sludge
- industrial plastic waste

Based on the data available from various sources of the government of India and reported in (4 & 5), it seems that about 62 million metric tons of solid waste is generated annually in the country, of which 5.6 million metric tons is plastic waste, 0.17 million metric tons is biomedical waste,

7.90 million metric tons is hazardous waste, and 0.15 million metric tons is e-waste. The per capita waste generation in Indian cities ranges from 200 to 600 g.

Municipal solid waste (MSW)

MSW is a heterogeneous mixture of daily items that people use and discard. It includes all types of biodegradable, inert, recyclable organic and inorganic waste materials. Although it has been mentioned above that 62 million metric tons of MSW is generated every year, which translates into 450 g of average per capita generation, the quantity actually collected is about 42 million metric tons, out of which only 12 million metric tons of MSW is treated by the municipal authorities. It is projected that by the year 2025 under the business-as-usual scenario about 140 million metric tons of MSW will be generated. MSW is made up of 9.0% plastic, 11.0% glass, 3.0% metal, 13.0% paper, 55.0% biodegradable materials, and 9.0% inert substance. Out of this, the entire quantity of plastic and paper and about 10.0% of the biodegradable materials constitute the recoverable fuel portion. Therefore, the key to achieve an efficient resource recovery is its segregation into individual fractions. The major problem in the above proposition is to find ways of using the segregated combustible fractions (SCF), which are not recyclable. These fractions may contain soiled paper, soiled cloth, contaminated plastics, packaging materials, leather, rubber, etc. Such fractions can be utilized as an AF only when they are converted into refuse-derived fuel (RDF) or solid recovered fuel (SRF) having defined quality parameters. The non-recyclable SCF is converted into SRF or RDF through a processing facility that can either be located outside or within a cement plant. A processing plant for RDF has complete facilities of appropriate screening, drying, air density separation, and further screening, blending, additive dispensation, press molding, final drying, and solidification, as shown in Figure 4.1.

In this context, it may be worthwhile to make a reference to the categories of RDF defined by ASTM for use in waste-to-energy plants (see Table 4.3) (6).

The above classification does not define the essential requirements of RDF suitable for cement kilns. Similarly, the classification of RDF into

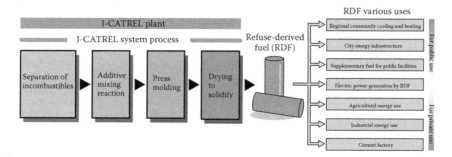

FIGURE 4.1 Conversion of combustible garbage into RDF by the J-Catrel process. (Courtesy of EBARA Corporation, Japan on "Zero Emission." Brochure.)

Table 4.3 Classification of RDF

Categories	Description
RDF-1	Discarded after processing
RDF-2	Processed to a coarse particle size with or without separation of ferrous metal
RDF-3	Processed to a particle size of 95% passing 50-mm square mesh, from which most metal, glass, and organics have been separated
RDF-4	Processed to a powder with 95% passing through no. 10 mesh (2-mm) screen, from which most metal, glass, and organics have been separated
RDF-5	Processed and densified into pellets, slugs, cubettes, or briquettes
RDF-6	Processed into a liquid fuel
RDF-7	Processed into a gaseous fuel

five categories in EN 15359:2011, based on CV, Cl–, and Hg contents in as-received samples is also not fully developed for the cement industry. An attempt has been made in India to specify the raw segregated combustible fraction (SCF) that could be converted into different grades of RDF for cement kilns. The recommended quality parameters are given in Table 4.4 and the desired specifications of the further processed fuels designated as SRF as distinguished from RDF are presented in Table 4.5 (6). It is expected that the efficacy of utilization of non-recyclable

Table 4.4 Recommended properties of SCF in raw state and for calciner use

Parameter	Raw SCF	SCF for inline calciner	SCF for separate line calciner
Size (mm)	< 400 (mm)	< 75	< 35
Gross calorific value (Cal/g)	> 1500	2500–3500	2500–3500
Moisture (%)	No limit	< 25	< 25
Ash (%)	No limit	< 20	< 20
Chlorine (%)	No limit	< 1.0%	< 1.0%

Table 4.5 Desired quality parameters of SRF

Parameter	SRF for inline calciner	SRF for separate line calciner	SRF for kiln burner
Size (mm)	< 75	< 35	< 20
Gross calorific value (cal/g)	4000–5000	4000–5000	4000–5000
Moisture (%)	< 15	< 15	< 15
Ash (%)	< 15	< 15	< 15
Chlorine (%)	0.8	0.8	0.8

combustible materials present in MSW can be enhanced by introducing proper specifications of RDF, SCF, and SRF.

Biomass residues

Biomass is the organic matter derived from agriculture, forestry, and living organisms. It is renewable, widely available, and carbon-neutral. Rice and paddy husk, mustard husk, bamboo dust, saw dust, food industry waste, wood chips, ground nut shells, coconut shell, tree bark, and straw are some examples of the materials commonly used or that can potentially serve as alternative fuel. It is estimated that about 500 million metric tons of biomass in the form of agricultural crop residues is generated in India. Since biomass is extensively used as domestic fuel, fodder, power generation, etc., the surplus available to serve as potentially alternative fuel in the cement plants may be in the range of 120–150 million metric tons per year. No prognosis has been made on its increased availability in future. A wild shrub called "juli flora," which grows uncontrolled in barren uncultivated land, is considered as a good biofuel. It is reported that power plants are operating on wood chips of juli flora. The composition and calorific value of certain varieties of biomass fuels are given in Table 4.6 (7).

From the above table, it is evident that the moisture content in the biomass residues varies in the range of 7 to 11% but the moisture content in the bamboo dust may be highly variable. Sometimes the as-received sample may show as high as 30% moisture, which comes down to less than 3% on atmospheric drying. A commonly observed value is shown in the table. Along with relatively high moisture content the agricultural waste materials have lower calorific value but due to their fixed carbon content and high volatile matter they turn out to be a group of easy-burning alternative fuel. The ash content in the rice straw and rice husk is much higher than that of the other biomass residues. In one particular study (8) it was shown that mustard husk contained higher alkalis compared to other agricultural residues. Hence it is always desirable to make a comparative evaluation of various biomass residues in respect of their sulfur, chlorine, alkali, and ash content before use. Some of the waste materials are known to be difficult to grind and, if ground together with coal, the mill output suffers. Hence their direct feeding may be preferred and the partial substitution of coal with these wastes does not have any adverse effects on the process and the product. In countries with high agricultural base, biomass and biomass residues can technically substitute 5–7% of the fuel requirement of cement plants.

Scrap motor tires

In the cement industry, scrap tires have been used as an auxiliary fuel for a long time. In older kiln systems without preheaters and precalciners, complete tires were fed into the kiln inlet through a simple mechanism to cover a relatively large proportion of the overall fuel requirement. However, in modern kiln systems with a separate precalciner vessel and a high calcination rate, the use of complete tires via the kiln inlet is restricted to a relatively small proportion of around 5% of the overall fuel requirement. To use higher proportions in the precalciner kilns, the shredding of tires into pieces of a maximum size of 50 × 50 mm becomes

Table 4.6 Fuel characteristics of some biomass fuels

Type of biomass	Gross CV (cal/g)	Net CV (cal/g)	Moisture (%)	C (%)	H (%)	N (%)	S (%)	Ash (%)	O (%)
Saw dust	3952	3798	10.81	43.65	5.83	0.52	0.11	2.05	47.84
Rice straw	3382	3228	11.01	38.69	5.78	0.59	0.11	14.54	40.29
Groundnut shell	4091	3937	10.40	44.55	6.02	0.02	0.09	5.16	44.16
Bamboo dust	4127	3973	9.90	46.89	6.77	0.34	0.10	2.09	43.81
Bagasse	4172	4018	8.38	46.73	6.24	0.44	0.01	3.49	43.09
Cane trash	4032	3878	10.33	43.76	6.19	0.93	0.10	5.91	43.11
Rice husk	4067	3912	12.24	34.52	5.55	1.35	0.15	24.34	34.09
Wheat straw	3980	3826	7.96	40.31	5.08	0.61	0.11	7.04	46.95
Coir waste	4312	4156	7.86	45.16	5.26	0.20	0.18	2.14	47.06

necessary. As fuel, the scrap tires are characterized by a high calorific value (> 6500 cal/g), a high volatiles content (> 60.0%), a low to moderate ash content (7–20%), and a moderate content of harmful substances, similar to most primary fuels.

According to the Rubber Manufacturers Association, the USA dispenses 379 million scrap tires each year, of which 52% or about 200 million pieces are burned as fuel (9). The cement industry in the USA is a very large user of tire-derived fuel. In India, according to the data of the All India Tire Manufacturing Association, 0.83 million metric tons of scrap tire was available in the year 2011–2012 and, considering a modest 5% increase per year, it is projected that the country would generate 1.32 million metric tons of scrap tires by 2025. After providing for the recycling of rubber and fuel requirement of small and medium enterprises, it is estimated that the cement industry in the country would in future have about 0.8 million metric tons of waste tire to be used as an alternative fuel.

Hazardous waste The national inventory of hazardous waste collated by the Central Pollution Control Board in India indicates that 41,523 manufacturing units generate 7.90 million tons every year, 42% of which is used for land filling, and 50% is considered recyclable, leaving 0.60 million tons to be disposed of through incineration. This mode of disposal of hazardous waste does not permit gainful use of the fuel value available in the waste materials and in addition it adds to the consumption of fossil fuel and emission of greenhouse gases. It is also projected that by 2025 the availability of hazardous waste necessitating incineration will cross 1 million tons. Its use as an alternative fuel in cement kilns has therefore been approved by the authorities concerned with the provision of appropriate emission norms.

The hazardous waste materials are mostly discharged in the form of slurry or sludge from petroleum refineries and manufacturing units of dye and dye intermediates, basic drugs, benzene-based organic chemicals, paint and coating materials, etc. The manufacturers of printing ink, printing companies, solvent recyclers, waste oil dealers, etc., also generate such wastes. Because of the proliferation of such waste materials, there has been some attempt by some pollution control authorities to draw up a guiding specification for such hazardous liquid waste fuels for cement kilns. An illustration of such a specification is given in Table 4.7.

The high calorific value related to their high volatiles content shows a great potential for this type of sludge to be used as an alternative fuel. The high sulfur content in the sludge and sludge ash may have to be taken care of through appropriate process steps.

Process-wise, the sludge samples available in semi-liquid state may be roll-pressed first to remove the water and other aqueous acidic contaminants. The cake so obtained can be easily handled and dried in a dryer. Sludge tests have indicated that it becomes soft at 75–80°C and becomes hard on cooling to the room temperature. The sludge turns into a brittle solid on heating to about 150°C for 3–4 h, followed by cooling. This processed sludge can be ground and used as a pulverized fuel.

Table 4.7 Tentative specification of hazardous waste fuels for cement kilns

Sl. no.	Parameters	Limiting values
1.	Heat content (Btu/lb.), min (kcal/kg)	10000 (5555)
2.	Suspended solids (%), max	30
3.	Sulfur (%), max	3
4.	Nitrogen (%), max	3
5.	Halogens (%), max	5
6.	pH	4–11
7.	Water as a separate phase (%), max	1
8.	Ash content (%), max	10
9.	Barium (ppm), max	5000
10.	Chromium (ppm), max	1500
11.	Lead (ppm), max	2500
12.	PCB (ppm)	None detected (< 50 ppm)

The sludge ignites at a temperature of 300°C, leaving behind a char-like material that burns at a temperature of 700–800°C. The sludge can be fed by screw conveyor along with the raw meal at the kiln inlet after drying with the preheater exit gas. Alternatively, the sludge can be kept in a fluid state at an appropriate temperature and pumped into the rotary kiln through an auxiliary burner.

Sewage sludge Of all varieties of sludge, sewage sludge, which is a by-product of urban wastewater treatment, has been used quite extensively in the cement industry and more particularly in the European countries and in the USA (10). Wastewater with a certain level of contamination and having about 5% solids in suspension, originating from residences, rainwater, and commercial establishments, is treated through processes of thickening, mechanical dewatering, conditioning, and drying. The main routes of sludge disposal are landfilling, agricultural application, incineration, and dumping into the sea. The generation of the sludge about a decade back in the European Union, barring the Eastern European countries, was more than 9.0 million tons (with 10% moisture). The situation in the USA was also similar and reached 6.3 million tons of dry biosolids in the year 1998 and showed a 16% per capita increase in generation. No reliable statistics of the generation of sewage sludge in India is readily available.

The disposal modes like landfilling, dumping, and agricultural application have their own environmental compulsions and limitations. Independent incineration is an energy-intensive process. Hence, burning in cement kilns has been receiving serious attention. Sewage sludge in its original liquid form has a very high water content. Mechanically dewatered sewage sludge with a water content of 55–75% would be combustible, but, from the process point of view, this would not be preferable.

Therefore, the sewage sludge is thermally dried to form a powder or a granulated mass with residual moisture of 5–25%.

The compositional details of the sewage sludge are given in Table 4.8, 4.9, and 4.10.

Generally, the use of sewage sludge for the production of cement clinker is an excellent option, since the high ash content of sewage sludge acts like a secondary raw material, while the organic matter is used as fuel. Although its calorific value is low, it is possible to achieve high flame temperature. As it has high oxygen content in its composition, the stoichiometric ratio of air (kg) to heat released is quite low compared to that of high-ash coal. However, for proper evaluation of combustion of

Table 4.8 Range of constituents of sewage sludge

Constituents	Typical range
Dry matter (DM) (g/L)	7–30
Volatile matter (%), DM	50–77
Fat (%), DM	10–18
Protein (%), DM	18–34
Fiber (%), DM	10–16
Calorific value (kCal/kg DM)	2580–3612

Table 4.9 Chemical composition of sludge

Parameters	Typical range (% DM)
Ash	31.0–44.6
Carbon	29.0–30.9
Hydrogen	3.8–4.6
Nitrogen	3.1–4.5
Sulfur	0.8–1.1
Chlorine	0.10–0.17
Oxygen	17.56–2650

Table 4.10 Typical composition of dry sludge ash samples

Oxides (%)	Ash 1	Ash 2
SiO_2	33.6	41.3
Al_2O_3	15.6	14.3
Fe_2O_3	3.6	6.8
CaO	22.2	19.8
MgO	2.3	3.2
P_2O_5	16.7	8.7
TiO_2	2.3	1.2

sewage sludge in cement kilns, the following parameters are important (10): P_2O_5, chlorine, nitrogen, and particle size. As a general rule, the P_2O_5 content in clinker should not exceed 0.5% as with higher concentration, the belite phase tends to increase. The chlorine content in the raw meal should be lower than 0.015% for avoiding the installation of a bypass system in the kiln. The total nitrogen content in the dried sludge in the form of nitrates and ammoniac can reach up to 8.0% of dry matter, which is significant and cannot be ignored for evaluating the possible effect on NO_x emission. The particle size of the sludge is important for both pneumatic transport and burning. Taking all these parameters into consideration, it is estimated that the theoretical substitution rate for sewage sludge would lie in the range of 0.10–0.17 kg sludge/kg clinker.

Sewage sludge with high mercury content may be problematic for the process. Since this element is highly volatile, it is not arrested in clinker. Instead, mercury is enriched in the external circuit between preheater and dust capturing facility. In this case, the emission of mercury is avoided by not recirculating the dust.

Industrial plastic waste

It is tentatively estimated that at least 100,000 tons of plastic waste are generated every day in India. Plastic waste that cannot be recycled can be used in cement kilns. The problem often encountered for use is the non-segregation of the plastic waste from other waste. Industrial plastic waste is particularly available from the waste paper-based paper mills, which have 1–2% plastic waste in wrappings, laminations, covers, and so on. It is also available in the form of discarded packaging bags, drums, etc. from various manufacturing units.

Miscellaneous alternative fuels

Apart from the categories described above, there are several other sources of alternative materials. A few such types are illustrated below.

- **Meat and bone meal.** Since in the European countries there is a prohibition of using meat and bone meal (MBM) in agriculture, new disposal methods have had to be found for the huge amount of material that is generated. Similar to other alternative fuels mentioned earlier, use of MBM as an auxiliary fuel in the burning process of cement clinker is a very good possibility for disposal and utilization. Animal meal is characterized by a relatively high calorific value and a high content of volatiles, but at the same time, the high levels of chloride and phosphate (see Table 4.1) have to be given due consideration in respect of the process and cement quality.

- **Spent pot-liners.** Spent pot-liners from the aluminum industry present another potentially valuable fuel resource. It is estimated that approximately 650,000 t/y of spent pot-liners are available globally. Their use had not been popular in the past due to the apprehension of fluoride and cyanide residues. The presence of sodium may also discourage use.

- **E-Waste.** Discarded electronic waste is the fastest-growing stream of waste in industrialized countries. This trend is also evident in

118　CEMENT PRODUCTION TECHNOLOGY

India and the opportunity to use this waste as an alternative fuel is being explored. The E-waste Rules were introduced in India in 2011 and were made effective from May 2012. There are 128 producers in 11 states, with 34 authorized e-waste collection centers in 19 states.

4.5 Systematic quality assessment

A systematic quality assessment scheme is essential to enhance the use of alternative fuels. Any such scheme is formulated with the following broad objectives:

- To handle a multiplicity of waste streams
- To differentiate hazardous from non-hazardous waste
- To obtain reliable test results with both precision and accuracy
- To know the ground state of contaminants and their environmental reaction potential
- To assess the risk of exposure to contaminants

The plan for essentially non-hazardous AFR assessment requires both a system and certain facilities. The system is focused on bulk sampling; the preparation of samples; proximate analysis; the measurement of chloride and sulfate contents, ultimate analysis, ash composition, calorific value, grinding behavior of solid fuels, particle size distribution, and ignition property; and the safety and precautionary measures to be adopted in handling the materials. For the sampling of agricultural waste, tire chips, MSW, industrial waste, rubber waste, etc., one may follow the procedure laid out in CEN/TS 15442. A minimum of 24 samples are to be drawn. For a typical supply of 60 trucks, it is preferable to plan sampling in two groups of 30 trucks each. One sample may be drawn from every two trucks, the selection of trucks being random. In order to prepare the analytical samples from MSW and agricultural waste it is essential to treat them in liquid nitrogen, to use a Fritsch mill, and to pass the material through a 500-µm screen. It is necessary to preserve the samples in sealed containers. For other fuels received in the form of powder or sludge it is essential to prepare the test samples in a dry state passing through a 212-µm screen. For chemical analysis of the properly prepared samples it is preferable to use thermo-gravimetric instruments and for ultimate analysis the facility for CHN&S determination is recommended. To determine the ash composition, it is advisable to use an elemental analytical instrument like the X-ray fluorescence spectrometer.

Measurement of organic pollutants

The measurement of the following eight organic pollutants is important for the cement industry:

- Non-methane volatile organic compounds (NMVOC)
- Dioxins and furans—polychlorinated dibenzodioxins (PCDD) and polychlorinated dibenzofurans (PCDF)—both expressed as a toxic equivalent 1-TEQ

- Polychlorinated biphenyls (PCB)
- Anthracene
- Benzene
- Naphthalene
- Bis (2ethylhexyl) phthalates (DEHP)
- Polycyclic aromatic hydrocarbons (PAH)

In some countries, it is mandatory to measure and report these pollutants in exhaust gases at least once a year. In many other countries they are not measured, neither do they have emission benchmarks. The European Pollution Release and Transfer (E-PRTR) data or the data from the US Environmental Protection Agency (EPA) are primarily relied upon. A study has been reported for the gas sampling and analysis of the above eight pollutants (11). The set-up used for isokinetic gas sampling is shown in Figure 4.2. Depending on the mode of capturing of the pollutants, some changes in the gas sampling set-up were made. The sampling methods and measurement procedures are summarized in Table 4.11.

The results of the measurements showed that the emission of NMVOCs, DEHPs, and benzene were close to E-PRTR annual emission limits, which are 100,000, 10, and 1000 µg. The rest of the pollutants comfortably met their annual emission limits, viz., 0.0001 µg for PCDDs and PCDFs, 0.1 µg for PCBs, 50 µg for anthracene, 100 µg for naphthalene, and 50 µg for PAHs. It should also be borne in mind that the benchmark emission parameters for the organic pollutants are not always available in different countries and they need to be developed for their effective monitoring and control.

Requirements of test facilities

Considering the multiple streams of alternative fuels, including the hazardous varieties that an assessment laboratory has to be equipped to

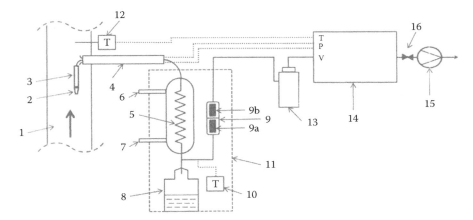

FIGURE 4.2 Typical set-up of facilities for the measurement of organic pollutants in a cement plant. (Reproduced from F. Sladeczek and E. Glodek, The measurement method of organic pollutants emitted into atmosphere by cement and lime industry, *Cement Wapno Beton*, No. 6, 2012, 423–442. With permission of Cement Wapno Beton, Poland.)

Table 4.11 Sampling and analysis of organic pollutants

Pollutant	Procedure/standard	Sorbent	Gas sampling method	Analytical method
NMVOC	EPA 0030 EN 13649:2001	Anasorb 747	Aspiration; measuring and control sorbent	GC/FID chromatography; HP 5890 II series flame ionization detector
PCDDs & PCDFs	PN EN 1948-1-3:2006	PUF	Filtration-condensation; filter + condensate + measuring and control sorbent	GC-MS/MS chromatography; Trace GC Ultra Gas Chromatograph with Thermo ITQ 1100 detector
PCBs	PN EN 1948-1-3:2006	PUF	Filtration-condensation; filtration + condensate + measuring and control sorbent	GC-MS/MS chromatography; Trace GC Ultra Gas Chromatograph with Thermo ITQ 1100 detector
Anthracene	ISO 11338-1-2:2003 EPA 0010	XAD-2	Filtration-condensation; filter + condensate + measuring and control sorbent	GC-MS chromatography; Trace GC Ultra Gas Chromatograph with Thermo DSQ detector
Benzene	EPA 0030 EN 13649:2001	Anasorb-747	Aspiration; measuring and control sorbent	GC/FID chromatography; HP 5890II series flame ionization detector
Naphthalene	ISO 11338-1-2:2003 EPA 0010	XAD-2	Filtration-condensation; filter + condensate + measuring and control sorbent	GC-MS chromatography; Trace GC Chromatograph with Thermo DSQ detector
DEPHs	EPA 0010	XAD-2	Filtration-condensation; filter + condensate + measuring and control sorbent	GC-MS chromatography; Trace GC Ultra Gas Chromatograph with Thermo DSQ detector
PAHs	ISO 11338-1-2:2003	XAD-2	Filtration-condensation; filter + condensate + measuring and control sorbent	HPLC/FLU chromatography; Merck HPLC C 18 PAH cartridge chromatograph with LiChroCART 250-3 column

ALTERNATIVE FUELS AND RAW MATERIALS

handle, its overall preparedness depends not only on the capability of on-site measurement of the critical organic pollutants as described above but also on the ability to conduct special tests on toxicity and several other parameters. The overall facility requirements can be tentatively summarized as follows:

- Proximate analysis
- Ultimate analysis
- Elemental determination (Cl, F, Pb, Cd, As, Hg, Cr, Co, Ni, Th, Cu, V, Sb, Mn, Se, Fe)
- Measurement of total organic carbon (TOC)
- Toxicity characteristics leaching procedure (TCLP)
- Total petroleum hydrocarbons (TPH)
- Organo-chlorine compounds
- Volatile organic compounds (VOCs) and semi-VOCs
- Polychloro-biphenyls (PCBs) and phenols (PCPs)
- Calorific value—gross and net
- Particle size determination (PSD) and particle flow characterization
- Viscosity of liquid waste
- Water content in liquid waste
- Solid content in liquid waste
- Hardgrove grindability tester

It should be borne in mind that environmental science and engineering are relatively young professions compared to other disciplines and depend heavily on both inorganic and organic chemistry. There have been rapid advances in analytical and characterizing tools for complex hybrid materials that one comes across in environmental chemistry. In moving towards reliable assessment of intrinsic quality of waste fuels and their emission impacts, it is important to be prepared with both unconventional expertise and advanced tools for sampling, sample preparation, analysis, and testing, as well as on-site measurements.

4.6 Co-processing of alternative fuels

Co-processing refers to the utilization of suitable waste material in the manufacturing process for the purpose of energy and/or resource recovery and resultant reduction in the use of conventional fuels and/or raw materials. The clinker and cement manufacturing system has certain operational features that make it eminently suitable for co-processing of AFR.

Advantages of cement kilns

Cement kilns have proved to be one of the best incinerators, which burns the waste at a high temperature of about 1400°C and which has long

Table 4.12 Combustion zone conditions in cement kilns versus hazardous waste incinerators

Sl. no.	Parameters	Cement kiln	Waste incineration
1.	Maximum gas temperature	> 2200°C	≤ 1480°C
2.	Maximum solid temperature	1420–1480°C	≤ 750°C
3.	Gas retention times at > 1100°C	6–10 seconds	0–3 seconds
4.	Solid retention time at > 1100°C	20–30 minutes	2–20 minutes
5.	Oxidizing conditions	Yes	Yes
6.	Turbulence (Reynolds' no.)	> 100.00	> 10,000

residence time in the combustion zone, unlike the normal incinerators. A comparison of combustion zone conditions in a cement kiln and an incinerator is given in Table 4.12.

In cement kiln systems, the ignition and burn-out behavior of alternative fuels is essentially influenced by particle size, moisture, ash content, and the content of volatiles. Particle size and moisture can be directly influenced by the kind and degree of processing, while the contents of ash and volatiles cannot be changed.

Cement kilns are generally suitable for the safe destruction of most types of hazardous and toxic organic material. It is generally considered that all organic compounds are adequately destroyed if exposed to a temperature of 1200°C for a residence time of two seconds under oxidizing conditions. The conditions in the burning zone of cement kilns exceed these requirements by a wide margin. Cement kilns further show a high degree of turbulence in the burning zone and a high thermal stability, which is a function of their large size and the amount of heated material in the kiln. Thus, even if a cement kiln is forced into an emergency shutdown, all hazardous material in the kiln should be completely destroyed, provided that automatic cut-offs prevent further injection of such material.

In addition, cement kilns operate under alkaline conditions. This means that, provided the rate of input of acidic precursors is controlled so as not to overload the system chemistry, all acidic gases such as HCl, SO_2, H_2SO_4, and HF generated during burning are either absorbed by the kiln charge or combined with the kiln dust in the form of harmless inorganic salts. A portion of these salts is generally acceptable in the clinker, while an excess may have to be discarded as kiln dust or bypass dust.

Metals are not destroyed in the burning process, although most of them are trapped in the clinker and in the collected dust. Materials containing significant quantities of heavy metallic constituents including lead, arsenic, mercury, cadmium, chromium, and thallium are not candidates for burning in cement kilns. Other metals, such as those typically found in soil (silicon, aluminum, iron, magnesium, calcium), constitute the bulk of the ash from the combustion of hazardous organic materials and are readily absorbed in the cement product since they are primary constituents of cement.

The hazardous materials, whether in gaseous, liquid, solid, or sludge form, must be introduced into the kiln at a point where the above-mentioned time-temperature conditions can be met. This means that the gas must be maintained at a minimum temperature, for a period of at least two seconds from the point of introduction. This minimum temperature is a function of the hazardous material; it is 1200°C for materials containing polychlorinated by-phenyls (PCBs); it can be lower for other materials. Emissions of carcinogenic substances should be limited to the largest possible extent by giving consideration to the principle of production.

It is therefore essential that the waste treatment and objectives of use must be coordinated with regard to process engineering and quality. Solid substitute fuels with an average lump size of 250–80 mm to be fed via the kiln inlet or a calciner of a rotary kiln are processed differently. In actual practice, there are two groups of alternative fuels. The first group is characterized by high to very high contents of volatiles and requires less grinding. Mostly, rough shredding (< 30 mm) is sufficient to ensure the desired combustion behavior. Unfortunately, these fuels are often inhomogeneous and difficult to handle. The other group of secondary fuels, like petroleum coke from oil refining or spent pot-liners from aluminum production, have lower reactivity (due to their low content of volatiles). Therefore, similar to anthracite, these fuels have to be ground very finely (< 5% residues on 90 µm) to enable sufficient burn-out at the calciner or sintering zone.

The grinding technology for many solid substitute fuels is still in an evolving stage. It seems that the operating principles of the Vertex mill, originally designed for grinding food articles, can also be used successfully for grinding solid substitute fuels. Depending on the grinding resistance, reduction ratios of up to > 100:1 are possible.

Co-processing is an approved activity in India and the Environment (Protection) 3rd Amendment Rules 2016 have defined the mandatory emission norms for cement plants operating with alternative fuels, which are briefly presented in Table 4.13, along with the EU Waste Incineration Directive (12).

Table 4.13 Emission norms of cement plants with co-processing of wastes

Parameters	EU directive	Indian norms
Particulate matter (PM) (mg/Nm^3)	30	30
SO$_2$ (mg/Nm^3 at 10% O$_2$)	Negotiable	100 (pyrite S < 0.25% in limestone); 700 (pyrite S 0.25–0.35% in limestone); 1000 (pyrite S > 0.5%)
NO$_x$ (mg/Nm^3 at 10% O$_2$)	500 (800 for existing plants)	600 for plants commissioned after August 25, 2014; 800 for prior plants with inline calciner; 1000 for prior plants with mixed streams of calciners
HCl (mg/Nm^3 at 10% O$_2$)	10	10
HF (mg/Nm^3 at 10% O$_2$)	1	1
TOC (mg/Nm^3)	Negotiable	10
Hg and its compounds (mg/Nm^3 at 10% O$_2$)	0.05	0.05
Cd + Tl and their compounds (mg/Nm^3 at 10% O$_2$)	0.05	0.05
Sb + As + Pb + Co + Cr + Cu + Mn + Ni + V (mg/Nm^3 at 10% O$_2$)	0.50	0.50
Dioxins + furans (ng/Nm^3 at 10% O$_2$)	0.1	0.1 (expressed in ng TEQ/Nm^3)

4.7 Systemic requirements for using alternative fuels

A pre-processing industry has sprung up in some countries to manage and broker various waste streams from numerous generating sources, which has boosted up the use of alternative fuels in those countries. Some of the cement plants have also set up integrated pre-processing plants in order to adopt the AFR strategy more effectively. The interface of the waste management and recycling industry with cement plants along with the facilities required there is shown in Figure 4.3 (13). The pre-processing industry has to be more alert and concerned with the mapping of waste material sources and needs to have an efficient infrastructure for collecting the waste materials, which can be treated at those units by undertaking operations like sorting, crushing, shredding, blending, pressing, etc. Mostly the business aim of this industry is to produce SRF or RDF with consistent quality parameters acceptable to the cement production units. It is also feasible to make the pre-processing plants capable of producing solvent fuels that are produced by blending different liquid wastes with strict specifications on water content, heating value, and viscosity. Once prepared to required specifications, the solid or liquid fuels are transported to cement plants for safe storing, handling, and burning without any adverse environmental and health issues for the employees or to the community.

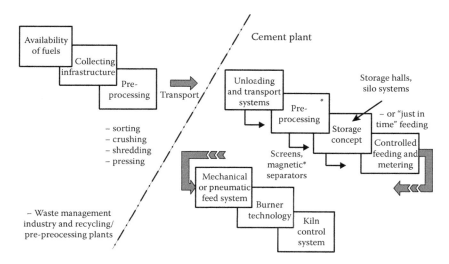

FIGURE 4.3 Interface between the waste management and recycling industry and the cement plants. (Reproduced from Cement International, Storing, metering and conveying solid secondary fuel—Design basis and equipment, *Cement International*, Vol. 5, 6/2007. 43–53. With permission from Cement International and Schenck Process GmbH.)

Storing, dosing, and conveying

Specially designed installations are required to control the use of waste from its arrival at the plant to its entry into the kiln. These comprise storage systems, dosing systems, and conveyance systems to the kiln. Designs vary depending on the characteristics of the alternative fuel used and the space available at each plant. The manner in which the product is received and stored on its arrival and the way in which it is put in the kiln will therefore have to be adapted to the plant in question. However, certain basic features are common to all plants (13,14). The AFR receiving system and the docking station are illustrated in Figures 4.4 and 4.5, respectively.

There are, however, two types of installations: fixed and mobile. Further, each installation, fixed or mobile, comprises four basic sections:

i. Waste storage
  - Storage yard with pit
  - Receiving hopper for unloading trucks with dump trailer
  - Receiving hopper for unloading trucks with moving floor trailors
  - Silo

ii. Conveyance to the dosing station
  - Single-rail hoist block with longitudinal movement or bridge crane with transverse and longitudinal movement, both with grab bucket
  - Mechanical conveyor

iii. Dosing station
  - Mobile or fixed station of modular design
  - Fixed station in one module

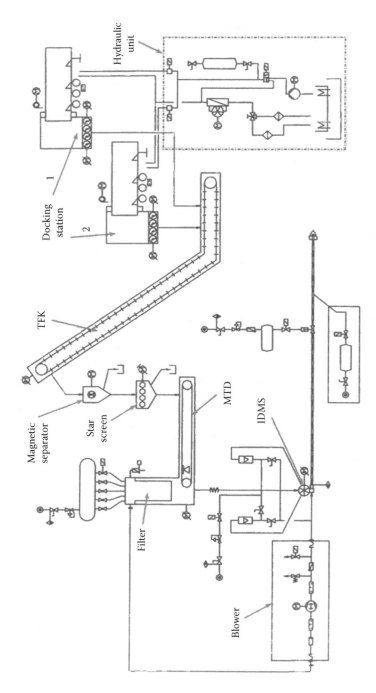

FIGURE 4.4 AFR receiving plant with walking floor container. (Reproduced from Cement International, Storing, metering and conveying solid secondary fuel—Design basis and equipment, *Cement International*, Vol. 5, 6/2007. 43–53. With permission from *Cement International* and Schenck Process GmbH.)

FIGURE 4.5 Docking system with walking floor container. (Reproduced from Cement International, Storing, metering and conveying solid secondary fuel—Design basis and equipment, *Cement International*, Vol. 5, 6/2007. 43–53. With permission from *Cement International* and Schenck Process GmbH.)

iv. Conveyance to the kiln
- Pneumatic conveyance
- Mechanical conveyance

Additionally, two other factors must be considered when designing an installation:

- Dosing of tires and whole large packages
- Automation and control

Some of these installations are elaborated below.

Storage yard with pit This is a rectangular yard approximately 4 or 8 m wide by 14 m long, with a pit 4 m deep, into which the trucks unload directly. This type of yard offers a large storage capacity that is clean and simple, as all the material remains inside the pit below the ground level of the plant.

For a storage yard 8 m wide, a bridge crane with longitudinal and transverse movement is used to allow the collection of material at any point on the yard. For a yard 4 m wide, a hoist block with only one direction of movement—longitudinal—is usually used. The operating results are the same as for the 8m yard, but only half the storage capacity is available. Both systems incorporate a hydraulically operated "octopus"-type grab bucket to carry the product in successive loads to the dosing unit.

The dosing station is situated at one end of the yard, usually under a receiving hopper into which the grab bucket drops its load and under which a belt weigh feeder is installed. The dosed product is discharged onto a conveyance system going to the kiln, which may be pneumatic for particle sizes less than 30 mm, and mechanical for larger particle sizes. If two alternative fuels are to be metered simultaneously, or one is to have a fixed supply and the other variable, a wholly symmetrical inline yard may be built, with a double-dosing system and sharing the same bridge crane and conveyance system to the kiln.

The storage yard and pit solution is ideal for all alternative fuel types, up to a maximum particle size of 100 mm.

Silo storage systems It is well-known that silos permit large storage capacities because of their vertical configuration. They are suitable for products with small particle size and also for powdery and easily extractable products, such as sewage sludge, animal meal, etc. However, for some materials, such as animal meal, the silo and its conveyance system need to be specifically designed to suit material characteristics, especially with regard to their efficacy of extraction.

One of the main advantages of using silos for the storage of powdery products or products with a very small particle size is that filling can be done by pneumatic conveyance. Another advantage is that the entire material can be fed to the dosing station completely in line by gravity, thus saving mechanical conveyance and simplifying the installation. This system is therefore preferred for powdery products or slippery products with a very small particle size. The flow sheet for silo storage is given in Figure 4.6.

Dosing and conveyance to kilns The best dosing system for alternative fuels is the belt weigh feeder, mainly because most of these fuels have a low density and so it is necessary to work with a large volume of load per meter. Depending on the type of fuel to be used, the belts may be made of rubber or synthetic material, with special characteristics such as greaseproofing if necessary. When pneumatic conveyance is used to send the dosed product to the kiln, a layer breaker is installed at the belt weigh feeder's outlet, which breaks up the layer of dosed material so that it does not fall into the conveying hopper in large lumps, thus ensuring a more even feed to the pneumatic conveyor.

The pneumatic conveyance to the kiln's burner usually includes the following components:

- Input hopper into the pipeline.
- Rotary valve, connected directly to the dosed product inlet hopper, for feeding the material to the pneumatic conveyance piping. Vertical pitch is generally preferred as conveying air cleans the rotor.
- Conveyor piping, through which the product is sent to the kiln's burner. The pipe dimensions are calculated with respect to the size of the dosed product, the flow rate, the length of the conveyance, and any bends that may be present on the route.

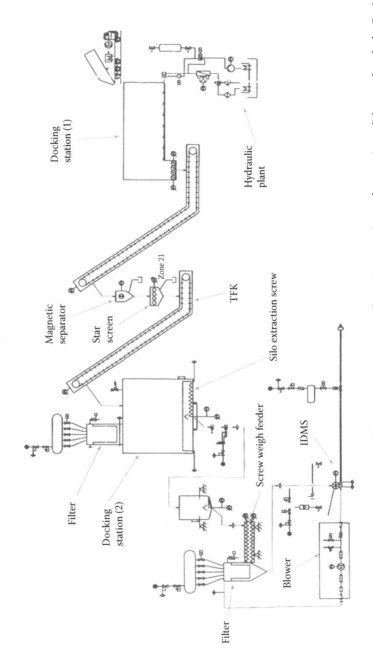

FIGURE 4.6 AFR storage silo in a plant. (Reproduced from Cement International, Storing, metering and conveying solid secondary fuel—Design basis and equipment, *Cement International*, Vol. 5, 6/2007. 43–53. With permission from *Cement International* and Schenck Process GmbH.)

- Instrumentation for controlling the pneumatic conveyance, such as pressure converters, pressure switches, and pressure gauges distributed at different points along the route for full pressure monitoring.
- Roots blower for pneumatic conveyance by thrust—triple-lobe for an air flow rate of about 10 m^3/min with a maximum thrust pressure of 1 bar.
- Soundproof cabin. The noise generated by a blower of these characteristics is 91 dBA, so it must be mounted in a closed room or a soundproof cabin. Using such a cabin made of steel and insulated with fiberglass, the perceptible noise level drops to 74 dBA. To prevent the blower overheating, the cabin usually includes a fan.
- A maximum particle size of 30 mm to permit problem-free pneumatic conveyance.

Mechanical conveyance of the dosed fuels to the kiln mainly depends on the particle size of the product and sometimes on the kind of material. As particle sizes of over 30 mm may cause problems in pneumatic conveyance, mechanical conveyance is in practice for larger particle size, up to approximately 100 mm. If the mode of conveyance is mechanical, the point of insertion in the kiln system normally changes to the precalciner or the preheater. This type of mechanical conveyance entails far higher investment cost than pneumatic conveyance, as it has to raise material several meters from the point of reception to the point of delivery. To avoid false air entering into the process, a double or triple gate is used at the material's entrance to the precalciner. When the dosing system is not in operation, the gate remains completely closed. The mechanical conveyance comprises different equipment depending on the specific needs of the plant, and may include belt conveyors, worm screws, apron conveyors, chain redlers, bucket elevators, and so on.

Dosing of tires and whole large packages For the installation to dose whole tires, or whole pressed bales of other products such as textiles, plastic, or any other waste with an approximate volume of 0.6 to 0.8 m^3, the facility that needs to be created for reception and storage of the fuels as well as for handling and conveyance up to the point of delivery in the precalciner or preheater turns out to be expensive. The dosing and kiln-feed system comprises two components. The first is a rubber belt or roller conveyor 6–8 m long and 1 m wide, which receives the package or tire from a preceding storage and conveyance system. The width of the belt can generally convey packages of up to 800 mm in width, and it has the necessary control devices to ensure that two packages do not enter the weighing system at the same time. The second is a weighing conveyor of approximately 1000 × 1000 mm with a weighing platform mounted underneath it, and a nominal capacity equivalent to the heaviest expected working weight. When working with whole tires or compact bales of fuel, the control system calculates the time in which a tire has to be unloaded into the kiln, depending on the weight measured beforehand and the flow rate set accordingly.

Automation and control The level of automation required in each installation will determine the rest of the control components, such as level reading and full mapping of the work area, grab load detection, communication between the fixed system and the mobile system, management of the safety interlocks for access to the warehouse, communication with the central control system, etc.

Precalciner design considerations

The design of precalciner systems for burning alternative fuels is essentially influenced by NO_x and CO formation within the combustion process. It is understood that the NO_x generation is related to the content of volatile matter and nitrogen in the fuel. There is a strongly decreasing trend of NO_x generation in precalciners when the volatiles content in the fuel increases. Further, it has also been observed from the comparative burning behavior of anthracite and petroleum coke that anthracite gives rise to lower NO_x generation despite its lower volatiles content, primarily due to its lower nitrogen content (12). Further, it is also known that precalciner design and atmosphere for burning significantly influence the NO_x emissions. Generally speaking, an inline calciner operated with a high-temperature oxidation zone yields lower NO_x emission than a separate-line calciner. It is also observed that with regular designs of precalciners, while it might be possible to meet the NO_x emission limit of 600 mg/Nm³ at 10% O_2 with high-volatile fuels, without secondary reduction measures, it is unlikely to achieve this performance when firing a low-volatile fuel like the petroleum coke. Hence, in order to deal with various streams of secondary and waste fuels, a multistage combustion system has become popular for NO_x control (Figure 4.7).

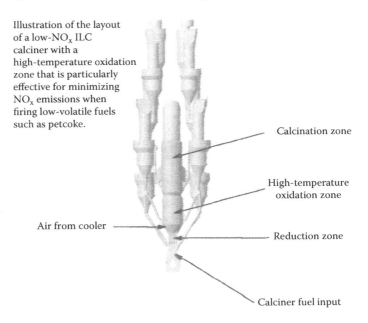

FIGURE 4.7 Adaptation of precalciner systems for AFR. (Courtesy of FLSmidth A/S.)

For some of the secondary fuels that are not sufficiently prepared and ground, the precalciner path needs to be extended for more retention time to ensure complete burnout. For very coarse secondary fuels, there are two other options. An ignition chamber is a special module for burning fuels with poor ignition properties like coarse petcoke and anthracite. It is integrated in the pure air side of the precalciner. A hot-spot burner creates a zone of high temperature (1000–1200°C) and the burnout happens in the extended calciner. A combustion chamber is the appropriate solution to burn very coarse fuels, for example, petcoke or anthracite ground to > 9% residue on 90 μm, as well as other lumpy secondary fuels. These fuels can be fed into the hot core zone of the combustion chamber at temperatures of greater than 1200°C. The partly burnt-out fuels are fed directly to the calciner, where they are lifted up with the gas stream and are fully burnt out in the calciner. A small amount of fuel particles may fall down in the inlet chamber and burn out in the kiln.

Use of a special combustion device "hotdisc," designed by major equipment supplier FLSmidth and already in operation in some plants, is an appropriate example in this regard. It is a refractory-lined special combustion chamber that is integrated with the precalciner system (Figure 4.8). The heavier and lumpy alternative fuels burn on the hotdisc table and residence time can be adjusted for different alternative fuels. Only the burnout ashes fall into the kiln inlet and are incorporated in the clinker. It is also energy efficient as it operates on recuperated process heat.

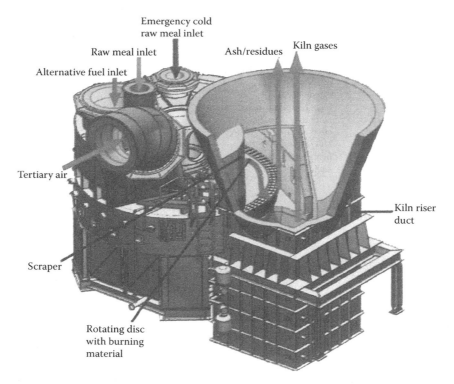

FIGURE 4.8 "Hotdisc" combustion for AFR. (Courtesy of FLSmidth A/S.)

4.8 Gasification technology

Modern gasification systems are designed to convert carbon-containing materials into a synthetic gas predominantly containing CO and H_2, which finds use in power generation or as a fuel for heating and burning purposes. Coal gasification processes differ widely but the three basic reactor technologies are the moving bed, the fluidized bed, and the entrained- flow gasifier (15). The moving-bed gasifier is operated counter-currently with steam and oxygen at temperatures much below the occurrence of slagging in coal (~ 550–600°C). Since this reactor design cannot gasify all types of coal, viz., the caking variety, a high-temperature slagging reactor has also been developed on similar design principles. A fluidized-bed gasifier is a back-mixed reactor in which coal particles in the feed are well mixed with coal particles already undergoing gasification. The gasifier operates at atmospheric pressure and moderately high temperatures (say, 1000°C). However, in this type of gasifier, significant amounts of unreacted carbon are lost both in the product gas and the discharged ash. Hence, this kind of gasifier is more suitable for highly reactive coal and in Germany it has been used for gasifying lignite. For higher operating temperatures, the fluidized-bed reactors produce less impurities than the conventional moving-bed gasifiers. The third type, i.e., the entrained-flow gasifier, is a plug-flow system in which coal particles react concurrently with steam and oxygen at atmospheric pressure. The reactor operates at higher temperatures than the fluidized-bed type and the residence time of coal particles is only a few seconds. The product contains almost pure CO and H_2 and hardly any methane and higher hydrocarbons, which are present in the moving-bed and fluidized-bed gasifiers. Hence, the cold gas efficiencies of the gasifiers are different.

While gasification technology in various forms has been in commercial use for quite some time, the first industrial-scale gasifier for a cement kiln precalciner system was successfully commissioned just a few years ago in Switzerland. The technology is now coming of age in the cement industry and shows high promise for the future. A gasifier design for the cement industry is shown in Figure 4.9 (16).

Use of tire-derived and other secondary fuels through gasification

Tires are already in widespread use in the cement industry but for operational convenience they are shredded before use. The key advantages for the gasifier system are that whole tires can be fed to the reactor without shredding and the proportion of substitution may rise to 40% of the overall fuel requirement.

The tires are taken from storage via a conveying system and metered into the gasifier through a special feeder/airlock (see Figure 4.10). A specific amount of tertiary air is diverted before the calciner into a bypass duct and injected into the reactor vessel. Inside the reactor, under reducing conditions the tires are decomposed to a gas, wire, and residual coke. The product gas leaves the gasifier via the product duct and is burnt in the calciner. The solids are transported out of the reactor via a discharge system. The heavier (wire) components fall through to the kiln

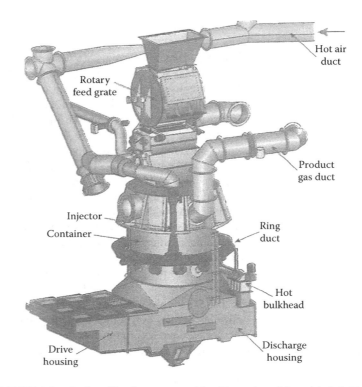

FIGURE 4.9 Coal gasifier for a cement kiln. (Reproduced from Mark S. Terry, *Indian Cement Review*, Part I, April 2005, 13–18 & Part II, May 2005, 15–20. With permission.)

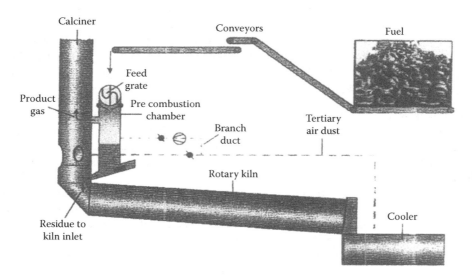

FIGURE 4.10 Waste tire gasifying system. (Reproduced from Mark S. Terry, *Indian Cement Review*, Part I, April 2005, 13–18 & Part II, May 2005, 15–20. With permission.)

inlet where they are fused and eventually incorporated into the clinker as they pass through the kiln. The lighter coke particles are entrained in the gas stream and flow into the calciner, where they burn out together with the product gas.

Although the gasifier had its first success with tires, it can be designed for a broad range of alternative or secondary fuels that are available in lump form. Depending on local availability and economics, it is technically feasible to utilize a variety of potential waste materials such as carpet scraps, vinyl upholstery, industrial rubber waste, or plastic car components. This development may also unfold an opportunity to supply synthetic gas to the main burner. It also opens up the opportunity of more efficient use of materials such as hazardous oil-bearing secondary materials from the refining industry, municipal sewage sludge, hydrocarbon-contaminated soils, and chlorinated hydrocarbon byproducts, all of which have already been used successfully in the gasification process. The challenge of the future is to further develop those systems and adapt them to the cement manufacturing process. This is especially encouraging since many of these materials already have acceptance within the cement industry.

4.9 Alternative raw materials

In recent years, cement manufacturing has successfully made use of a number of wastes and byproducts as supplementary raw materials. Typically, these materials are lime-bearing carbonates, paper sludge, lime waste from water purification plants, sludge from sugar and fertilizer industries, slag from metallurgical plants, coal ash from power plants, red mud from the aluminum industry, mineral and ore tailings from the mining industry, catalyst fines, sludge from sewage and effluent treatment plants, foundry sand, etc. Depending on their composition and characteristics they supplement different natural raw mix components. As a result, they help conserve natural raw material resources on one side and recycle the polluting industrial wastes on the other. A detailed treatise on the subject is given in (17).

Some of the major industrial wastes in India, which could find use in cement industry, are:

- Lime-rich sludge from paper pulp, sugar, ammonium fertilizer, and calcium carbide plants, estimated at about 5–6 million metric tons per year.
- Fluoride-bearing sludge and mining rejects amounting to about 0.5 million metric tons per year.
- Red mud from the aluminum industry, estimated at about 0.2 million metric tons per year.
- Blending components for cement grinding, such as fly ash (about 200 million metric tons per year) and blast furnace slag (about 10 million metric tons per year).
- Steel slag, amounting to about 20 million metric tons per year.

Of the above wastes, the lime-rich sludge from the paper, sugar, fertilizer, and carbide industries generally has CaO content as high as 75–77% on a loss-free basis and, consequently, has the potential of substituting expensive higher grades of limestone in raw mix; further, due to the fine powdery nature of these wastes they contribute to increased throughput of the raw mill and improved burnability of the kiln feed. However, the carbide sludge and the fertilizer sludge contain high sulfate, which limits their use as a raw mix component. The fluorine-bearing wastes may act as effective mineralizers, if the compatibility and benefits of using them with the main raw materials of the cement plant are established through trials in the laboratory and in the plant. Wastes like red mud, fly ash, iron and steel slags, etc. can be used as corrective materials, although fly ash and granulated blast furnace slags are more effectively used as mineral additives in producing finished blended cements. The major areas of concern in using the alternative raw materials are their economic and pollution-free transportation, storage, and feeding in the production process.

4.10 Environmental aspects

During the combustion of alternative fuels, national emissions standards are applied by the concerned authorities and mostly implemented by permits in each case. Within the given standards, the technical specifications for co-processing and the waste to be used may vary from country to country or even from one cement plant to another. Special attention is necessary for reliable emission control and monitoring systems.

Dust emissions come from the following four sources in a plant:

- kiln/raw mill main stack
- clinker cooler stack
- cement mill stack
- dust-capturing outlets of material transfer points

By providing bag filters or efficient electro-static precipitators it is technically feasible to achieve particulate emission as low as 10 mg/Nm^3. Hence, implementation of the widely stipulated limit of 30 mg/Nm^3 does not appear to be difficult.

Sulfur dioxide results from the oxidation of sulfides or elemental sulfur contained in the fuel and raw materials in a favorable temperature and oxygen regime. The alkaline nature of cement particles, however, allows SO_2 to be absorbed, reducing its emission to the atmosphere. The emission is generally below 100 mg/Nm^3 but is strongly related to the pyrite contained in limestone. If the emission exceeds 1000 mg due to high pyrite content in raw material, it may be necessary to take recourse to secondary measures for emission reduction such as the addition of absorbent or wet scrubbing. The desulfurization of flue gas may lead to generation of $CaSO_4$, which will have to be disposed of in a gainful manner.

Thermal NO_x results from the oxidation of molecular nitrogen in air in the kiln's burning zone where the temperature is greater than 1200°C. The fuel NO_x results from the oxidation of nitrogen in fuel at lower combustion temperatures. The range of total uncontrolled emission ranges from 300 to 2000 mg/Nm³. The NO_x emission may be reduced by applying the following measures:

- use of low-NO_x calciners
- staged combustion by feeding raw materials with a high content of volatile organic compounds
- selective non-catalytic reduction (SNCR)
- selective catalytic reduction (SCR)

The primary feature of low-NO_x calciners is an oxygen-deficient combustion zone to form molecular nitrogen and other products including CO. The initial combustion zone is followed in the gas flow path by a secondary combustion zone in which the residual CO is oxidized to CO_2. The function of staged combustion is to develop a reducing zone in the flue gas path after the burning zone. The nitrous oxides formed in the burning zone break down to molecular nitrogen in the reducing zone. The most prevalent location for staged combustion is in the riser duct between the kiln inlet and the calciner vessel. The SNCR process is based on feeding anhydrous ammonia or ammonium hydroxide to the preheater tower in the temperature window of 870°C–1090°C to form water and molecular nitrogen. The SCR technology involves the use of a catalyst that accomplishes the reaction between ammonia and nitrous oxide in a lower temperature window of 300°C–450°C. While the SNCR technology has been successfully used in Europe and the USA, the SCR technology, although proven, has not been tried in the cement plants.

So far as the organic emissions are concerned, the vapors are emitted mainly by the cracking of the constituents of petroleum and kerogens present in the raw materials. Uncontrolled emission is usually below 50 mg/Nm³ and can be kept low by avoiding the feeding of raw materials containing high-volatile organic compounds. For PCDD and PCDF the available data are limited but the present observation is that the emission levels are below 1-TEQ.

It has, however, been the practice that an operating unit using alternative fuels should have continuous control for dust, SO_2, and NO_x and yearly monitoring of all other parameters including heavy metals and dioxins/furans. Thus, on the whole, the parameters that need to be monitored are NO_x, SO_2, CO, CO_2, H_2O, HCl, HF, NH_3, C_6H_6, O_2, TOC, and dioxin/furan. In addition, the heavy metals that need to be monitored are: antimony (Sb), arsenic (As), cadmium (Cd), chromium (Cr), cobalt (CO), copper (Cu), lead (Pb), manganese (Mn), mercury (Hg), nickel (Ni), thallium (Tl), and vanadium (v).

4.11 Summary

From the perspectives of both energy conservation and pollution abatement, the use of alternative fuels and raw materials (AFR) has emerged as an important operational strategy in the cement industry. The alternative fuels and raw materials can be classified as incombustible and combustible. The latter is further classified as industrial and agricultural. The industrial wastes are subdivided into non-hazardous and hazardous varieties. The AFRs are available in different physical states and forms. They can be gaseous, liquid, and solid ranging from powders to lumps.

The hazardous wastes are identified on the basis of their corrosivity, ignitability, reactivity, toxicity, and certain biological criteria. There are standard procedures and guidelines for testing and characterizing the hazardous wastes.

For effective use of AFRs it is essential to study the feasibility of the project which includes source mapping, quality evaluation, operational issues like odor, noise, emission, and several other parameters, engagement with stakeholders, processing options, and viability.

The waste streams are numerous but the more abundant ones on which the implementation of the strategy depends in India are the municipal solid waste (MSW), biomass or agricultural residues, scrap tires, hazardous wastes that are neither recycled nor used in landfilling, sewage sludge, organic solvents, spent pot-liners, meat and bone meal, plastic wastes, electronic wastes, etc. All the AFRs have different properties and composition. Hence, a systematic quality assessment of the materials is essential, for which proper sampling, adoption of standard test procedures, and the use of appropriate analytical instruments are necessary. For the emission of organic pollutants, the benchmark values need to be collated by conducting field measurements in plants.

Co-processing of alternative fuels in cement plants has been approved by the environmental authorities in India and the emission norms for the plants using alternative fuels have been enforced. The emission norms are comparable with those in some of the other countries.

There are special infrastructural requirements for using alternative fuels. These include collection and supply, storing and dosing, burning in precalciners and kilns, and special environmental protection measures. The basic design considerations for such installations have been briefly discussed in the chapter.

The potential of adopting the gasification technology for more effective use of alternative fuels has been dealt with. The technology has had a successful trial in Europe and holds high promise as a viable option.

There are a large number of industrial wastes that are being used or have the potential of being used as supplementary raw materials in cement plants. This kind of application is obviously different from the use of wastes as fuel. The practice and potential have been indicated in this chapter. The use of granulated blast furnace slag and fly ash has not been covered here, as these byproducts have been dealt with elsewhere in the book.

The use of alternative fuels is a progressively increasing measure in the industry but the strategy has been to respect the "waste hierarchy" as defined below:

- Waste should be used in cement kilns if and only if there are no better methods for its ecological and economical recovery and recycling.
- Additional emissions and negative impacts have to be avoided.
- The quality of cement must remain unchanged as compared to products made with conventional fuels.
- The use of alternative fuels should be an integral part of waste management and should be in line with the national and international norms.

References

1. ROBERT MATHAI, Calciner technology for secondary fuels, *Indian Cement Review*, April 2005, 21–26.
2. DANIEL A. VALLERO, *Environmental Contaminants: Assessment and Controls*, Elsevier Academic Press, UK (2005).
3. Resource Company (Australia), Successful AF implementation, *International Cement Review*, January 2015, 66–68.
4. N. A. VISWANATHAN, S. K. HANDOO, AND K. K. ROY CHOWDHURY, Resource conservation, waste utilization and alternative fuel usage in Indian cement industry, *Special Publication of Cement Manufacturers' Association on Alternative Fuels and Raw Materials*, 3rd International Conference on AFR, New Delhi March 23–24, 2017.
5. ULHAS PARLIKAR, P. V. KIRAN ANANTH, K. MURALIKRISHNAN, AND V. KANNAN, Waste mapping and forecasting for alternative fuel usage in cement plants, *ibid*.
6. ULHAS PARLIKAR, Recommended quality standards for SRF/RDF derived from MSW to improve their resource recovery for cement kilns, *ibid*.
7. D. SIVAGURUNATHAN, Alternate fuel usage – an update on the experience of the India Cements Ltd, 2nd International Conference on enhanced Usage of Alternate Fuels and Raw Materials in Cement Industry, February 19–20, 2015, A CMA-IIP Special Publication, Cement Manufacturers Association, New Delhi (2015).
8. T. BISWAS, Minimizing build-up problems through understanding alternative fuels compatibility with regular fuels, *Cement, Energy & Environment*, Vol. 16, Nos 1 &2. CMA, New Delhi (2016)
9. https://en.wikipedia.org/wiki/Waste tires, retrieved on July 29, 2017.
10. RENATO GRECO, Combustion of unusual fuels, *World Cement*, February 2006, 33–38.
11. F. SLADECZEK AND E. GLODEK, The measurement method of organic pollutants emitted into atmosphere by cement and lime industry, *Cement Wapno Beton*, No. 6, 2012, 423–442.
12. LARS S. JENSEN, Fulfilling emission guarantees, *World Cement*, December 2005, 87–94.
13. Cement International, Storing, metering and conveying solid secondary fuel—Design basis and equipment, *Cement International*, Vol. 5, 6/2007. 43–53.

14. JOSE RODA, The alternative movement, *World Cement*, February 2006, 39–51.
15. JACOB A. MOUIJIN, MICHIEL MAKKEE, and ANNELIES VAN DIEPEN, *Chemical Process Technology,* John Wiley & Sons Ltd, UK (2001).
16. MARK S. TERRY, Future trends in cement manufacture, *Indian Cement Review*, Part I, April 2005, 13–18 & Part II, May 2005, 15–20.
17. JAVED I. BHATTY AND JOHN GAJDA, Use of alternative materials in cement manufacture, in *Innovations in Portland Cement Manufacturing* (Eds Javed I. BHATTY, E. M. MILLER, and S. H. KOSMATKA), Portland Cement Association, USA (2004).

CHAPTER FIVE

Pyroprocessing and clinker cooling

5.1 Preamble

Over the span of nearly two centuries of existence of Portland cement the fundamentals of the manufacturing chemistry have not undergone any significant changes, while considerable engineering advances have been made in the equipment and software for cement manufacture. These advances have resulted in the operating capacity of a single kiln system to exceed as much as 12,000 tons per day, supported by automation, instrumentation, and computer-aided controls.

It should be borne in mind that in the total cement manufacturing process in general and in pyroprocessing in particular the application of human discretion has progressively reduced, increasing dependence on electronic gadgets and precision in hardware design. Needless to say, the characteristics of materials, including impurities present in them, play a critical role in the design of equipment and process.

The pyroprocessing stage in a cement plant extends after the point of extraction of kiln feed materials from its storage, through the steps of weighing, transportation, and feeding to the kiln, to the point of discharge of clinker from the cooler, going up to the clinker silo (Figure 5.1). This part of the plant is regarded as the heart of the cement making process. The main chemical reactions take place in this section of the production process and most of the operating costs are incurred here. This stage is also important for the process improvement opportunities. This chapter deals with the unit operations in this part of the process.

5.2 Clinker formation process

The clinker formation process can be represented in a simplified form by the reaction steps given in Figure 5.2 (1). It is evident from the summary diagram that the process depends on two basic parameters: reactivity and burnability of the raw mixes. Conceptually, the reactivity of a given

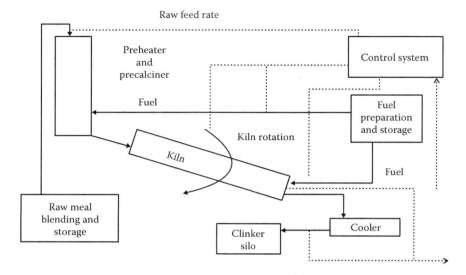

FIGURE 5.1 A simplified outline of the pyroprocessing stage.

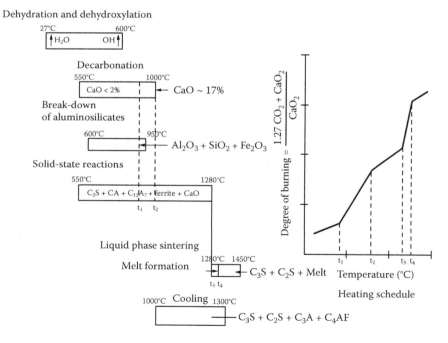

FIGURE 5.2 Approximate reaction sequence in clinker making. (From A. K. Chatterjee, Chemico-mineralogical characteristics of raw materials, in *Advances in Cement Technology* (Ed. S. N. Ghosh), Pergamon Press, UK, 1983.)

PYROPROCESSING AND CLINKER COOLING 143

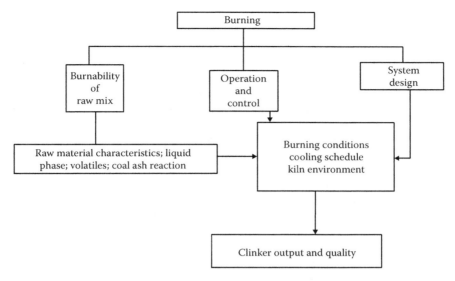

FIGURE 5.3 Major factors controlling the burning operation. (From A. K. Chatterjee, Chemico-mineralogical characteristics of raw materials, in *Advances in Cement Technology* (Ed. S. N. Ghosh), Pergamon Press, UK, 1983.)

raw mix signifies the quantitatively achievable rate of different clinker forming reactions at respective temperatures within practical time limits. The chemical composition including minor and trace constituents, the phases or compounds present, and the fineness and homogeneity of the given raw mix control its reactivity.

Burnability, unlike reactivity, is a qualitative parameter and signifies the overall measure of ease or difficulty of burning of raw mixes under practical operating conditions of a kiln system. In this context, one has to bear in mind that the burning process has several independent and interrelated controlling factors, as outlined in Figure 5.3.

Clinkering reactions and kiln systems

Barring the continuance of some old wet-process or semi-dry process kilns, the new and modern plants almost exclusively operate dry-process five-stage preheater-precalciner kiln systems with efficient grate coolers. Such kiln systems are also provided with multi-channel burners. For all practical purposes this system has to take care of the following four major reaction steps:

- decarbonation of limestone
- solid-state reactions
- liquid-phase sintering
- reorganization of the clinker microstructure through cooling

The above chemical transformations and the corresponding processing parameters and relevant equipment performance occur synergistically, to the extent it is practically feasible. The basic aspects of the limestone decarbonation reactions are summarized below, while the other three reactions will be discussed in the relevant sections later.

The removal of CO$_2$ from limestone by heating, also known in practice as "limestone calcination," is dependent on the physical and chemical characteristics of limestone on one hand and preheater–precalciner system on the other. The relation of material characteristics of limestone with its dissociation had been dealt with by the author in one of his earlier publications (2). The relevant aspects are recapitulated here.

The influence of the partial pressure of carbon dioxide on the dissociation rate and temperature of calcium carbonate is shown in Figure 5.4.

In the decarbonation reaction there is a definite relation between the reaction temperature and the partial pressure of CO$_2$ and its concentration. If the temperature and pressure are in equilibrium, dissociation is static, but if there is a minute change in one of these variables, dissociation proceeds. In other words, in a mixture of 10 g CaO and 10 g CaCO$_3$ kept in equilibrium at 898°C, 760 mm pressure, and 100% CO$_2$, there would be no change. If the temperature shifts towards 890°C, the reaction will move towards calcite, while a rise in temperature to 910°C will push the reaction towards formation of CaO. As already stated, at 898°C the carbon dioxide dissociation pressure reaches one atmosphere and between 800 and 900°C the pressure–temperature relationship is almost linear. Another important factor of calcination kinetics is the concentration of carbon dioxide in the gas stream. Apart from the influence of partial pressure and concentration of carbon dioxide, the dissociation rate and temperature of limestone

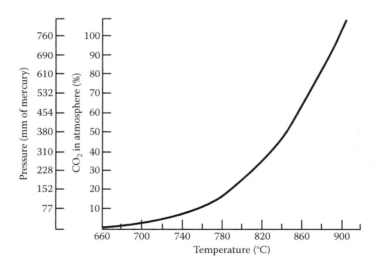

FIGURE 5.4 Influence of CO$_2$ concentration and pressure on dissociation of CaCO$_3$. (From A. K. Chatterjee, Chemico-mineralogical characteristics of raw materials, in *Advances in Cement Technology* (Ed. S. N. Ghosh), Pergamon Press, UK, 1983.)

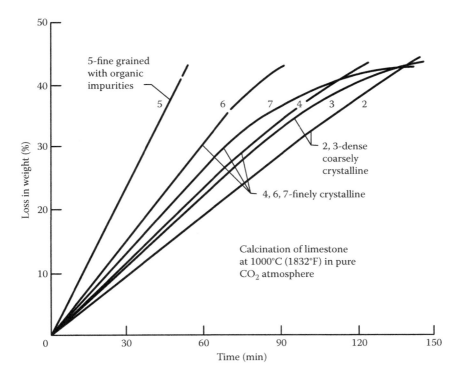

FIGURE 5.5 Dissociation of different types of limestone. (From A. K. Chatterjee, Chemico-mineralogical characteristics of raw materials, in *Advances in Cement Technology* (Ed. S. N. Ghosh), Pergamon Press, UK, 1983.)

vary with the presence of impurities as well as the grain size of calcite and quartz phases (Figure 5.5). Although the dissociation temperature of pure calcite is given as 898°C, it has been found that depending on grain size of the mineral phase and solid solubility of CaO in $CaCO_3$ the dissociation temperature may vary widely. The effect of the associated minerals and impurities is basically to lower the decomposition temperature. Further, it should be understood that the decarbonation reaction is a solid-state diffusion process in which the decomposition of calcite starts with the formation of pseudomorphs of two-dimensional CaO crystals. Only after some time, known as the induction period, the three-dimensional nuclei of CaO crystals appear. The coarser the calcite crystals and more perfect their structure, the longer the induction period.

The associated minerals and microstructure also influence the activation energy for dissociation of calcium carbonate. For a given set of limestone samples the activation energy for dissociation was observed to vary from 30–60 kCal/mole (Figure 5.6).

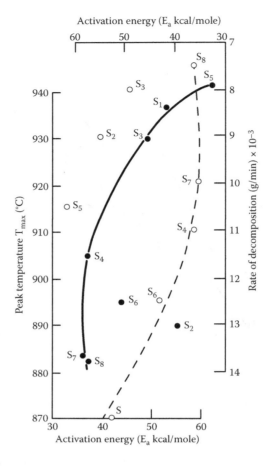

FIGURE 5.6 Activation energy versus rate of dissociation (dashed line) and peak temperature (solid line) of different types of limestone. (From A. K. Chatterjee, Chemico-mineralogical characteristics of raw materials, in *Advances in Cement Technology* (Ed. S. N. Ghosh), Pergamon Press, UK, 1983.)

5.3 Preheater-precalciner systems

Historically, it is known that the cyclone preheater kiln was first patented in 1934 in Czechoslovakia but the first preheater kiln was built and commercialized in 1951. The cyclone preheaters are also known as suspension preheaters and the basic principle of the system is the counter-current heating of cold kiln feed by hot kiln gases (Figure 5.7). Suspension preheaters facilitated the construction of kilns of capacities up to 4500 tons per day with two preheater streams working in conjunction with one rotary kiln. The system progressively proliferated in the industry as a more thermally efficient process.

Advances made in the various facets of preheater-precalciner systems have been published from time to time (3–7). Consolidation of the chemistry and engineering of the system is attempted here. A preheater system can have any number of stages, although for all practical purposes

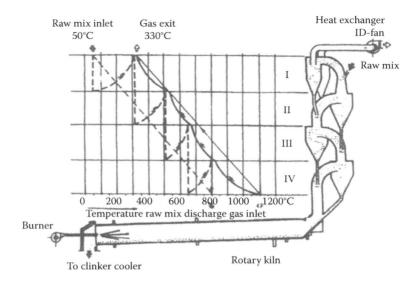

FIGURE 5.7 Four-stage cyclone preheater. (Reproduced from W. Kurdowski, Cement burning technologies, in *Advances in Cement Technology* (Ed. S. N. Ghosh), Tech Books International, New Delhi, 2002. With permission.)

the number is restricted to four to six stages. In a four-stage preheater kiln system the material entering the kiln is at 800°C and is generally 20–30% calcined. Gases from the kiln system leave at around 350°C. The residence time of the material in the preheater is approximately 30 s and in the kiln about 40 min. The kiln rotates at two revolutions per minute and the pressure drop in the preheater is 300–600 mm of water. The gas velocity through the cyclone ducts is maintained typically around 20 m/s. The kiln capacities range up to 3500 tons per day with specific fuel consumption of 750–800 kCal/kg of clinker. Since in the preheater stages the total material residence time does not exceed 30 s and the degree of decarbonation of limestones does not exceed 30%, a part of the horizontal kiln volume is used up to complete the decarbonation reaction.

In order to enhance the decarbonation capacity of the preheater, attempts were made to fire additional fuel in the riser duct, which eventually led to the development of preheater-precalciner kiln systems with special vessels added to the preheater string. A typical example is given in Figure 5.8, in which an MFC (Mitsubishi Fluidized-bed Calciner) is in the preheater line.

The maximum heat consumption is in the raw meal calcination, occurring at about 900°C, and amounts to about 2000 kJ/kg of clinker. At the same time, as the temperatures of the gases and material increase, the properties of the latter change, and change more radically with the appearance of the liquid phase inside the kiln. Figure 5.9 shows the Q-T diagram of the clinker burning process, which demonstrates that the temperature difference between the gases and material keeps on widening during the calcining stage. Keeping the temperature difference in view and the nature of solid-state reaction up to this stage, the most energy-efficient solution for the clinkering process was to divide

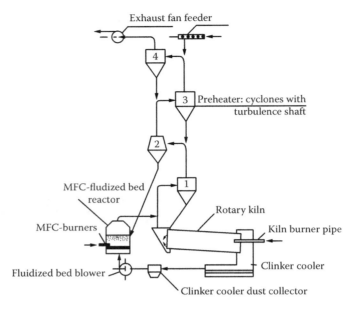

FIGURE 5.8 Schematic diagram of a preheater kiln with MFC precalciner. (Reproduced from W. Kurdowski, Cement burning technologies, in *Advances in Cement Technology* (Ed. S. N. Ghosh), Tech Books International, New Delhi, 2002. With permission.)

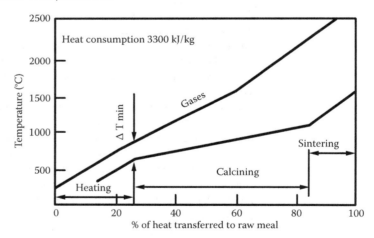

FIGURE 5.9 Q-T diagram of clinker burning. (Reproduced from W. Kurdowski, Cement burning technologies, in *Advances in Cement Technology* (Ed. S. N. Ghosh), Tech Books International, New Delhi, 2002. With permission.)

it into two sub-systems: stage of preheating extended up to completion of calcining of raw meal in particle suspension mode, and the stage of sintering the calcined raw meal in the presence of a liquid medium.

The precalciner process, based on the above theoretical foundation, was conceived to be accomplished with a second stage of firing, and it greatly expanded the systems capability. It became possible to generate and absorb over 60% of the total heat input in the calcination zone where the heat consumption was high due to the endothermic nature of the decarbonation reaction. Fuel combustion and, to an even greater extent, solids residence time depend on the gas flow regime. The calculated gas residence time varies in different designs from 1.4–1.7 seconds in systems with a separate precalciner vessel to 4–5 seconds in the extended duct system. Some precalciner vessels are designed with a swirling or cyclonic motion of the gas stream inside them, which gives solids relatively longer residence time. The modern preheater-precalciner kiln systems are designed for at least 85% calcination, as computed by the following equation:

$$C = 100 \times 1 - \left(LOI_{sample} \times (100 - LOI_{raw\ meal}) / LOI_{raw\ meal} \times (100 - LOI_{sample}) \right)$$

(5.1)

Where
  C = degree of calcination
  LOI_{sample} = loss on ignition of the sample analyzed
  $LOI_{raw\ meal}$ = loss on ignition of the raw meal

The degree of calcination in the precalciner depends on:

- inside temperature
- residence time of the raw meal
- gas/solids separation
- effect of dust circulation
- kinetic behavior of the limestone, as discussed earlier

One should bear in mind that it is the control of the calcination reaction that determines the stability of the kiln operation. From the experience gained with different types of precalciner kilns there appears to be a relationship between the variations in degree of calcination and the residence time of the material in the precalciner. The systems with shorter solids residence time tend to demonstrate wide fluctuations of calcination.

The basic configurations of precalciner systems are related to the flow of secondary air and tertiary air from the kiln and cooler respectively into the precalcining system. The different configurations in operating plants can be summarized as follows:

  I. The secondary air for combustion from the kiln flows entirely into the precalciner vessel without any separate tertiary air duct from the cooler. This type of precalciner, therefore, operates with combustion gas, impoverished in oxygen (10–14%), and is suitable only for a lower proportion of fuel feeding (20–30%).

  II. The kiln gases do not pass through the precalciner, which uses only the hot tertiary air from the cooler. The kiln gases mix with the product of the precalciner prior to entering the preheater cyclone. This combustion in the precalciner is accomplished with hot air containing 21% oxygen. The fuel proportion in the precalciner can be in the range of 50–65%.

III. The tertiary air from the clinker cooler and the kiln gases are mixed prior to the precalciner and then added to the stream entering the precalciner. The ratio of tertiary air:secondary air is controlled in accordance with the process requirements. This configuration can also provide for a bypass for the kiln gases, if required.

IV. If there are two preheater streams for a kiln, one is fed by the kiln gases and the other by the gases from the precalciner. The tertiary air from the clinker cooler provides the combustion air to the precalciner.

Without going into the design details, one may broadly distinguish two types of precalciners: inline calciners (ILC) and separate-line calciners (SLC) (Figure 5.10). The main difference is that in ILC the kiln gases pass through the precalciner and in SLC they do not. Another important aspect of precalciner operation is the temperature of the tertiary air. The higher the temperature, the lower the heat consumption and more stable the kiln system. Thus, the precalciner kiln systems get strongly linked with the design and efficiency of the clinker coolers.

The original equipment suppliers have developed different designs of precalciners for better energy efficiency and other operational advantages. The more widely used commercial systems include the FLS design, KHD design, Mitsubishi Fluidized-Bed Calciner, Reinforced Suspension Preheater of Onoda design, Suspension Preheater with Flash Furnace of IHI design, and others. It is relevant to mention here that the plethora of design variations of preheater-precalciner systems are aimed towards achieving the following objectives:

- Meeting the contradictory demands of minimizing the temperature and oxygen levels and complete burn-out of fuel.

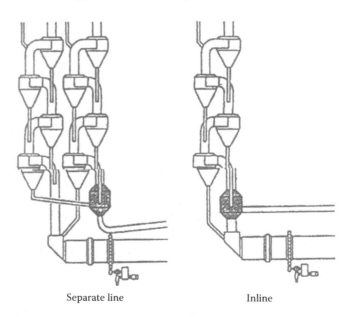

Separate line   Inline

FIGURE 5.10 Arrangements of inline and separate-line calciners.

- Relief of the thermal load in the rotary kiln
- Achieving a controllable degree of calcination
- Abatement of NO_x emission
- Use of alternative or hard-to-burn fuels

So far as the above objectives are concerned, and more particularly the last two, the cement machinery industry has come up with system designs for reduced NO_x emissions (Figure 5.11) and for firing different types of fuel in the precalciner (Figure 5.12).

Progress in preheater cyclone design

As a result of the design improvements in the preheater cyclones, the material separation efficiency has increased and the pressure drop across the preheater has reduced substantially. Pressure drop in cyclones has been reduced by providing larger inlets, larger outlets, and sloped inlet shelves, and by eliminating any inside horizontal surfaces. The development of high-efficiency, low-pressure-drop cyclones has enabled the installation of additional stages of heat-exchanging cyclones to reduce the exit gas temperature and effect fuel saving with marginal increase in the cost of installation. Exit gas temperatures, static pressures, and specific fuel consumption for modern preheater-precalciner kilns are given in Table 5.1 (5):

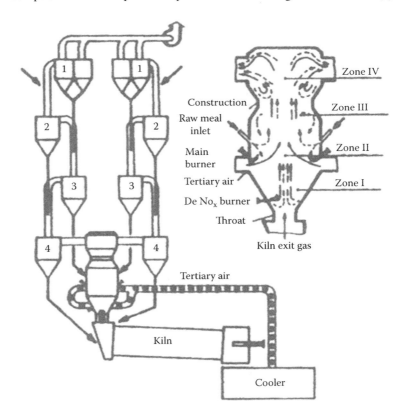

FIGURE 5.11 De-NO_x calciner of Kobe Steel and Nihon Cement Co., Japan. (Courtesy of Trade bulletin of the company.)

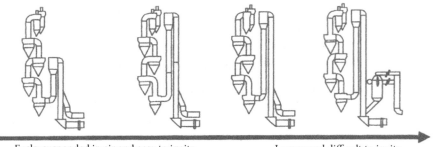

Fuels: suspended in air and easy to ignite   Lumpy and difficult to ignite

FIGURE 5.12 Fuel-compliant flash tube calciners of KHD design. (From Mathias Mersmann, KHD Humboldt Wedag International, presentation at IEEE-IAS/PCA conference, Charleston, USA, May 2007.)

Table 5.1 Comparison of the operating parameters of preheaters with increasing stages of cyclones

No. of stages	Exit temperature (°C)	Pressure drop (mm H_2O)	SHC (kCal/kg)
4	350	350	800
5	320	500	775
6	260	550	750

It may be relevant to mention here that the latest designs of preheater-precalciner kilns aim at achieving specific heat consumption of less than 700 kCal/kg clinker.

5.4 Rotary kiln systems

Since the innovation of Portland cement by Joseph Aspdin in the mid-nineteenth-century, burning was carried out in batch-type shaft kilns. The throughput rates were very low and heat consumption was high. Towards the end of the nineteenth century, wet-process rotary kilns appeared on the scene. The semi-dry Lepol grate pre-heater, developed in 1928, and the cyclone suspension heater, introduced in 1951, were two past innovative steps in pyroprocessing to make the clinker production less energy-consuming than in the wet-process kilns. In the last three decades, all the energy-intensive kiln systems have steadily declined in importance. The development of precalciners has probably been the biggest break-through in the process engineering of cement manufacture since the innovation of four-stage cyclone suspension preheaters. The permutation and combination of different preheater-precalciner configurations with even a single type of cooler, e.g., the grate cooler, can result in a wide range of pyroprocessing systems. As an illustration, the system designs of one of the machinery suppliers are shown schematically in Figure 5.13. The modern preheater-precalciner kiln systems may

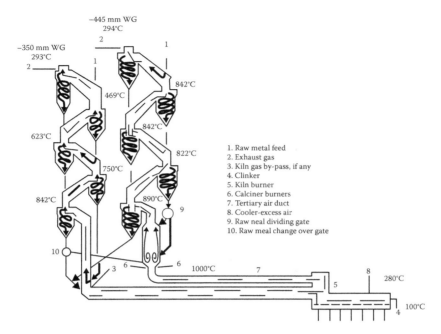

FIGURE 5.13 Typical schematic representation of a modern large kiln system of FLS design. (Courtesy of Trade catalogue of FLSmidth A/S, Denmark. Trade catalogue.)

have up to three strings of preheater and up to two precalciner vessels. Single kilns with multiple preheater strings are now operational, with over 12000 tons per day of clinker production.

Another significant development in rotary systems is the design of short two-support kilns with a length:internal-diameter ratio of 10–14. This is considerably lower than the typical ratios of 16–18 of three-support systems. Two-support kilns are larger in diameter than three-support kilns for the same production rate and volumetric loading, and therefore have lower thermal loading than a traditional kiln (Figure 5.14) (1).

The success of two-support kilns has been accompanied with quite a few mechanical innovations like tangential tire suspension, double-lamella seals, etc., which are not being dealt with in this chapter. While these short kilns offer substantial economic and mechanical advantages, they meet all the process technology requirements for the clinker making process, despite the fact that the average residence time in such kilns is of the order of 21 min or about half the time of conventional kilns. Depending on the burnability of raw materials and the clinker quality desired, the volumetric loading ranges from 4.5 to 5.3 tpd/m^3. The kiln diameter does not exceed 6 m to ensure reasonably long refractory lining life. Under plant conditions the entire pyroprocessing of raw meals in these short kilns is complete in about 20 min.

Heat balance of kiln systems

The measurement of heat balance in a rotary kiln system is an effective tool in the hands of operators to judge the productivity and thermal efficiency of the system. To determine the balance, it is necessary to

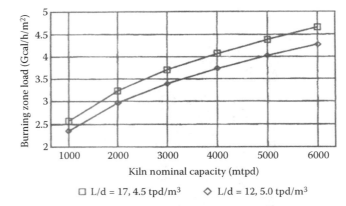

FIGURE 5.14 Comparison of burning zone thermal loads of traditional kilns (L/D = 17) and short kilns (L/D = 12). (From Information Brochure of FLSmidth on Rotax 2 kiln components, December 2005.)

measure the quantities and temperatures of all materials entering and leaving the system. In addition, surface temperature measurements for the equipment are taken and standard formulae used to calculate direct heat losses by radiation and convection. The heat content of each solid and gas, referred to as its sensible heat, is calculated as a difference from its value at ambient temperature. The calorific value of a fuel is used to calculate the energy input and the theoretical heat required to carry out the chemical reactions is calculated by the following formula:

$$H_{total} = 2.22\,Al_2O_3 + 6.48\,MgO + 7.66\,CaO - 5.116\,SiO_2 - 0.59\,Fe_2O_3$$
(5.2)

The gases leaving the kiln or pre-heater, drawn through the system by a large ID fan, are analyzed so that the sensible heat of each component can be calculated. In fact, the exhaust gas temperature and the oxygen content are measured with the help of instruments along with the fuel and raw mix feed to obtain a continuous record of the air–fuel–material balance in the kiln.

Based on the above measurements, the following heat balance is worked out:

Heat input	Heat output
Combustion of fuel	Theoretical heat of clinker formation
Sensible heat of fuel	Exit gas losses
Organics in feed	Evaporation of moisture
Sensible heat in kiln feed	Dust in exit gas
Sensible heat of cooler air	Clinker discharge
Sensible heat of primary air	Cooler stack losses
Sensible heat of infiltrated air	Kiln shell losses
	Calculation of waste dust
	Unaccounted losses

For reference purposes, typical parameters of heat balance for a six-stage preheater-precalciner kiln can be indicated as follows:

a. surface heat loss (kiln): 4.3%
b. surface heat loss (preheater): 5.0%
c. cooler heat loss: 16.4%
d. exit gas heat loss: 20.7%
e. heat of reaction: 57.1%
f. sensible heat from raw material and fuel: −3.6%

Based on actual heat balance it has been observed that different kiln systems operate generally with the following ranges of heat consumption:

Type of kiln	kCal/kg clinker
Wet kiln	1380–1700
Long dry kiln	950–1150
Suspension preheater kiln	750–850
Preheater-precalciner kiln	680–750

5.5 Kiln burners and combustion

Kiln burner technology has undergone substantial improvements and burners with multiple channels for air and fuel flow are in universal use. The multiple channels provide flexibility in changing the flame shape and its intensity by the control of air flow in the axial and swirl channels, air velocities, and adjustment of burner position. Three-channel and five-channel burners are in operation. While the three-channel solid fuel burners have the provision of additionally feeding a gaseous or a liquid fuel, the five-channel burners are designed for simultaneous firing of three fuels in proportioned quantities (Figure 5.15) (3).

Further improvement in burner technology has been reported. The essence of the new burner design is a sandwich concept, in which the

FIGURE 5.15 Schematic diagram of multi-channel burners. (From A. K. Chatterjee, *Cement & Concrete Research*, Vol. 41, Issue 7, July 2011.)

fuels are captured between two air streams and exposed to high oxygen accessibility and turbulence. As an aid for need-based flame shaping, the outer jet streams can be directed in different angles, forming either a convergent or a divergent or a rotational flow pattern. For this purpose, the jet nozzles are individually mounted on air supply stubs, which can be rotated over 360 degrees (Figure 5.16) (8).

Unstable flames adversely affect kiln operation. They are also unsafe to plant and personnel. An unstable flame is one that has a varying ignition point and a variable stand-off distance from the burner tip. Ensuring early ignition of fuel entering the combustion zone enhances the flame stability. In rotary kilns this is best achieved by ensuring recirculation of hot gases as the ignition source for new fuel. External circulation from properly designed burners brings hot combustion gases back into the combustion region to help stabilize the ignition point. Internal circulation can even be more effective.

Another factor that requires consideration in the context of burner technology is "flame momentum." If the burner jet momentum (fuel and primary air) is less than that required for complete entrainment of the secondary air into the fuel jet, then fuel-air mixing will be inadequate for good combustion and a long, lazy flame with high CO is the likely result. If the burner momentum is greater than that required for complete secondary air entrainment, then the excess momentum of the fuel jet is dissipated in pulling back exhaust gases from further down the kiln into the flame. This "recirculation" has a positive effect in stabilizing the flame, although too much recirculation can be detrimental to fuel efficiency. These aspects are discussed in Chapter 3.

It should also be borne in mind that high flame momentum generally increases heat consumption and electricity demand. When there are high chloride and sulfur inputs without bypass, there could be enhancement of cyclone clogging tendency with high flame momentum. Still, when high-sulfur fuels like petcoke are used in the kiln, the flame momentum is kept on the higher side along with high rotational speed of the kiln to minimize the entrapment of sulfur in clinker.

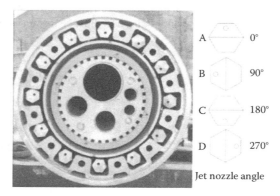

FIGURE 5.16 Rotational jet nozzles. (From Mathias Mersmann, KHD Humboldt Wedag International, presentation at IEEE-IAS/PCA conference, Charleston, USA, May 2007.)

5.6 Clinker coolers

Cooling is an integral and essential part of the clinkering process to serve two basic functions: first, to reduce the clinker temperature so that it can be conveyed, stored, and processed further and, second, to recover a substantial part of the clinker heat for reuse as a secondary or tertiary air stream for combustion. In the pyroprocessing system, clinker cooling occurs in two steps:

  a. inside the kiln between the burning zone and the kiln discharge end;
  b. inside the cooler.

In the first step the clinker is cooled from the burning zone temperature to about 1200–1250°C and in the second step the cooling process continues to about 100°C. The clinker coolers are also necessary for freezing the final clinker phase assemblage and microstructure. However, in the global cement industry one comes across a wide variety of coolers in operation, viz.,

  a. tube coolers
   1. rotary
   2. planetary
  b. grate coolers
   1. traveling
   2. reciprocating
    – inclined
    – horizontal
    – combination
    – dual-stage with duotherm air
  c. gravity coolers
   1. secondary "g" type
   2. shaft

The development history of the above types of coolers is given in (9). Barring the gravity coolers for which enough data are not available, the salient features of the other types are summarized in Table 5.2.

In both planetary and rotary coolers, cooling is based upon cascading the clinker through the air flow. A planetary cooler is expected to cool the clinker faster than in a rotary cooler as in the planetary cooler the clinker is distributed in a number of smaller tubes, thereby ensuring better air-clinker contact. In grate coolers, cooling is based on cross–current cooling air, which passes through the bed of clinker on the grate, leading to a very high air-clinker contact area and faster cooling. Grate coolers have undergone several stages of modification. Needless to say, although the reciprocating grate coolers were developed to overcome the deficiencies of rotary and planetary coolers as far back as 1951, this is the system that has become

Table 5.2 Average data for different types of coolers

Parameters	TG	RG-I	RG-H	RG-C	RG-D	Planetary	Rotary
Max. operating capacity (tons/day)	3500	2500	2500	12000	3500	5000	3000
Clinker temperature (°C)							
Inlet	1400	1400	1400	1400	1300	1200	1300
Outlet	135	105	120	105	65	165	210
Cooling air (Nm³/kg Cl)	2.1	1.6	1.8	1.3	1.2	0.90	0.90
Secondary air temp. (°C)	870	880	795	880	925	740	870
Exit air temp. (°C)	240	260	265	260	345	–	–
Thermal efficiency (%)	67	68	63	68	72	72	73

Note: RG-C: reciprocating grate-combined; RG-D: reciprocating grate- duotherm; RG-H: reciprocating grate-horizontal; RG-I: reciprocating grate-inclined; TG: traveling grate.

suitable today, with further mechanical improvements to match clinker capacities up to about 12,000 tons/day production. Thus, grate coolers have become the preferred system for the modern large-capacity cement plants. A schematic diagram of a typical grate cooler with heat recuperation for the precalciners is given in Figure 5.17. In the common design the clinker bed is moved on the reciprocating grates, which are inclined in the hot zone and horizontal further down the cooler. There may be up to eight under-grate compartments with separate fans to provide the cooling air and three grate sections for moving the clinker bed. Grate coolers with "air beam" designs are now available, in which there is direct ducting of

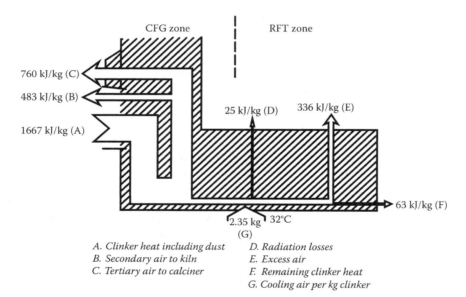

A. Clinker heat including dust
B. Secondary air to kiln
C. Tertiary air to calciner
D. Radiation losses
E. Excess air
F. Remaining clinker heat
G. Cooling air per kg clinker

FIGURE 5.17 A schematic diagram of a grate cooler with heat recuperation in modern plants. (Reproduced from W. Kurdowski, Cement burning technologies, in *Advances in Cement Technology* (Ed. S. N. Ghosh), Tech Books International, New Delhi, 2002. With permission.)

cooling air to hollow grate support beams. In this variant air is directed into the clinker bed more efficiently than what is possible in the conventional design. Further, the grates used with air beams, called "controlled flow grates," allow air to pass horizontally through slots into recesses in the grate surface. This largely eliminates fall-through of clinker fines and renders the air flow less dependent on the bed resistance. Other refinements in the grate cooler design include a pendulum frame for the moving grate, which is claimed to minimize drive maintenance. A recent development of cooler design is the "cross bar cooler," which consists totally of static grates with clinker transport effected by reciprocating pusher bars above the grate surface. This cooler also incorporates an ingenious flow regulator on each grate, which maintains constant air flow through the clinker bed regardless of bed porosity. Innovative design features are also seen in the "polytrack," "pyrofloor," and "eta-cooler" coolers supplied by different manufacturers. Mention may also be made of the new revolving-disc cooler (RDC) (1), which is installed below the discharge end of the kiln and is provided with a clinker distribution system (CDS) above a slowly rotating disc. A special removing device discharges the clinker onto a roller crusher, located at the outer diameter of the RDC. The diameter of the RDC is synchronized with the kiln diameter and kiln capacity. The total residence time in this cooler is estimated at 30 min, which is somewhat lower than a reciprocating grate cooler (~ 45 min). The geometry of this cooler as conceived is quite innovative and different from the previous and modern grate coolers but its commercial success is yet to be known.

In a modern grate cooler, the heat loss can be reduced to about 400 kJ/kg of clinker. The recuperation of heat as secondary and tertiary air is shown in Figure 5.17, which also shows two zones of the cooler: controlled flow gate (CFG) and rapid fall-through (RFT). In the heat recuperation zone the cooling air is supplied directly to the grate plates through a system of ducts and hollow beams instead of through an undergrate compartment, as in conventional coolers. By rapid fall-through of clinker, the operational problem of discharge of streaks of red-hot clinker, often called "red river," is avoided.

Cooling effects on clinker quality

The effect of cooling conditions on clinker quality is well known but the correlation of clinker quality with a specific cooling schedule has proved difficult to prove due to interactive effects of many parameters. It appears that slow cooling up to, say, 1200°C with rapid cooling to the discharge temperature is expected to maximize the formation of alite, the stabilization of the desired polymarph of belite, tricalcium aluminate, and alumino-ferrite phases. With slow cooling in the range of 1200–1000°C, often there is a tendency for decomposition of alite with release of free lime. It may also alter the polymorphic form of belite. In addition, it results in coarsely crystallized clinker with hard grinding characteristics. Slow cooling, as is well known, may result in the crystallization of coarse periclase grains, if the clinker has high magnesia content, which is detrimental to the quality of cement.

In view of these observations it is evident that the total effect of cooling is the result of cooling inside and outside the kiln. As a result, the differences in the cooling schedule obtained in different types of coolers assume a lower degree of significance in modifying the clinker quality. In fact, it has been proven in practice that no significant difference in the quality characteristics of clinker is detected when different types of coolers are used, if they are designed and operated properly.

The process importance of the cooler design is more for heat recuperation than clinker quality. As mentioned earlier, an important aspect of precalciner operation is the temperature of the tertiary air for both fuel economy and kiln stability. This demands that the cooler design and operation should ensure a consistently high efficiency. It may be borne in mind that small differences in cooler throat, kiln hood, tertiary air off-take, secondary air temperature, and velocity can have an enormous effect on the air flow patterns for combustion of fuel.

If the raw material has a strong tendency to form coatings and rings in the rotary kiln, the rate of clinker discharge from the kiln is likely to fluctuate. As a result, in both the rotary and planetary coolers the exit temperature of discharged clinker will vary greatly, as the radiation losses are constant and the rate of cooling of air cannot be altered very much. However, in the rotary cooler the rate of movement of the clinker can be varied independently of the kiln. Variation in clinker discharge rate from the kiln can most effectively be compromised in the grate cooler.

Irrespective of the cooler systems, it is desirable that the clinker that enters the cooler is mainly nodular in form and not dusty. For example, the operating experience has shown that, other parameters remaining the same, if the dust content is limited to about 15% (−0.5 mm), the clinker is expected to possess better grinding behavior. Apart from dust

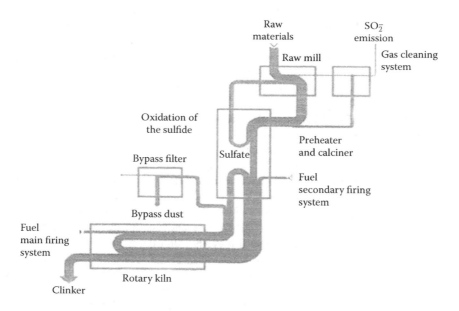

FIGURE 5.18 A typical sulphur cycle in a preheater-precalciner kiln. (From *The Cement Environmental Yearbook of the British Cement Association*, UK, 1997.)

circulation, consequent increase in clinker discharge temperature, transport difficulties, etc., the dusty clinker causes overheating of grate steel parts, wear of burner pipe, poor visibility, dust ring, etc.

5.7 Volatiles cycle in preheater-precalciner kiln systems

The volatile constituents in the kiln feed, and particularly sulfates, alkalis, and chlorides, pass through condensation-volatilization cycles in the preheater-precalciner kiln systems (10). The system volatility has two components: primary volatility (E_1) and secondary or circuit volatility (E_2). The primary volatility is the vaporization loss of K_2O, Na_2O, SO_3, and Cl^- (and others, if present), when the kiln feed passes through the burning zone of a kiln for the first time. The secondary or circuit volatility is the vaporization loss of these oxides/radicals after the vaporized phases have undergone condensation reactions in the cooler parts of the preheater cyclones and retraced their path through the kiln in subsequent movements. A typical sulfate cycle is shown in Figure 5.18.

The primary volatility of the chemical constituents is normally determined with the help of specially designed laboratory facilities but the circuit volatility is difficult to estimate as it is measured on the basis of analysis of circuit samples properly collected over a period of investigation. An illustration is given in Table 5.3. These values cannot be considered universal as they are specific to materials and systems.

From the above data, it appears that the circuit volatility of alkalis and sulfates increases substantially from their initial primary volatility, which is not so for the chlorides. It is important to note in this context that the actual volatility in the kiln system will also depend on the loss of volatiles from the system and the amount passing into the clinker.

In the preheater-precalciner kiln systems volatiles practically have no escape route and consequently the circulating volatiles may either lead to cyclone plugging due to the formation of low-melting phases or excessive entrapment in the clinker. This situation is partially overcome by diverting a portion of the kiln exit gases through a bypass duct. A modern bypass system consists of an air-quench chamber, a shut-off valve, a water-quench chamber, and a dust collector. The air-quench chamber is used to mix ambient air with the kiln gases to rapidly cool

Table 5.3 An example of volatilization factors

Oxides/radicals	Kiln feed (%)	Fuel (%)	Primary volatility	Circuit volatility
K_2O	0.820	–	0.67	0.90
Na_2O	0.160	–	0.53	0.80
SO_3	0.710	2.200	0.55	0.90
Cl^-	0.007	0.010	0.99	0.99
Loss on ignition	34.200	99.980	–	–

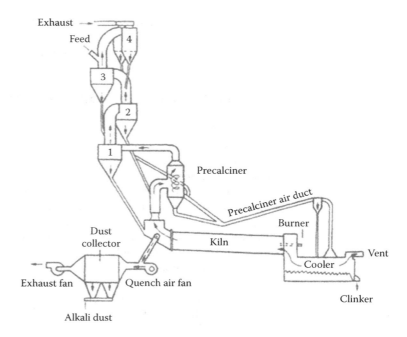

FIGURE 5.19 A schematic layout of a bypass system.

the objectionable compounds. The effect is to cause quick transition from vapor to solid state and prevent liquid formation. The water quench chamber is used to cool the gases to lower temperatures for dust collection. It is obvious that the bypass systems add to energy consumption and material losses. At 30% bypass the fuel consumption increases by approximately 15% and the material losses by about 7.5%. The bypass concept and arrangement are shown in Figure 5.19. The order of volatility in a kiln system is different for different compounds and for the common volatile constituents the order is as follows:

Cl > K > S > Na

The volatility of alkalis, and particularly of potassium oxide, is related to the presence of SO_2/SO_3, water vapor, and Cl^-. The alkali volatility can be decreased by the sulfates and increased by the halides.

5.8 Refractory lining materials in the kiln system

The refractory lining of rotary cement kilns is an essential process requirement to protect the machinery and personnel from high-temperature activities, to reduce heat losses to the surrounding environment, and to provide the right temperature inside the kiln to achieve the clinkering reactions. The operating conditions of the lining inside the rotary kilns are different from the static furnaces. Over almost the entire length of the kiln the lining periodically comes into contact with charge material being burned. During this process, the hot material gets mixed and remixed on the refractory surface, causing impact and abrasion on it. In the burning zone a melt phase appears and forms a coating on the refractory surface. On each rotation of the kiln, the lining is subjected to chemical and thermal actions of the material and the gas stream. Even during the stable operation of the kiln, the lining is subjected to considerable temperature fluctuation as it is exposed by turn once to the gas stream and again to the charge material during each rotation. The temperature fluctuation can range from ± 40–100°C. The refractory requirement in the kiln, therefore, is for dense, strong, spalling-resistant and reactive shaped bricks, the property demands from which may often be contradictory.

The suspension preheaters with multistage cyclones, precalciner vessels, ducts, etc. are also lined with refractory materials but the functional requirements of these lining materials are significantly different as these parts are static but of large surface area and complex geometry. The precalciner, the kiln inlet, and the lowest cyclone stages are particularly sensitive to plugging and material build-up. The lining requirement in this section is more for unshaped refractories with high insulation properties, which at the same time should be non-clogging in nature.

Premature failure of the refractory lining leads to unplanned shutdown of kilns and production interruptions. The preheater clogging constitutes nearly 50% of all kiln shutdowns. In case of excessive build-up of materials in the riser duct or the lowest cyclone, the pressure drop through the preheater may become high and be a production-limiting factor. In most instances, unproductive hours due to destabilization of operations could be of greater economic significance than the actual shutdown duration. Ultimately, the production losses determine the crucial role of the refractory lining, rather than the cost of materials and installation.

Refractory materials and kiln zonation

The refractory materials used in lining the rotary cement kilns and the preheater-precalciner tower can be placed in three broad categories: alumina bricks, basic bricks, and monolithic products. Four types of alumina bricks find use, viz., firebricks of 35–40% Al_2O_3, lightweight insulating bricks of low-alumina content, high-alumina bricks of 50–70% Al_2O_3, and another variety having 75–85% Al_2O_3. The basic bricks are made from either magnesite or dolomite and the types that are used include magnesia-chromite, periclase-spinel bricks (magnesia bricks containing spinel phase), fired dolomite (lime and magnesia), and stabilized magnesia with no chromite phase. Of all the varieties of monolithic refractories, the most commonly used one is the castable of alumina-silica group which can be cast in situ. The castables are a mix of suitable aggregates mostly with calcium aluminate cement (CAC) as a binder. Depending on the binder content, they are classified as:

- conventional/dense and insulating: > 2.5% CAC
- low-cement castable: 1.0–2.5% CAC

- ultra-low-cement castable: 0.2–1.0% CAC
- no-cement castable: the hydraulic bond is replaced by chemical bonds such as water glass, silica sol, phosphates, etc.

These varieties of refractories are used in different zones of the kiln system, as illustrated in Table 5.4 and Figure 5.20.

The burning zone refractories are the most critical lining materials in the kiln systems as they undergo serious thermo-chemical stresses. In order to obtain the optimum efficiency of lining, its thickness must be adopted in accordance with kiln shell diameter as indicated below:

Lining thickness (mm)	Kiln shell dia. (m)
250	> 5.5
220	4.5–5.8
200	3.5–5.0
180	2.8–4.2
160	≤ 3.5

Magnesia-chromite bricks have been the standard lining material for several decades. It has, however, been established that that the clinker phases and alkali oxides react between 800 and 1400°C with chromite to form hexavalent chromium compounds that are undesirable from an environmental point of view. The periclase-spinel and pure magnesia bricks were developed as alternatives. However, it has been observed that the magnesia-based refractories do not always favor the formation of coating in the burning zone. Hence, dolomite bricks with higher reactivity with clinker have been used in the kilns. Notwithstanding various

Table 5.4 Zones with different stresses in the preheater-precalciner kiln systems

Zonation	Approximate length (expressed as multiplier of kiln diameter (D))	Types of refractories
Discharge end	1–2 D	75–85% Al_2O_3
Transition zone (discharge end)	1–2 D	Basic bricks
Burning zone	3–4 D (preheater kiln) 5–7 D (precalciner kiln)	Basic bricks
Transition zone (feed end)	2–3 D (preheater kiln) 3–4 D (precalciner kiln)	Basic bricks
Safety zone	2 D	50–70% Al_2O_3
Preheating zone	5–8 D (preheater kiln) 2–4 (precalciner kiln)	Lightweight fire bricks
Feed zone	1–2 D	35–40% Al_2O_3
Cyclones, ducts, etc.	Static vertical tower section	Castables

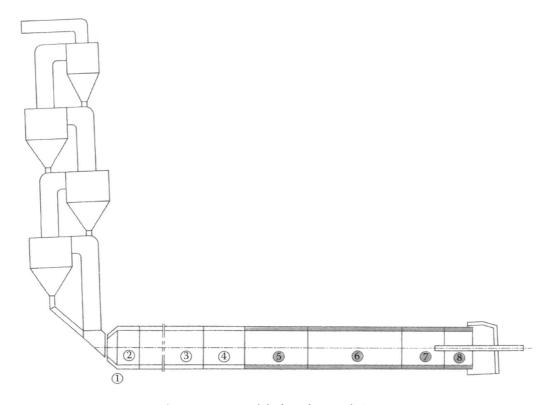

FIGURE 5.20 Zonation in the cement rotary kiln for refractory lining.

options in practice, the life of the burning zone lining has mostly been 10–13 months. In other sections of the kiln, viz., the preheating zone, the lightweight bricks and fireclay bricks give a life of 3.0–3.5 years. On the whole, the specific consumption of refractories in the preheater and precalciner kilns were reported to be 0.89 and 0.56 g/ton clinker respectively in a survey conducted in the past (11). In fact, the specific consumption of refractories is a minor concern for the cement plants, the major concern being the downtime of the kiln due to premature failure of the lining.

No burning zone lining can survive without a clinker coating over its surface, which is formed during the course of the kiln campaign by chemisorption and adsorption of the charge material. Subsequently the coating formation process is accentuated by infiltration of liquid phase of the clinker and its reaction with the lining material. Maintenance of a stable coating over the refractory surface is a dynamic process of coating thickness growth, falling off of the overgrown coating and reformation of new coating layers. The prime requirement for the stable coating is the steady-state operation of the kiln and compositional compatibility of the kiln feed and the lining material. The presence of excess alkalis, sulfates, and chlorides in the kiln atmosphere are detrimental to the formation of a stable coating.

Lining of suspension preheater section

As already mentioned the preferred materials in this section are the monolithic refractories in place of brick linings. Cyclones, pipelines, ducts, and inlet chambers are installed with low-cement castables. Additionally, for some of the more complex portions even precast shapes are sometimes installed. Material build-up in in the lower cyclones and kiln inlet is likely to be more frequent in the case of use of alternative fuels in the precalciner and in the kiln. It has already been mentioned that volatile constituents like alkali oxides from the raw meal and sulfate and chloride from fuel are responsible for the formation of volatile cycles between the preheater and the upper transition zone of the kiln. Although the primary measures for dislodgement or reduction of such build-up are operational, some relief can also be achieved by using special monolithic refractories providing an impervious vitrified layer on the refractory surface to prevent build-up.

Degradation of refractory lining in service

A comprehensive range of case studies on used bricks and castables from various zones of the total rotary kiln system has provided ample evidence of change in the course of wear of refractory lining following the introduction of industrial wastes and secondary fuels (12). Significant salt infiltrations without, or in combination with, other wear factors have been the major factor for refractory degradation. The influence and interaction of salt components with the refractory lining are dependent on the prevailing temperatures, the partial pressure of oxygen, and the alkali:chlorine:sulfur ratio in various zones of the kiln. The formation of salts, such as K_2SO_4 (M.P. 1040°C) and KCl (M.P. 800°C), leads to densification of the brick structure, accompanied by a reduction in elasticity and thermal shock resistance of the affected lining area. In addition, significant concentrations of heavy metal compounds like PbS can occur primarily in the upper transition zone. The net result of such reactions is deep spalling in portions of the lining. If there is excess alkali in the kiln atmosphere and the lining is of magnesia-chrome brick, there can be formation of the chromate phase, which can also lead to premature wear of the affected part of the brick or castable lining. The simultaneous concentration of K_2O, Al_2O_3, and SiO_2 in stoichiometric proportions during the process may lead to the formation of feldspathoid phases, which is accompanied with significant volume increase (up to 30%) resulting in structural weakening. In the case of dolomite bricks CaO is reformed into $CaSO_4$ or CaS, or even $CaCO_3$ in an excess-sulfur environment and if CO is present due to organic impregnation in dolomite bricks. These phase formations lead to volume increase and a weakening of the affected portion of the lining.

Occurrence of corrosion in the kiln shell is well known in cement plants and is related to the kiln gas atmosphere. The use of alternative fuels may cause corrosion of the shell and also of the metal anchors used in the castable lining. This occurs due to the scaling of metals caused by the migration of the volatiles in kiln gases through open joints and pores in the refractory lining to the cooler side of the metal components.

5.9 Energy consumption and kiln emissions

The specific heat consumption in clinker production varies from 700–750 kCal/kg in dry process preheater-precalciner kilns with capacity ranging from 4000 to 7000 tons/day. There are production units with better or worse consumption of heat, depending on their age, design, raw materials, fuel, and local conditions. For the pyroprocessing stage from preheater to cooler the following factors prevail in determining the specific heat consumption:

a. Preheater
- pressure drop across preheater
- efficiency of top cyclones
- dust load in gases leaving preheater
- temperature of gases leaving preheater
- efficiency of preheater fan
- specific power consumption of preheater fan

b. Precalciner
- percentage of fuel in the precalciner
- degree of calcination achieved
- retention time in the precalciner
- temperature at the precalciner exit
- pressure drop across the precalciner
- CO at the precalciner outlet
- specific volume of the precalciner, including ducts
- specific heat load in the precalciner (million kCal/h/m^3)

c. Rotary kiln
- output in tons/day
- capacity
- thermal loads in the burning zone
- degree of filling
- retention time
- mean time between failures of refractories
- rotational speed and slope

d. Clinker cooler
- specific output per grate area in tons/day
- temperature of clinker leaving the cooler
- recuperation efficiency
- temperatures of the secondary and tertiary air flows
- temperature of the vent air
- cooling air in Nm^3/kg Cl

- bed thickness of clinker
- specific power consumption of the cooler with its auxiliaries, such as grates, fans, clinker breaker, etc.

As already explained, the heat economy of different kiln systems depends mainly on the size, number of preheater stages, rate of kiln gas bypass, if any, the raw mix composition and the fuel type. Further, a precalciner kiln and a preheater kiln without a precalciner, with the same number of cyclone stages, may not show large difference in specific heat consumption. This is because of the fact that in the preheater kiln system the lower radiation losses from the cyclones are balanced by the significantly higher radiation losses from the kiln. This relation is reversed in the precalciner kilns.

So far as the specific power consumption is concerned, the pieces of equipment demanding most power in a kiln system are the exhaust gas fan motors, the cooler fan motors (for a grate cooler), and the kiln drive motor. The power consumption of the exhaust fans is mainly dependent on the total pressure loss in the kiln system. The major part of this pressure drop occurs in the preheater. This pressure drop can be reduced by increasing the preheater cyclone dimensions, but, for a given preheater geometry, a certain minimum pressure drop is essential for stable operation. Generally speaking, the total specific power consumption of a precalcining system with grate cooler will be higher than that of the preheater system without the precalciner. For a preheater system having a pressure drop of 400 mm WG, the specific power consumption is estimated at 9 kWh/t, and for a precalcining system of comparable configuration but with pressure drop of 520 mm WG, the specific power consumption is estimated to be 16 kWh/t. Notwithstanding this difference, the precalciner kiln system is the preferred option for modern large-capacity cement production units because of its multifarious advantages.

Kiln emissions

The major emissions from the cement kilns are CO_2, dust, NO_x, and SO_2—all other emissions being negligible and occasionally plant-specific. The industry is primarily concerned with the emission of CO_2, which occurs due to limestone decomposition (50–55%), coal combustion (40–45%), and electricity consumption (up to 10%). In the present manufacturing process, it is estimated that about 535 kg CO_2 is released per metric ton of clinker from limestone calcination and about 330 kg CO_2 per metric ton of clinker from fuel combustion, resulting in direct emissions of 835 kg CO_2 per metric ton of clinker. The industry at large has adopted the following key levers to reduce CO_2 emission:

- Minimizing the use of clinker in cement
- Enhancing use of alternative fuels
- Making unit operations more energy-efficient
- Generating electricity with waste heat
- Gradually adopting renewable energy

In addition, the possibility of capturing and recycling CO_2 is being explored.

So far as dust emission is concerned, the global standards specify the limits of dust loading in the exit gases to 30–50 mg/Nm³. With advanced electrostatic precipitators and efficient fabric filters the limiting norms are being met by the industry. The formation of nitrogen oxides (NO_x) is related to the high flame temperature in the kiln and NO is the main constituent (at about 95%) in what is called the thermal NO_x. The amount of NO_x emission increases if the fuel that is fired in the precalciner contained a high amount of nitrogen. As most of the NO is converted to NO_2 in the atmosphere, emissions are given as NO_2 per cubic meter of exhaust gases. The measured values of NO_x emission have shown a range of 300–2000 mg/Nm³, but in most cases the emission exceeds the global emissions limit of 500–800 mg/Nm³. The adoption of reduction measures is important in this case and the technologies being tried include the installation of low-NO_x burners, staged combustion, and selective non-catalytic reduction (SNCR). SO_2 emission is generally below 300 mg/Nm³ but may in some cases be higher. High SO_2 emissions are often attributable to the sulfides contained in the raw material, which oxidizes between 370 and 420°C, prevailing in the kiln preheater. In those cases, adoption of technology for the desulfurization of flue gases by injecting calcium hydroxide may be required.

Apart from the above major emissions, it is possible that minor emissions of several organic compounds such as dioxins and furans, polychlorinated biphehyls, polycyclic aromatic hydrocarbons, the BTEX group (benzene, tolouene, ethylbenzene, and xylene), and inorganic acids like HCl and HF may occur from the kiln system. Emission of heavy metals is also encountered in the clinker production process. These emissions seldom cross the norms globally specified. When they are detected in excess, the examination of raw materials and fuels becomes important in order to identify the sources and to eliminate them accordingly. The emission of organic compounds has become more important with the firing of alternative fuels and compliance with the emissions norms is essential.

5.10 Kiln control strategies

Kiln operation and control requires both instruments and expertise. The following parameters are critical:

- burning zone material temperature
- clinker temperature
- kiln shell temperature
- feed-end gas temperature
- feed-end oxygen level

High-temperature color cameras and optical pyrometers are commonly used to provide the operator with continuous images of the kiln.

Since the kiln and conveyors central to the process are both moving, non-contact temperature measurement using infrared techniques has long been recognized as a more effective way of meeting the objectives. It is a standard practice to use a ratio thermometer that is suitable for temperatures ranging from 1000–2600°C. The ratio thermometer produces two outputs at slightly different wavelengths, which are averaged by the signal processor to give the true product temperature regardless of any temporary obstruction in the sight path. The signal processors of the latest generation are multichannel-type and are capable of accepting signals from multiple infrared thermometers.

Clinker en route from the cooler to storage should be below 300°C but the larger lumps may be even hotter. An infrared thermometer can be set up to continuously measure the clinker temperature and activate the water spray system. Damage to the kiln shell and the inside refractory lining due to excessively high temperatures and thermal shock can be avoided by monitoring the shell temperature with an infrared scanner along with a rotating mirror and associated electronics. The system is installed at a distance of 30–40 m and up to 1000 measurements can be taken. The output signals are transmitted to a computer containing data-acquisition electronics and software, which displays a colored thermal image representing the distribution of temperatures along the kiln shell and location of cold and hot spots.

Several years back, NO_x generated in the burning zone was taken as the indirect measurement of the burning zone temperature. With the use of indirect firing systems and low-NO_x burners it was found that NO_x could not always be relied upon to give good indication of the burning zone conditions. Therefore, there is more and more reliance on pyrometer or infrared thermometer measurements, as explained above. There is another approach to determine the kiln condition by measuring the kiln torque with the help of kiln ampere. The kiln ampere gives an indication of the fluxing state of the material inside the kiln, while the infrared thermometer or any other pyrometer measures the material temperature. In recent years, a combination of kiln ampere and pyrometer measurements has been used to ascertain the burning zone conditions more realistically. Further, there are several proprietary expert systems that have proven valuable in optimizing the burning process and they make use of measured values in addition to many other process inputs. Ultimately, control of the process is accomplished by adjustments to kiln feed, fuel rate, and induced draft fan speed. The simultaneous control of kiln feed and speed ensures proper volumetric filling (say, 7–12%). While the cyclone preheater kilns rotate at 2–2.5 rpm, the precalciner kilns rotate much faster (3.5–5.0 rpm), both the systems having more or less similar material residence time (20–30 min) in the kiln system. The feed-end gas temperature and the feed-end oxygen level are targeted at 1000°C and 2% respectively. The entire kiln control is of course linked to the composition of the raw meal, fuel, and clinker, as determined by the plant laboratory.

5.11 Summary

Pyroprocessing is at center stage of a clinker making plant. This part of the process extends from the feeding of raw meal to the discharge of clinker into the storage silo. Clinker formation takes place through four major steps: decarbonation of limestone, solid-state reactions of lime and other oxides, liquid-phase sintering, and consolidation of microstructure by cooling. These reaction steps occur in preheater cyclones, precalciner vessels, kilns, and coolers.

The cyclone preheaters are a static counter-current heat transfer system in which the hot kiln gases are used to heat up the raw meal from the ambient conditions to about 800°C before the material enters into the rotary kiln for further processing. The cyclone preheaters, also known as suspension preheaters, generally have four to six stages of cyclones and the raw meal gets calcined to an extent of about 30% before it enters the kiln. Hence, a significant part of the decomposition of limestone takes place in the rotary tube.

Precalciners were developed as an additional stage in the cyclone preheater system to enhance the degree of calcination or decarbonation of raw meal to at least 85%, and preferably up to 95% outside the rotary tube. There are various designs of precalciners available and their dispositions in the preheater systems also vary. All these variations in system design are aimed at meeting the contradictory demands of minimizing the temperature and oxygen levels and completing the burn-out of the fuel, providing relief of thermal load to the rotary kiln, and achieving the required degree of decarbonation prior to burning in the rotary kiln. The rotary kilns can either be three-support or two-support systems with their merits and demerits. There have been significant advances in the design of burners to the kilns and precalciners. Three- or five-channel burners are generally used in the kilns. The flame shape, its stability, and its momentum are some of the important characteristics that define the course of burning.

Cooling is an integral and essential part of the clinkering process. Amongst the various types of coolers, advancements have most often been made in the design of grate coolers. These improvements include grate design, air flow regulators, sealing devices, etc. The developments have been prompted by the process requirements of high and consistent temperature of the secondary and tertiary air flows from the cooler to the kiln and precalciner and also by the demand of easy maintenance. The effects of cooling conditions on clinker quality have been widely studied. It appears that slow cooling up to, say, 1200°C with subsequent rapid cooling up to the discharge temperature is expected to maximize the formation of alite, stabilization of desired belite polymorph, tricalcium aluminate, and calcium alumino ferrite phases. It may also be borne in mind that the cooling process is more for heat recuperation than clinker quality.

A system configuration with five stages of cyclones, a precalcining reactor, a three-pier rotary kiln, and a grate cooler is common these days. Depending on the output required, one may have more than one string of preheater cyclones with different arrangements of secondary and tertiary air flows. The capacity of kiln systems may go up to 12,000 tons/day, although the majority of kilns today are in the output range of 4000–7000 tons/day. In such complex kiln systems, the specific heat and power consumption depends on a large number of design and operational parameters, as discussed in this chapter. So far as the emissions norms are concerned, compliance for dust emission is satisfactory; the NO_x and SO_2 emissions may require additional technological measures but they are known. CO_2 emission continues to be an active area for abatement research in the industry. The emission of organic and inorganic pollutants is not generally an area of concern, although greater attention is being paid due to the increasing use of alternative fuels.

The refractory lining materials are important from the clinker production angle. The burning zones of the kilns are lined with basic bricks of different composition and quality, while the rest of the system, including the preheater cyclones, precalciner vessels, and coolers are lined with alumina bricks and monolithic refractories, the details of which have been provided in this chapter.

References

1. A. K. CHATTERJEE, Chemistry and engineering of the production process – incremental advances and lack of breakthroughs, *Cement & Concrete Research*, Vol. 41, Issue 7, July 2011.
2. A. K. CHATTERJEE, Chemico-mineralogical characteristics of raw materials, in *Advances in Cement Technology* (Ed. S. N. Ghosh), Pergamon Press, UK (1983).
3. A. K. CHATTERJEE, Modernization of cement plants for productivity and energy conservation, in *Cement and Concrete Science and Technology* (Ed. S. N. Ghosh), Vol. 1, Part 1, ABI Books Pvt. Ltd., New Delhi (1991).
4. CLANS BECH AND ALEX MISHULOVICH, Preheaters and Precalciners, in *Innovations in Portland Cement Manufacturing* (Eds J. I. Bhatty, F. M. Miller, and S. H. Kosmatka), Portland Cement Association, Skokie, Illinois, USA (2004).
5. PHILIP A. ALSOP, *The Cement Plant Operating Handbook*, Tradeship Publications Ltd., UK (2007).
6. P. C. SOGANI and KAMAL KUMAR, Technological trends, in *Modernization and Technology Upgradation in Cement Plants* (Eds S. N. Ghosh and Kamal Kumar), Vol. 5, Akademia Books International, New Delhi (1999).
7. W. KURDOWSKI, Cement burning technologies, in *Advances in Cement Technology* (Ed. S. N. Ghosh), Tech Books International, New Delhi (2002).
8. MATHIAS HERSMANN, *Pyroprocessing technology – continuous improvement*, IEEE-IAS/PCA Conference, Charleston, USA (May 2007).
9. A. K. CHATTERJEE, S. KRISHNAN, AND B. S. RANGNEKAR, Clinker coolers – types, efficiency and effect on clinker quality, ALL India Seminar on Cement Manufacture, NCB, New Delhi (1981).

10. A. K. CHATTERJEE, Role of volatiles in cement manufacture, in *Advances in Cement Technology* (Ed. S. N. Ghosh), Pergamon Press Ltd., Oxford, UK (1983).
11. P. BARTHA, Present technology of the refractory lining in cement burning plants, RefraSymposium'86, Refratechnik GmbH, Gottingen, Germany (1986).
12. P. BARTHA AND J. SODJE, Degradation of refractories in cement rotary kilns fired with waste fuels, *C N Refractories Special Issues*, Vol. 5, Verlag Schmid GmbH, Freiburg, Germany. 2001.

CHAPTER SIX

Clinker grinding and cement making

6.1 Preamble

From the process steps narrated earlier, an intermediate product, called clinker, is obtained, which is subjected to the next step of grinding with gypsum and other additives, as required, to obtain the final product: Portland cement. This step in the cement manufacturing process involves various grinding systems. The differences in the design, layout, and operation of these systems reflect in the grinding of clinker. It has also been widely observed that the composition and crystal structure of the individual phases and their resultant microstructure control the overall grinding response of clinker.

Even now, the cement grinding process is overwhelmingly based on ball mills, which are used either in isolation (open-circuit) or in conjunction with separators (closed-circuit). A third variant of the ball mill system that has emerged in the last decade or so is the hybrid system, having separators and hydraulic roll presses. In parallel, vertical roller mills have also been deployed for cement grinding. In recent times, the clinker grinding process seems to be moving increasingly from ball mills to vertical roller mills, hydraulic roll presses (without ball mills), and horizontal roller mills. These changes are apparently dictated by the demands of energy conservation and productivity.

The demand for cement production has been continuously increasing all over the world. The grinding process being highly energy-intensive, increasing cement demand has a strong bearing on economics and environment. Meeting customer needs with minimum energy costs and maximum environmental benefits is of paramount importance. The present chapter is devoted to understanding the material properties of clinker and the multifarious grinding systems deployed to convert clinker into cement.

6.2 Clinker characteristics

Phase composition

Portland cement clinker is a dark-gray nodular material. The nodule size ranges from less than 1 mm to 25 mm or more and nodules are composed mainly of calcium silicates (typically 70–80%). In the industrially produced clinker there are four principal phases, the composition of which is given in Table 6.1 (1).

In addition to the four primary phases detailed in Table 6.1, an industrial clinker is likely to contain minor quantities of periclase (MgO), free lime (CaO), and alkali sulfates. The chemical aspects of transformation and stabilization of the above four major phases have long been known (2,3) and the issues that are important in the present context are highlighted below.

- Alite or C_3S phase

    The pure C_3S at temperatures ranging from ambient to about 1100°C has seven polymorphs:

620°C	920°C	980°C	990°C	1060°C	1070°C	
$T_1 \leftrightarrow$	$T_2 \leftrightarrow$	$T_3 \leftrightarrow$	$M_1 \leftrightarrow$	$M_2 \leftrightarrow$	$M_3 \leftrightarrow$	R

Three of the polymorphs belong to the triclinic system (T), another three to the monoclinic system (M), and the seventh belongs to the trigonal rhombohedral system (R). However, industrial clinkers generally show the presence of M_1 or M_3 or their mixture. Rarely, the T2 variety has been observed. The alite phase generally has an impurity level of 3–4% and its density increases to 3.15 from 3.12 of pure C_3S. It seems that the changes in hydraulic properties of

Table 6.1 Primary phase composition of clinker

Phase	Basic composition	Foreign elements present	Crystallochemical state
Alite	Tricalcium silicate (C_3S)	Al, Fe, Mg, Cr, Ti, S, P, Ba, Mn, Na, K	7 polymorphs
Belite	Dicalcium silicate (C_2S)	Al, Fe, Mg, Cr, Ti, Mn, V, Ba, S, P, Na, K	5 polymorphs
Calcium aluminate (celite)	Tricalcium aluminate (C_3A)	Fe, Mg, Si, Ti, Na, K	No polymorphs, foreign elements as solid solutions; alkalis cause change in crystal symmetry
Calcium aluminoferrite (brownmillerite)	Tetracalcium alumino-ferrite	Mg, Si, Ti, Mn, Cr	Solid solution series between C_2F and hypothetical phase C_2A up to C_4AF

alite are due more to structural defects, disorders, etc. than to its polymorphic form.

- Belite or C_2S phase

  Unlike the alite phase, the properties of belite are more sensitive to different polymorphic modifications. From ambient to 1500°C, this phase may change its polymorphic state, as shown in Figure 6.1. The high-temperature modifications have high reactivity but their stabilization at ambient temperature is not easy. Industrial clinkers predominantly show the presence of the beta polymorphic phase, sometimes together with small proportions of the high-temperature forms. The beta form possesses good hydraulic properties, while the gamma form is of poor hydraulic behavior.

  The incorporation of foreign elements in the belite phase is of the order of 4–6%, resulting in a change of density from 3.30 to 3.326 of pure C_2S.

  One should note that stabilization of the beta phase of belite is quite important for achieving the desired hydraulic properties of cement.

- C_3A phase

  As mentioned in Table 6.1, the C_3A phase does not have any polymorphic transformation but foreign elements can occur as solid solutions in C_3A up to 10%, but only the alkali elements cause changes in its crystal symmetry to orthorhombic, and monoclinic forms (see Table 6.2). Pure C_3A is cubic in symmetry.

  The different crystal forms of C_3A show variations in their properties. This phase is particularly sensitive to the use of chemical admixtures.

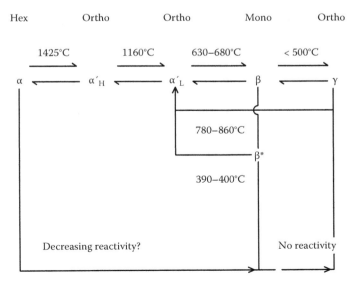

FIGURE 6.1 Polymorphic transformations of the belite phase.

Table 6.2 Modifications of C_3A structure with alkali content

Approximate Na_2O (%)	Value of x in the general formula $Na_{2x}Ca_{3-x}Al_2O_6$	Designation	Crystal systems
0–1.0	0–0.04	C_I	Cubic
1.0–2.4	0.04–0.10	C_{II}	Cubic
2.4–3.7	0.10–0.16	C_{II} + O	–
3.7–4.6	0.16–0.20	O	Orthorhombic
4.6–5.7	0.20–0.25	M	Monoclinic

- Calcium aluminoferrite phase

    This phase in clinker is basically represented by C_2 (A, F) in which there is complete miscibility between C_2F and hypothetical C_2A up to C_4AF. More precisely, this is represented as a solid solution series of the following composition: $Ca_2 (Al_x Fe_{1-x})_2 O_5$, in which for C_4AF, x = 0.5. The C_4AF phase has a density of 3.732 and possesses orthorhombic symmetry. Numerous studies have established that the ferrite phase in clinker may not necessarily be C_4AF in composition. In Indian clinkers, the composition of this phase is close to C_6AF_2 as determined in a large number of clinkers studied in the author's laboratories. It has also been established by various researchers that Mg, Si, Ti, Mn, and Cr may be incorporated in the lattice of C_2 (A, F).

    The ferrite phase is regarded as having low hydraulic properties. Nevertheless, it does contribute to the strength development in cement. It has been observed that C_6AF_2 hydrates faster than C_4AF.

Proportions of the major clinker phases and their microstructure

The precise quantitative correlation between the proportions of major phases present in clinkers and their properties is still a basic problem of cement technology. By and large, the optimum content of alite is taken as 50–55% and that of belite as 20–30%; the C_3A and C_4AF are targeted at 8–12% and 12–14%, respectively (4).

The basic microstructural feature of polycrystalline materials is the individual grain. Each grain has size, shape, and orientation parameters. In a mono-phase material the above parameters can substantially alter its properties. In a poly-phase material like clinker the properties are controlled jointly by all the grains of the constituent phases, where, apart from the size and shape, the relative distribution of the grains and the nature of matrix become important.

The microstructure of clinker is best studied under a petrographic microscope. For more accurate and magnified observations, scanning electron microscopes (SEMs) are used. The microstructure of clinker for all practical purposes is determined by

   i. size and shape of alite grains
   ii. size and shape of belite grains

iii. characteristics of C_3A and C_4AF as the interstitial phases
iv. size and clustering of periclase
v. nature of occurrence of free lime
vi. clustering of the silicate phases
vii. clinker nodule porosity
viii. presence of relicts of coarse grains of calcite and quartz

The size of the alite crystals generally ranges from 10 to 25 μm. It may often become coarser—up to 50–60 μm. The crystals are prismatic and sometimes pseudohexagonal, having clearly defined faces and being yellowish-brown in color, as observed under the optical microscope (Figure 6.2a). Less regular with rounded corners and faces with re-entrant angles are also seen, signifying textural deterioration (Figure 6.2b). Internal structures observed in alite crystals are lamellae due to twinning as well as inclusions of belite. Twinning in alite is evidence of polymorphic transformation during the clinkering process. A triple cyclic twinning indicates trigonal-monoclinic inversion and polysynthetic twinning shows monoclinic-triclinic inversion.

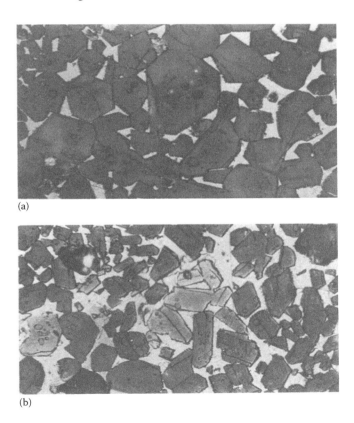

FIGURE 6.2 Microstructural views of alite and belite crystals in clinker; (a) alite crystals with good development of faces; (b) alite crystals of different types with imperfect faces. *(Continued)*

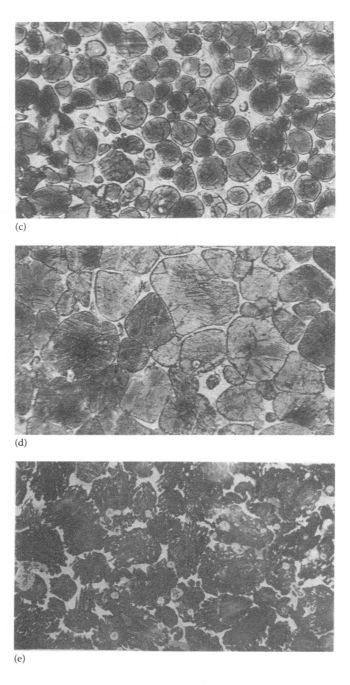

FIGURE 6.2 (CONTINUED) Microstructural views of alite and belite crystals in clinker; (c) perfect rounded crystals of belite; (d) striated belite crystals of varying degree of roundness; (e) subrounded and lamellar belite crystals.

Belite is typically observed as 25–40-μm rounded crystals, which show blue or red color in the etched sections under the microscope; sometimes the grains display multi-dimensional lamellae or striations. These striations indicate twinning resulting from polymorphic transformations. According to the microscopic features, belite is classified into three types:

Type I: At least two sets of striations are present and normally beta phase is seen.

Type II: One set of striations is seen and the polymorph may be either ά or a mixture of α and β forms.

Type III: Free of striations, often clear and mainly ά polymorph is present.

The microstructural variations of the belite grains are shown in Figure 6.2 (c, d, and e).

For a good microstructure, the distribution of alite and belite crystal should be homogeneous, as shown in Figure 6.3. If there are problems of homogeneity or grinding in the raw mix, they manifest in the clinker microstructure. For example, the belite crystals may occur in clusters and occasionally the large clusters indicate that the site was formerly occupied by silica particles (Figure 6.4).

The interstitial areas between the silicate grains contain small crystals of C_3A and C_4AF. The aluminate phase appears as angular crystals, size being a function of the cooling rate. The alkali-modified aluminates are observed in prismatic or cigar-like shape. The ferrite grains appear intermixed with C_3A, as bright reflective laths (Figure 6.5).

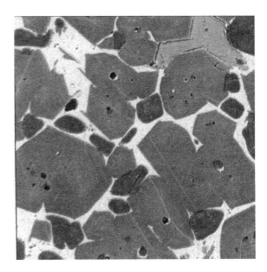

FIGURE 6.3 Photomicrograph of clinker showing uniform distribution of silicate phases.

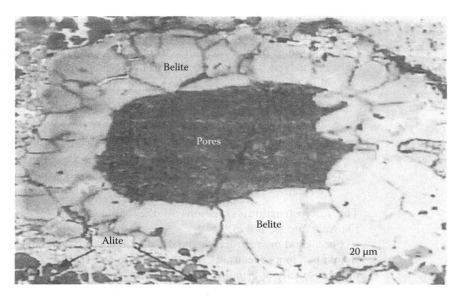

FIGURE 6.4 Photomicrograph showing a large pore inside a belite cluster.

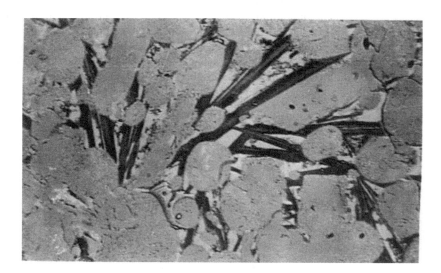

FIGURE 6.5 Clinker microstructure showing well differentiated matrix of ferrite and aluminate.

Free lime (CaO) grains are observed in water-etched polished sections as brightly colored drop-like phases 10–20 μm in size, present in a dispersed condition (Figure 6.6) or in clusters (Figure 6.7). It is important to note that the cluster of free lime crystals often retains the shape of the original coarse calcite grain.

Periclase (MgO) occurs as needle-shaped crystals, sometimes in clusters. In order to achieve soundness in cement hydration it is important to have crystals of small size, even when they are dispersed. If the crystals

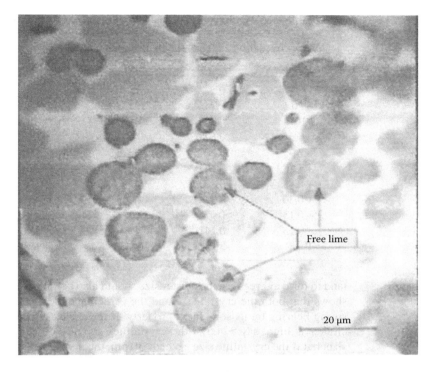

FIGURE 6.6 Photomicrograph of clinker showing free lime in dispersed state.

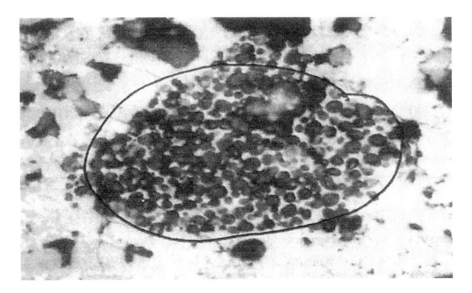

FIGURE 6.7 Free lime nest as a result of coarse calcite particles.

Table 6.3 Computation of fineness modulus (FM) of clinker

Sieve fraction (mm)	Weight (g)	Percentage of the total weight	Cumulative percentage
+ 80	0	0	–
+ 40	0.248	7.64	7.64
+10	0.687	21.15	28.79
– 5	0.679	20.91	49.70
+ 2.35	0.500	15.39	65.09
+ 1.18	0.347	10.68	75.77
+ 0.88	0.123	3.79	79.56
+ 0.30	0.346	10.65	90.21
+ 0.15	0.077	2.37	92.58
– 0.15	0.241	7.42	100.00
Total	3.248	100.00	589.34

tend to form aggregates, the cluster size should also be small. Experience shows that both the crystal size and the percentage of periclase have a strong bearing on its soundness property. If the periclase percentage in clinker is high, say > 4.5%, the crystallites should be below 5 μm in size, but if the crystallite size exceeds 30 μm, the percentage of periclase should be as low as 1% for similar soundness property.

Clinker granulometry

In addition to the phase composition and microstructure, the clinker grindability depends on its granulometry. While a clinker with 50–60% passing 15 mm is often considered to be coarse, the passing level of 80–90% through 15 mm characterizes the clinker as fine. The size distribution of clinker nodules can be expressed by a parameter called "fineness modulus" (FM), which is calculated from its sieve analysis (Table 6.3). Clinkers are generally classified on the basis of FM as follows:

Normal clinker: 4–7 FM

Dusty clinker: < 4 FM

Coarse clinker: > 7 FM

From the above table, FM can be calculated as follows:

FM = 589.34/100 = 5.89

and the value signifies that the given clinker can be categorized as of normal granulometry.

Broad characteristics of clinkers

From various studies carried out in the author's laboratories, the broad characteristics of Indian clinkers can be summarized as follows:

- Alite (%): 46–62
- Alite crystal size (μm): 15–30
- Belite (%): 25–40

- Belite crystal size (μm): 12–20
- $C_3A + C_4AF$ (%): 16–18
- Granulometry: 45–90% passing 15 mm
- Porosity (%): 12–20
- Bond's grindability index: 11–18 kWh/t cl

Experience has shown that a clinker becomes harder to grind when

i. its granulometry becomes finer;
ii. its microstructure is ill-formed;
iii. its total porosity comes from only a few large pores, instead of from well-dispersed small pores;
iv. its belite content is high;
v. the crystal sizes of alite and belite are large;
vi. the sulfate content in clinker is high.

Conversely, a clinker containing less than 2% minus 1 mm size fraction or higher alkali contents (say, 0.5% R_2O) or relatively high free lime tends to show softer grinding tendency.

6.3 Clinker grinding systems

The clinker grinding system was once predominated by ball mills, used either in isolation (open-circuit) or in conjunction with separators (closed-circuit). But due to the inherent problem of low milling efficiency, more energy-efficient systems such as ball mills in combination with separators and high-pressure roll presses, vertical roller mills, direct grinding high-pressure roll presses, and horizontal roller mills (known as horomills) appeared on the scene. Their broad performance comparison is given in Table 6.4 (5).

From the table, it is evident that the ball mills can yield products with very high specific surface area and wide size distribution of particles but with 30–40% higher energy consumption as compared to other grinding systems. The newer grinding systems generally have a limitation of achieving high specific surface area and the particle size distribution is also rather steep. However, in some of the more recent installations

Table 6.4 Performance comparison of different cement milling systems

Sl. No.	Parameter	Ball mill	Vertical roller mill	High-pressure grinding roll	Horomill
1.	Specific surface area (Blaine), cm²/g	>6000	4500	4000	4000
2.	Rosin-Rammler plot for particle size distribution	0.85–1.1	0.85–1.1	1.0–1.1	1.05–1.1
3.	Relative energy demand for the entire circuit (%)	100	60–70	50–60	70

of VRM higher product fineness has been achieved at much lower energy consumption. But they may not still match the high specific surface area that can be achieved by ball mills. Similarly, the high-pressure grinding rolls and horomills continue to have the limitations of reaching the limits of both the particle size distribution and specific surface area that are achieved by ball mills. Perhaps due to these reasons, notwithstanding the high specific energy demand, ball mills are still the most widely used grinding system. A broad description of the constructional and operational features of these milling systems is given below.

A. Ball mills

Ball mills are cylindrical shells designated by their inside diameter, length, and connected drive motor power. The shell is protected by steel liners and is usually divided by a diaphragm into two compartments. High-chrome alloys are now almost exclusively used for balls and liners and the ball usage should not exceed 50 g/t of cement. The first compartment is primarily to break the clinker nodules. The differences in the features of the first and second compartment are given in Table 6.5.

While coarse grinding benefits from a range of ball sizes, the efficiency of fine grinding, which is attrition-based, is achieved with single-size (20–25 mm) balls. If a range of ball sizes are charged, classifying liners are employed to retain the larger balls close to the diaphragm and the smaller balls at the discharge. The discharge screen slots are at least 3 mm wider than the diaphragm slots and 5 mm less than the smallest ball.

A typical schematic flow-sheet of a two-compartment closed-circuit ball mill system is shown in Figure 6.8. The ball mill internals are shown in Figure 6.9.

The ball quantity, the type and condition of the shell liners, and the mill speed determine the power drawn by the mill. The critical speed of a ball mill is the speed at which the centrifugal force just holds the charge to the shell during rotation. With relation to this critical speed, ball mills typically operate at 75% capacity. The volumetric charge loading is typically kept at 25–35%. The normal air velocity is 0.8–1.2 m/s related to the open mill cross-section above the ball charge. Fully air-swept mills operate with an air velocity of 5–6 m/s.

Table 6.5 Comparison of the features and functions of the two compartments in a ball mill

Features	First compartment	Second compartment
Typical lengths	30–44%	67–70%
Ball dimensions	90–50 mm	50–15 mm
Clinker top size	30 mm	2–3 mm
Power drawn	9–10 kWh/t	24–25 kWh/t
Charge specific surface	9.5–10.5 m^2/t charge	35–38 m^2/t

For yielding a finished product, mainly cement, with large surfaces
1. Feeding bin
2. Feeding bin
3. Proportioning belt weightmeters
4. Belt feeder
5. Tube mill
6. Discharge housing
7. Airslide
8. Bucket elevator
9. Airslide
10. Cyclone air classifier
11. Grits return belt with weighing device
12. Airslide for finished product
13. Preliminary separator
14. Dust removal plant
15. Dust removal blower

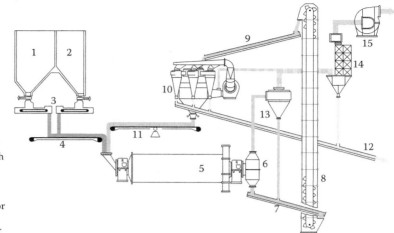

FIGURE 6.8 A closed-circuit ball mill system.

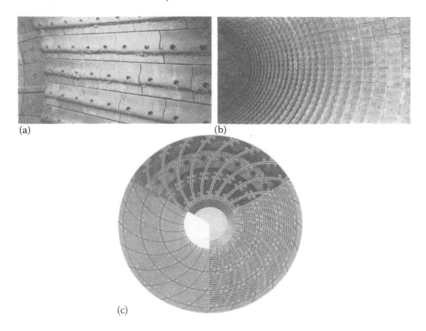

FIGURE 6.9 The ball mill internals: (a) mill shell lining of the first chamber; (b) classifying lining of the second chamber; (c) the diaphragm wall between the two chambers. (Courtesy of KHD on grinding plants. Trade catalogue.)

The discharge temperature of cement is maintained between 90 and 115°C. Water sprays are used to control mill temperature but all care has to be taken to ensure that the water is evaporated without causing the hydration of cement. It is easier to add water to the first compartment, but the best effect is achieved by spraying at the diaphragm and the second compartment. Maintenance of the temperature within a given range is essential to avoid the dehydration of gypsum, which affects the cement quality. Air flow through the mill serves to remove water from the mill.

Another parameter that is quite important to assess the mill performance is the mill filling or mill hold-up or power loading (see Figure 6.10) (6). For very low powder loading (say, less than 60% void filling) it is reasonable to expect inefficient grinding since there will be a large proportion of energy spent on media to media impacts without any grinding of the material. Conversely, in the case of high power loading (say, more than 120% void filling) there would be a cushioning effect and absorbance of energy without any energy optimization. Mill filling is monitored with the help of microphones located externally approximately one meter from each end of the mills. Decreasing microphone sound amplitude and increasing mill drive power indicate that a mill is filling up. Conversely, increasing sound and decreasing mill power indicate the emptying of the mill.

Several types of separators, also known as air classifiers, are employed in mill circuits. The function that a classifier is required to perform is simply to separate as effectively as possible the mill product at the desired cut size so that the fewest possible number of fine particles below the size will get into the tailings and vice versa. The mixed particles are introduced into a classifying zone of the separator where centrifugal force of a rotating system and drag force of air currents act upon the particles; depending upon the direction of the resultant force, the particles get into coarse or fine streams. There are numerous design variations of each type and the status of separator technology has been reviewed in (7). In a broad sense, the following three types of separators representing different stages of development are relevant in the present context:

a. Static separators, also called grit separators, do not have any moving parts and effect separation by the cyclonic air flow

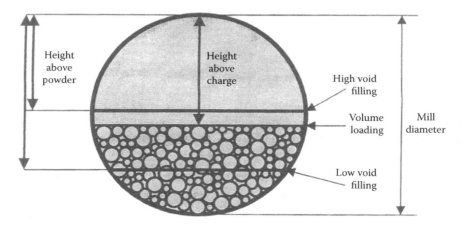

FIGURE 6.10 Schematic representation of powder filling in ball mills.

induced by guide vanes, which are so adjusted as to make the dust collectors collect 45-μm particles that can go as product. Figure 6.11 shows the basic twin-cone design on the left-hand side and an improved feature of a bottom-impact cone on the right-hand side. The conveying air, along with solids, enters the bottom of the grit separator and expands into the annular space of the outer cone. Due to a sudden expansion and drop in velocity the coarser particles move to the tailings discharge chute. In some designs, the tailings cone is provided with an impact cone at the bottom, which is suspended by strings from the top frame of the separator. The gap between the impact cone surface and the inner cone can be adjusted. After passing from this gap and before reaching the rejects discharge chute, the coarse material gets reclassified by the incoming air stream. Even with such improved features the static separators do not reach an efficiency level of 70%.

b. Mechanical separators are first-generation dynamic separators that have been extensively used to classify the mill product. They contain a rotating distribution disc for dispersing feed particles into the classification zone, a centrally driven fan for creating air circulation necessary for classification, an auxiliary fan, return air vanes for tailings reclassification along with a tailing cone, and a product discharge cone. Variations in the design of the above features resulted in several types of separators achieving varying degrees of separation efficiency. The material is fed onto a rotating dispersion plate and is thrown up into a rising air stream. Coarse particles either fall from the dispersion plate or are rejected between the auxiliary fan blades and the control valve. Fine dust remains entrained through the main fan and gets detrained as the gas flows

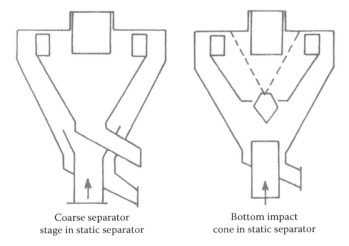

Coarse separator
stage in static separator

Bottom impact
cone in static separator

FIGURE 6.11 Coarse separation stage and bottom impact cone in a static separator.

downwards with decreasing velocity and due to diversion through the return vanes. The operating adjustments are done with the number of auxiliary blades, the clearance between auxiliary blades and control valve, and the radial position of the main fan blades. Figure 6.12 shows a sketch of a mechanical separator and its layout in closed-circuit grinding. After the size classification, the coarse fraction, which is also called rejects or tailings, is returned to the mill at the feed end and the fines with product size is sent to the product silo.

c. High-efficiency cyclone separators are an improvement over mechanical separators. The improvement came through the adoption of spiral air classification and the introduction of caged rotors for compact classification zones. O-Sepa, developed by Onoda Cement Co. Ltd., Japan in 1983, is considered to be the first successful separator in this category. There are many other versions of high-efficiency separators, such as Sepax of FLSmidth & Co., Sepol of Krupp Polysius group, Sepmaster-SKS of KHD-Humboldt-Wedag, among others. The schematic diagram of the O-Sepa model is given in Figure 6.13 (7). Material is fed onto a rotating dispersion plate to be dispersed into the classifying air stream, which is sucked from tangential inlet ducts through fixed guide vanes. A horizontal vortex is formed by the rotor, which classifies particles between the centrifugal force and the inward air flow. The fine fraction exits upwards with the exhaust air for subsequent dust collection while the coarser fraction descends and is discharged from the bottom of the vessel. Fineness can be increased by increasing the rotor speed; air flow is separately controlled by the separator ID fan; effective material dispersion is assured by buffer plates around the periphery of

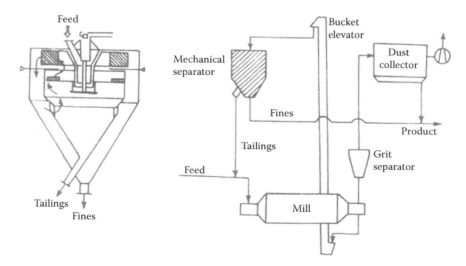

FIGURE 6.12 Conventional mechanical separator and its layout in a grinding circuit.

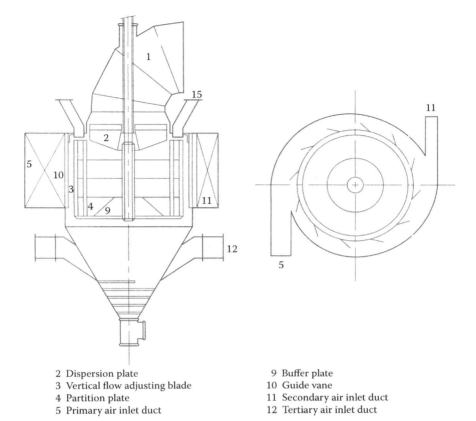

2 Dispersion plate
3 Vertical flow adjusting blade
4 Partition plate
5 Primary air inlet duct
9 Buffer plate
10 Guide vane
11 Secondary air inlet duct
12 Tertiary air inlet duct

FIGURE 6.13 Schematic diagram of a high-efficiency separator (O-Sepa). (Reproduced from J. Yardi, *World Cement*, October, 87–113, 2005. With permission of World Cement, UK.)

the dispersion table; uniform distribution of incoming air is ensured by the design of the air ducts and guide vanes; and, finally, the height:diameter ratio of the rotor controls the retention time of particles in the separating zone.

A convenient means of describing the performance of a separator is to determine the grade efficiency curve (known as the Tromp curve or the selectivity curve). The Tromp curve is determined from the particle size distributions for the separator feed, fines, and rejects (a, f, r). The average circulating load (A/F) can be derived from the sum of three particle size distributions. The Tromp curve is defined as follows:

$$\frac{\text{Mass of material at size I in rejects}}{\text{Mass of material at size I in feed}} \times 100$$

i.e., $\text{Tromp}_i = \dfrac{R_{ri}}{A_{ai}} \times 100$

The cut point (equiprobable size), imperfection (sharpness of cut), and bypass can then be derived from the Tromp curve (see Figure 6.14). The bypass indicates the amount of fines being incorrectly placed into the rejects stream. Many separators display a noticeable "fish-hook," i.e., the grade efficiencies increase after the minimum. This happens since the particles are treated as coarser ones on account of agglomeration or due to adhesion and entrainment with the coarser particles.

As the bypass increases, the amount of fines returned to the mill increases. Obviously, the bypass is larger for a reduced separator efficiency and a higher circulating load. It may be relevant to mention here that a circulating load is typically 200–300% with mechanical separators and 150–200% with high-efficiency separators.

On the whole, the following parameters are found to significantly influence ball mill performance:

- mill ventilation
- media grading
- diaphragm design and condition
- number of chambers
- separator type, efficiency, and circulating load
- volume loading
- mill speed
- material characteristics

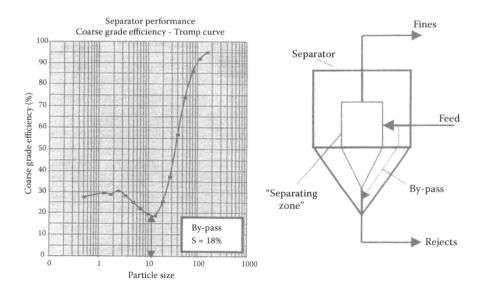

FIGURE 6.14 Concept of Tromp curve.

B. Vertical roller mills

Vertical roller mills (VRM) are air-swept grinding mills with built-in separators. Roller mills grind in a closed circuit. All types of VRMs contain grinding tables rotating on thrust bearings mounted on a right-angle gear box. Roller mills employed in the cement industry have grinding elements of various shapes. Some have cylindrical rollers and flat or inclined tables, and some have convex-shaped tires and grooved tracks on the table. A few mills contain big hollow balls as grinding elements. The grinding force is exerted by hydraulic pressure. A schematic diagram of a vertical roller mill is given in Figure 6.15. All the major equipment suppliers have their models of VRMs, viz., Loesche, Raymond, FLSmidth (Atox), Peiffer (MPS), Krupp Polysius, and others.

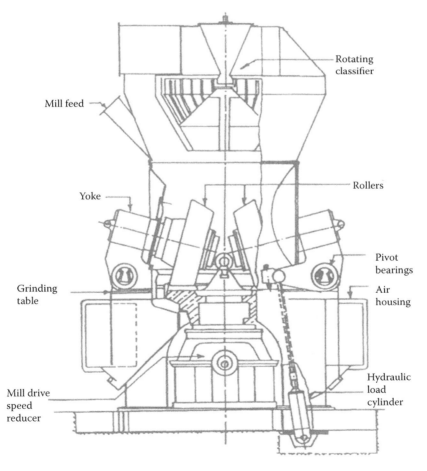

FIGURE 6.15 Sketch of a typical vertical roller mill.

In the vertical roller mills of today, the material is ground on a pan or table rotating about a vertical axis onto which rollers are placed by spring or hydraulic pressure. These mills are air-swept, using hot gas for drying while grinding, with a classifier incorporated in an airtight casing within the body of the machine. The vertical roller mills are designed with two, three, four, or six rollers. Although, to start with, roller mills were used for raw materials and coal, i.e., for relatively softer materials, in recent years, improved metallurgy and modified roll configurations have led to their increasing use for cement and slag grinding. The significant engineering advances in vertical roller mill technology have been reported in (8–13).

The main requirement in grinding cement and mineral admixtures in vertical roller mills is the stabilization of the grinding bed. Compared to the grinding of raw materials, the internal circulation between the grinding table and the classifier is substantially higher in cement grinding. Without a well-stabilized grinding bed, mill vibrations can not be avoided.

After a period of intensive research work based on CFD simulation, one of the machinery manufacturers, Loesche, introduced the modified module concept for cement. Two of the four full-size grinding rollers have been replaced by smaller rolls (see Figure 6.16). The larger rollers, called the master rollers, are for material grinding, while the smaller rollers, called the support rollers, are used for compaction, de-aeration, and bed building.

A simple clinker grinding circuit with a vertical roller mill is shown in Figure 6.17. Feed materials are transported from the storage bins via belt feeders and rotary air locks into the mill. In the

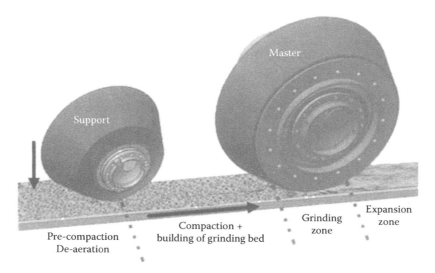

FIGURE 6.16 Working principle of master and support rollers in a VRM. (Reproduced from Caroline Hacklander-Woywadt, *World Cement*, September, 71–77, 2005. With permission of World Cement, UK.)

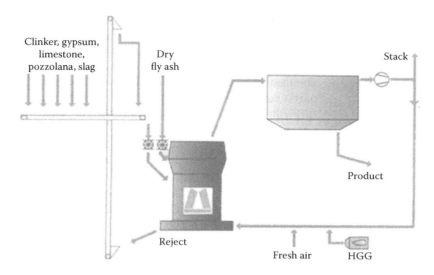

FIGURE 6.17 Typical flow chart of a VRM. (Reproduced from Caroline Hacklander-Woywadt, *World Cement*, September, 71–77, 2005. With permission of World Cement, UK.)

mill the material is ground, classified, and dried by exhaust gas or an auxiliary hot gas generator. The coarse material from the grinding table is returned via the reject conveyor to the feed material. The finished product from the inbuilt high-efficiency separator leaves the mill in a gas stream and is captured in a bag filter. The fan behind the filter provides the required draft. In general, the capacity range of VRMs for cement grinding is 60–300 t/day, while for slag grinding the capacity ranges from 45–200 t/day. The specific power consumption ranges from 15–35 kWh/t, depending on the products and their fineness.

When the vertical roller mills were first introduced for cement grinding, there were concerns about the product quality due to steep particle size distribution curves. Today it is observed that the particle size distribution curves of VRM products can be changed by manipulating grinding pressure, grinding bed thickness, grinding table speed, and working pressure.

C. High-pressure grinding rolls

High-pressure grinding rolls, or hydraulic roll press, as they are otherwise known, have been extensively used for pre-grinding in a variety of circuits and they have also been used as standalone cement mills. Many early roll presses suffered from the rapid wear of roll surfaces and bearing failures but, progressively with improved metallurgy and lower operating pressures, roll press grinding has been adopted for clinker grinding in many plants (11).

The most common circuits are for pre-grinding with slab recirculation and other modes with de-agglomeration. Clinker to be ground is compressed in a gap between two counter-rotating grinding rolls with circumferential speed of 1.0 to 1.8 m/s to form compacted cake with a volumetric proportion of solids of over 40%, which may contain up to 40% fines (< 90 μm). The fines are obtained by de-agglomeration of compacted cake and the residual part is sent for further grinding. Three different circuits are shown in Figure 6.18. In the hybrid mode, the roll press is not close-circuited, while the ball mill is in closed loop. In the combined mode both the roll press and the ball mill are closed-circuited. In the finish grinding mode the roll press itself performs the function of finish grinding without any ball mill in the circuit.

It is important to consider that the performance of the roll press systems depends on how the compacted slab discharge is handled downstream. It can be split between recirculation of 80–100% relative to the fresh feed, and transfer to ball mill for finish grinding. Another alternative is to pass the compacted cake through a hammer mill for de-agglomeration. These systems generally allow a production increase of up to 40%. Greater capacity increase can be achieved if fines from roll press products are removed by installing a V-separator. The V-separator both de-agglomerates and classifies in a static configuration of stepped plates down which the material cascades through

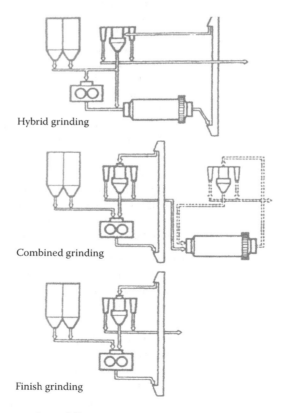

FIGURE 6.18 Three different modes of using high-pressure grinding rolls.

a cross flow of air. A further development of this separator is the VSK model that combines the functions of the V-separator with an integral high-efficiency separator. The VSK separator has a dynamic cage wheel in the same housing as the static part. The cage wheel is mounted on a horizontal shaft, which is supported on a roller bearing (Figure 6.19). The cut size of this assembly is between 25 and 150 μm.

Finish grinding of materials in a hydraulic roll press operated in a closed circuit with a classifier is practically feasible with an

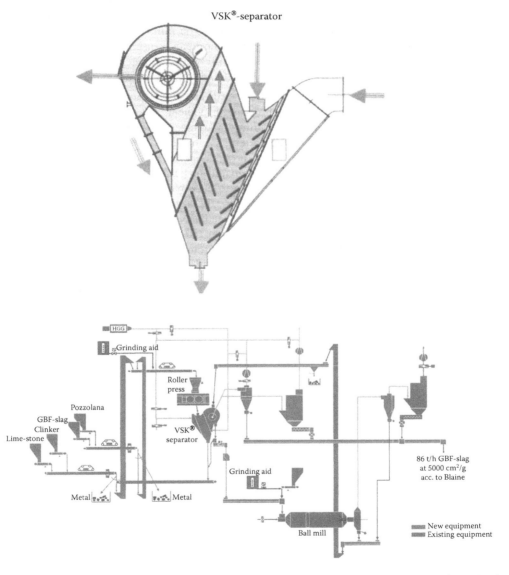

FIGURE 6.19 A schematic diagram of a VSK separator and its installation in a grinding circuit. (Reproduced from Hans Peter Klockner and Henrich Hose, *World Cement*, September, 85–88, 2005. With permission of World Cement, UK.)

almost-50% saving in power consumption as compared to a ball mill. However, due to various operational issues this has not been a very popular system, although some installations have been working in different countries, including India.

D. Horizontal roller mill (horomill)

Horomill has been designed and manufactured by FCB with a view to combining the benefits of both vertical roller mills and ball mills (12). This is a cylindrical mill shell, rotating above critical speed, with a single idler roller and internal fittings to control the flow of material. The roller is in free rotation but is hydraulically pressed against the shell. The mill may operate either in closed circuit with a separator or it can be used for pre-grinding in a ball mill circuit. The horomill is approximately one-third the length of an equivalent ball mill and offers a power saving of 30–50%.

Horizontal roller mills are suitable for use as a single-stage finish grinding process for cement. After the initial problems, the scaled-up horomill is now regarded as technically mature. More than 15 years after its introduction to the market, more than 40 horomills are now in operation, mostly for cement grinding (Figure 6.20), Compared to ball mills, as mentioned earlier, an energy saving of up to 50% has been reported. Even compared to vertical roller mills, the saving is up to 20%. The horomill's present limitation is relatively low throughput rates, the largest capacity reportedly being 180 t/h for normal Portland cement (see Table 6.6). A modified horomill, called a BETA mill, designed for 250 t/h capacity, has recently been installed (Figure 6.21).

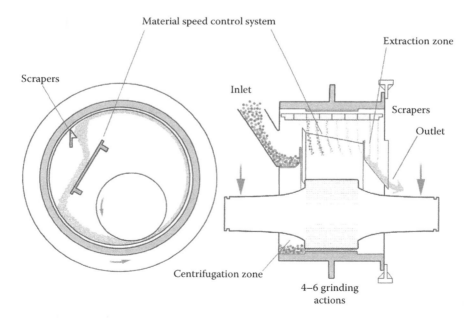

FIGURE 6.20 Working principle of a horomill.

Table 6.6 Typical illustration of horomill sizing and output[a]

Mill Size Designation	CEM I 300 m²/kg (t/h)	CEM I 400 m²/kg (t/h)	CEM-II/A-P 350 m²/kg (t/h)	CEM-II/A-P 450 m²/kg (t/h)	Slag 350 m²/kg (t/h)	Slag 450 m²/kg (t/h)	Raw meal 10% +90 µm (t/h)
3600	95	60	115	75	60	42	225
4000	125	80	150	100	78	56	295
4600	180	115	215	145	114	80	425
2 × 4200	280	180	340	230	180	128	680

[a] As provided by fives FCB based on mean material hardness and moisture.

FIGURE 6.21 Industrial installation of a horomill. (Reproduced with the permission of the Cement Manufacturers' Association, India, *Cement, Energy and Environment*, Vol. 9, No. 4, 2010, p. 7.)

6.4 Energy conservation and material characteristics

The largest amount of electrical energy in the entire cement production process is consumed in grinding the clinker. The power requirements of the above major grinding systems have already been compared in Table 6.4. In high-pressure grinding rolls the material to be ground is comminuted by high compression stressing in a gap between two counter-rotating grinding rolls and is pressed into flakes. With cement grinding the greatest energy savings can be achieved by using high-pressure grinding rolls in a finish grinding configuration. The electrical energy requirement can be lowered by up to 50% compared to a ball mill. However, the changed grinding principle in relation to a ball mill can lead to substantial changes in the finished cements. A semi-finish grinding system is therefore often employed for cement grinding, in which the high-pressure grinding rolls are used for primary comminution. The following three process variants have proved successful for cement grinding:

- Primary comminution in high-pressure grinding rolls followed by fine grinding in a ball mill, which is operated in closed circuit with a classifier.
- Hybrid grinding, i.e., primary comminution in high-pressure grinding rolls followed by the grinding in a ball mill-classifier circuit with partial return of the classifier tailings to the high-pressure grinding rolls.
- Combined grinding, i.e., primary grinding in a circuit comprising high-pressure grinding rolls and classifier followed by fine grinding in a ball mill that may possibly be operated in combination with another classifier.

Energy savings of up to 35% when compared with a plain ball mill are possible with the configuration of high-pressure grinding rolls in combination with a ball mill and the cement quality parameters remain unchanged. The larger the fine fraction generated in the high-pressure grinding rolls, the higher is the energy conservation.

In addition to the high-pressure grinding rolls, the vertical roller mill, as already stated, has also become established in recent years, particularly for grinding blast furnace slag for making Portland slag cement. In VRM the mill feed is comminuted by pressure and friction between a rotating grinding table and two to four grinding rolls pressed hydraulically down on it. The horizontal roller mill has also recently come into use; this has a lining on the grinding zone against which a grinding roller is pressed hydraulically. Roller mills can also achieve energy savings of up to 35%.

While conserving energy in the grinding process, one should not lose sight of the properties of finished cements. For a given material composition of the mill feed the properties of the cements from the different grinding processes are influenced essentially by the particle size distribution obtained and by the differing degrees of dehydration of the set-regulating gypsum in the mill.

It may be recalled that in the RRSB plot the characterization of cement is done by two parameters, namely the slope (n) and the position parameter (d), defined as the average passing particle size of 63.3% of cement. The slope of the graph is a measure of the ratio between the coarse and fine constituents of the cement while the position parameter describes the fineness level of the cement. In order to achieve the same specific surface area, it is necessary to grind a cement having a steep particle size distribution more finely than a cement with a flatter particle size distribution that already contains a higher proportion of fines.

As far as the cement properties are concerned it is generally true that for a given specific surface area cements with steep particle size distributions always have higher water demands for a given consistency than cements of the same specific surface area with flatter particle size distribution. This is shown schematically in Figure 6.22 (14). The amount

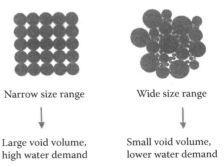

FIGURE 6.22 Effects of particle size distribution of cement on the intergranular voids. (Reproduced from H. M. Ludwig, Influence of process technology on the manufacture of market-oriented cements, in *Proceedings of 2002 VDZ congress on Process Technology for Cement Manufacturing*, Verlag Bau+Technik GmbH, Dusseldorf, Germany, 2003. With permission.)

of water needed for filling the void volume between the cement particles plays an overriding role for the water demand of cement for consistency. The larger void volumes found in cements with narrower particle size distribution always lead to higher water demands. On the other hand, in cements with wider particle size distribution the voids occurring between coarser particles can be filled by the finer constituents of the cements. The resulting reduction in void volume leads to a lower water demand.

A comparison of the different mill systems shows that ball mills alone and high-pressure grinding rolls used in conjunction with ball mills generate the widest particle size distributions. On the other hand, roller mills and high-pressure grinding rolls used in a finish grinding configuration generate relatively narrow particle size distributions. For example, a cement, ground in a ball mill, had specific surface area of 390 m^2/kg, n = 0.86 and d = 19 µm and the same cement having identical specific surface area but ground in a high-pressure grinding rolls system had n = 1.10 and d = 14 µm. With roller mills and high-pressure grinding rolls it is possible to widen the particle size distribution, although within narrow limits only. This is controlled by using the recirculating loads and also, for the roller mill, the applied pressure. Also interesting are developments in which two selective classifiers with cuts at different particle sizes are placed downstream of the high-pressure grinding rolls so that the mixture from them provides a significantly wider particle size distribution. However, it should be mentioned that all the measures to control these mill systems towards flatter particle size ranges always lead to a reduction in mill output.

Dehydration of gypsum

In addition to the particle size distribution, the extent of dehydration of gypsum is of great importance in cement grinding. The nature and quantity of gypsum to be added to the milling system is linked primarily to the aluminate phase in the clinker so that its rate of hydration can be regulated. If there is a gross mismatch between the quantity and reactivity of the aluminate phase and the sulfate in solution, the cement may display either "false set" or "flash set." Normal setting occurs when the sulfate ion concentration in the solution is low for a low quantity and low reactivity of the aluminate and high for a high quantity and high reactivity of the aluminate. "False set" is caused by the formation of secondary gypsum due to an excess of sulfate; this bridges the interstitial spaces between the particles in the cement paste and therefore interferes with normal setting. "Flash set" happens when there is inadequate sulfate in the solution, causing the formation of sulfate-deficient or sulfate-free hydrate phases such as the monosulfate hydrate or calcium aluminate hydrate. Here, again, the interstitial spaces between the particles are bridged by such phases altering the setting behavior of the cement.

As explained in an earlier chapter, gypsum may occur in dihydrate, hemihydrate, and anhydrite forms. The hemihydrate form has the highest solubility at 25°C (6 g/L), compared to the other two forms (2.4 g/L for the dihydrate and 0.1 g/L for the anhydrite). Since gypsum loses its water molecules due to the temperature inside the mill and since different mill

Table 6.7 Comparison of gypsum dehydration between ball mill and high-pressure grinding rolls

Parameters	Ball mill	High-pressure grinding rolls
Mill temperature (°C)	115	60
Anhydrite (%)	3.0	2.7
Hemihydrate (%)	1.9	0.4
Dihydrate (%)	0.1	2.3

systems have different atmosphere, the formation of hemihydrate and anhydrite differs from mill to mill.

Compared to ball mills, significantly lower material temperatures prevail in vertical roller mills and high-pressure grinding rolls in the finished grinding configuration, with the result that conversion of the dihydrate to hemihydrate occurs only partially or not at all. Table 6.7 shows the differing degrees of gypsum conversion that can occur when using ball mills and high-pressure grinding rolls.

In roller mills, which are air-swept systems, the gypsum dehydration can be controlled by the air temperature, but this control option is not available with high-pressure grinding rolls. The only remaining option is to add the gypsum in already partially dehydrated form as hemihydrate so that clinkers containing high levels of reactive aluminate can still be processed properly.

Grinding of mineral admixtures and fillers

So far as the grinding of mineral admixtures and fillers is concerned, it is important to ensure that there is reasonable compatibility between the clinker being ground and mineral admixtures being added. The admixtures include granulated blast furnace slag, pozzolanic materials such as fly ash, volcanic ash or trass, calcined clay, micro silica or condensed silica fume, reactive filler like the limestone powder, and so on. It is desirable that the selection of the admixture is done, depending on the final product application and the existing clinker properties. Two examples of appropriate combination based on clinker reactivity are given in Table 6.8.

As a rule, clinker of fairly low reactivity leads to Portland cements that are characterized by a lower early strength and by high 28-day strengths. The upper limits for the 28-day compressive strengths specified in the cement standard are sometimes exceeded. By adding finely ground, less-reactive filler material, such as limestone powder, it is possible to produce a significant increase in the early strengths and reduce the 28-day compressive strengths. In contrast, highly reactive clinker produces Portland cements that reach high early strengths but only moderate 28-day strengths. These clinkers also have a tendency to accelerated stiffening, especially at fairly high temperatures. In this case there is much to recommend combining them with a latently hydraulic admixture, such as granulated blast furnace slag, which hydrates slowly and

Table 6.8 Clinker reactivity-based admixture selection

	Very reactive clinker	Very reactive clinker	Less reactive clinker	Less reactive clinker
Cement type	CEM I	CEM II/A-S	CEM I	CEM II/A-LL
Blaine specific surface area (m^2/kg)	420	400	380	490
Water demand for consistency (%)	30.0	29.0	29.0	28.0
Initial setting time (min)	150	180	160	170
Compressive strength after three days curing (MPa)	38.5	33.4	28.4	32.8
Compressive strength after 28 days curing (MPa)	55.8	59.2	61.4	59.0

contributes to the strength development at a later stage. This improves the workability properties and significantly increases the final strength.

In addition to the compatibility of clinker with the mineral admixtures as explained above, the grinding process adopted has its own effect. The varying grinding characteristics of the additive components play a crucial role in deciding whether these components should be simultaneously ground with clinker or they should be ground separately and then mixed with ground clinker. It is well known that the blast furnace slag is harder to grind than even clinker, while the limestone rock is much softer to grind; many other additive components come in between these extremes in terms of their grinding characteristics. In a specific study (13) it was found that for the same position parameter d of 16 µm the RRSB slope for the blast furnace slag and for the clinker was 0.9, while for trass it was 0.7, and for the limestone it was 0.5. Such differences in grinding characteristics must be taken into account, while selecting the grinding system and deciding on the operational parameters. These differences in grinding characteristics of input materials also affect the overall particle size distribution of the finished cement.

Manufacture of cements containing blast furnace slag and limestone powder

When clinker and slag are ground together to produce the Portland slag cement, it results in enriching the finer fractions of the product with clinker particles and the coarse fractions with slag particles because of the different degrees of grindability of clinker and blast furnace slag. In separate grinding followed by mixing of the components a more uniform distribution of clinker and blast furnace slag is observed over the entire particle size range (see Table 6.9).

What is important to note from the table is that in the intergrinding process the quantity of slag particles below 12 µm was only 2%, which increased to 5% in separate grinding and blending process for the same cement with similar Blain surface. The particle distribution pattern in Portland limestone cement is quite the opposite of the above pattern. This variety of cement is produced generally by the intergrinding process and the limestone particles tend to concentrate in the finer fractions of the cement. It is seen that the water demand for consistency decreases

Table 6.9 Distribution of clinker and slag particles in cements obtained from intergrinding and separate grinding processes

Size fraction of particles (μm)	Percentage of the fractions	Slag: Clinker in intergrinding process	Slag: Clinker in separate grinding process
> 32	20	12:8	6:14
12–32	40	9:31	9:31
6–12	18	2:16	4:14
< 6	22	0:22	1:21

with increasing proportion of limestone, despite grinding of the cement to higher Blaine surface for achieving early strength. This apparently contradictory trend in Portland limestone cement is explained by two factors—one, the finer clinker that picks up more water is substituted by the limestone powder, and second, the entire particle size distribution of the Portland limestone cement becomes flatter with increasing proportion of limestone, which causes the volume of water-filled voids to reduce.

6.5 Grinding aids in cement manufacture

The grinding aids are synthetically prepared chemicals or processed or refined industrial wastes that help the grinding process to achieve the following objectives:

- Increase the cement output without increasing the clinker capacity
- Enhance the incorporation of supplementary cementitious materials in the blended cement
- Reduce the power demand in grinding
- Derive the benefit of lower carbon dioxide emission due to reduced use of clinker and electricity

There can be three types of cement grinding aids. The simplest type is aimed at increasing the mill output. The second type increases the output but also enhances the strength of cement. The third type is functionally the most complicated as it increases the output, enhances the strength, and improves the flow of cement. Chemically the grinding aids are complex in formulation but mostly based on water-soluble organics, such as alkaloamines. Experience and test data indicate that the most effective grinding aids are polar in nature. In simple terms, the polar compounds have a more positive portion on one part of the molecule and a more negative portion in another part. They develop a dipole moment in a force field, which causes the dipole to orient itself with the direction of the force field. The polar compounds therefore react at the unbalanced valence points as the new surfaces are generated by grinding. The negative part of the molecule is attracted to the positive valences and vice versa. By this action, the agglomeration tendencies of the cement particles are reduced effectively. The grinding aids function by coating the new particles, thereby shielding or satisfying the unsatisfied valence forces that are produced in the grinding process. The mechanism of adsorption of the grinding aid molecule onto the surface of the cement particles, including the micro-cracks, which reduce the energy needed to break the particle and reduce the surface charges, is also known as "Rebinder's effect." Further, it is well established that these additions contribute to the reduction of the "Van der Waals" forces between cement particles, thus preventing agglomeration and hence increasing fluidity. Grinding aids, being organic polar products, react during the gaseous phase by adsorbing onto the surface of cement particles during grinding. They form a very thin mono-molecular film (10–20 μm) around the charged particles, which reduces or neutralizes the electrostatic charge. However, it may be noted that the effects of grinding aids such as triethnolamine are noticed in grinding products to high Blaine surface, say, 350 m^2/kg, as the grinding behavior of clinker at lower surfaces is governed more by the basic clinker characteristics such as its porosity and alite and belite contents.

Effect of grinding aids on cement hydration

Alkaloamine-based grinding aids have a positive effect on cement hydration and leads to an increase in mechanical strengths. It is known that during the first stages of hydration of a cement paste, both crystalline and amorphous phases appear. The amorphous phase is an indication of formation of a complex family of calcium silicate hydrates. It has been seen that the amorphous phase is higher in the cement hydrated with grinding aids, which could be the reason for better mechanical strength. It has also been seen that the chemical effect of a grinding aid is to accelerate the reactivity of the aluminate phase. Certain specially formulated alkaloamines can have a catalytic effect on cement hydration by increasing the hydration of the ferrite phase.

Grinding aids for improving the flow properties of cement

A large number of plants are affected by the problems of cement flow. The cement flow problems can be three-fold. The first problem is known as "pack-set" and is defined as the condition that inhibits the start of flowing; the second problem is the lack of "flowability," which refers to the ease with which the cement flows once it is set in motion; and the third is the problem of "silo-set," which is the absorption of moisture in storage or formation of flow-inhibiting complex sulfate phase like syngenite. The first two problems are solved to a large extent by using grinding aids, which may not have any impact on the third problem.

Operational features of a ball mill and effect of grinding aids

Rittinger's law, expounded in 1867, had shown a linear correlation between the evolution of specific surface area and the energy spent on creating a new surface in ball mill grinding. In general, this is true at the beginning of the grinding process when the Blaine surface area is still low. However, in practice it has been proven that this correlation is no longer true beyond certain fineness and the energy used is lost in heat. Hence, it has been observed that as the mill exit fineness increases, the efficiency decreases (Figure 6.23). The reduction in grinding efficiency

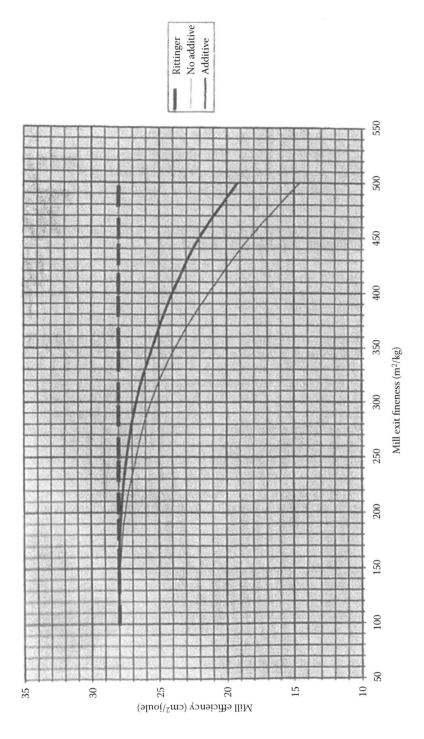

FIGURE 6.23 Typical effects of grinding aids on ball mill efficiency and product fineness.

is a combined effect of agglomeration, subsequent de-agglomeration, and the occurrence of coating. It has been seen in practice that cement particles can coat the grinding media, block the mill internals, and agglomerate and form small plates that absorb the impact. These phenomena are characterized by an increase in energy consumption whilst maintaining a constant specific surface area. The extent of agglomeration depends on:

- the specific characteristics of the materials to be ground,
- the operating parameters of the mill,
- the efficiency and distribution of the grinding media,
- the fineness of the cement particles, and
- internal conditions of the mill (humidity, temperature, ventilation, condition of the lining plates, etc.).

The above problems are overcome in the ball mills with the use of suitable grinding aids. In addition, when the efficiency of the mill separator gets reduced for increased loading of the separator as a result of poor dispersion, particle agglomeration and fines entrainment, the improved particle dispersion resulting from the grinding aids helps in improving the separator efficiency.

Grinding aids for VRMs

The main requirement in grinding cement in VRMs is the stabilization of the grinding bed. The internal circulation between the grinding table and the classifier is quite high in VRMs, which results in a large amount of fine-grained and highly aerated particles moving over the grinding table. With a well-stabilized grinding bed, mill vibrations are avoided and the operation becomes smooth and efficient. In order to achieve the desired particle size distribution, different adjustments are made, such as higher grinding pressures, change of grinding bed thickness, variation in grinding table speed, and working pressure. In addition to the above operational steps, water spraying is done to stabilize the bed in VRMs. Grinding aids often turn out to be a better option, instead of simple water spraying, to stabilize the bed, particularly when products are ground to a higher fineness.

Grinding aids for roller press systems

In order to improve the throughout in a given ball mill system, an emerging practice is to add a roller press to the ball mill circuit. The efficiency of such a system obviously depends on the effective operation of the roller press. For maximizing the amount of work to be done by the roller press, it is necessary to recycle a portion of the pressed material back to the roller press for further grinding. In such hybrid systems, the output from the roller press finally comes to the ball mill and its particle-related problems as described earlier often arise under unfavorable circumstances. The ball mill in this configuration responds to grinding aids as positively as it does in the independent system. It may be relevant to mention here that the hybrid grinding system can be converted into a finish grinding mode by eliminating the ball mill section but often with a limitation of very steep particle size distribution or relatively lower specific surface area in the finished product. The material, passing through the roller press, becomes quite aerated and then compaction of this material becomes difficult, resulting in instability of the roller press. In order to tackle these problems, surface-active performance-enhancing additives such as grinding aids may be helpful.

Dosing of grinding aids

One ton of cement is said to contain about 4000 million sq. cm of surface area. A grinding aid must have to be spread uniformly over this area or at least over the surface points at which the electrovalent bonds have been broken. If the affected surface area is considered only 10% of the total surface area, it is necessary still to cover at least 400 million sq. cm surface area per ton of cement. This demonstrates how important it is to dose grinding aids accurately and uniformly for effective results. The dosing arrangement generally consists of the following equipment:

a. a dilution facility, mainly in an aqueous medium, along with a hydrometer to monitor the specific gravity

b. adjustable pumps, a flow-metering instrument, and consistent flow to the pump

c. a feeding aspirator with an air pressure regulator to spray into the mill without vaporizing

d. an interlocking arrangement with the mill feed

The dosing system may vary from situation to situation and should be engineered based on the layout and specific needs of a plant. What is important to bear in mind is the accuracy and consistency of dosing so that the following overall benefits can be derived:

- significant reduction of the energy used during grinding
- gain in productivity
- improvement of flow and pack-set characteristics of the powder, eliminating lump formation in silos and bulk tankers
- reduction in wear and tear of grinding media
- optimization of particle size distribution in powders
- improvement of cement performance strength at 3, 7, and 28 days

6.6 Storage, dispatch, and bagging of cement

Storage of cement

From the grinding mills the finished cement is conveyed pneumatically to multiple storage silos that are mostly made of concrete. The total capacity of cement storage should match with the requirement of holding the cement stock for the mandatory early-age strength measurement, in accordance with national standards, before releasing to market. In the majority of countries, this may mean holding the stock for a minimum production period of two or three days. The number and capacity of cement silos are also decided, depending upon the types of cement produced, need for blending of each type of cement for uniformity before supply, and demand fluctuations in the market fed by the concerned cement producing unit.

As a basic requirement of the process of manufacture, the cement conveyed to the storage silos should not be hot. The temperature should generally be less than 70°C as cement conveyed to and stored in silos with higher temperatures may lead to dehydration of gypsum inside the silos, reaction of water thus released with surrounding fine cement particles, and promotion of false set in cement, lump formation, and build-up in the silos. Because of the setting characteristics and technical difficulties of complete evaluation of silos, they are prone to developing dead stocks over periods of time. Periodically, the silos should be cleaned from inside either manually or by remotely controlled mechanical equipment. For tracking the extent of filling and discharge, the silos are equipped with level indicators; alternatively, lasers are also used. For consistent flow of cement during discharge the silos are designed and installed with aeration arrangements at the bottom.

There are various designs of silos seen in operation in cement works. The designs are related to the vintage of the plants on one side and the capacity requirement on the other. The design varieties include:

a. Cone-bottom silos: they serve smaller capacities, generally constructed up to 14 m diameter, and are equipped with air fluidization at the bottom of a steel cone along with air-slide conveyors.

b. Flat-bottom silos: they are of larger capacities, generally constructed up to 30 m diameter with fluidized floors having central or side discharge points and air-slide conveyors.

c. Central inverted-cone silos: they are designed with central distribution boxes under the cone and have the advantage of lower floor area and fluidization area.

d. Dome silos: they are installed for large capacities in excess of 30,000 tons, are suitable for ship loading, with fully fluidized floors; such silos require low pack-set cements to avoid problems arising from poor discharge flows.

e. Multiple-compartment silos: they are suitable for plants producing several varieties of cements that need to be stored and supplied separately.

Dispatch of cement

Shipment of cement is country-specific. In some countries, the preferred mode is in bulk form and, in others, in packed bags. But in almost all countries the consignments are transported in all the three ways: road, rail, and waterways, including sea routes. For road consignments, pneumatic trucks are the preferred carriers and truck capacities may range from 9 to 25 tons. For railway transportation, specially designed wagons for bulk supplies or closed wagons for carrying packed cement bags are used. The consignment weight is approximately 2000 tons. Automatic loaders are employed for bagged cement consignments in road and rail supplies. For ocean routes large cargo ships are used, although for shorter distance in some countries barges are preferred.

Bagging of cement The bagging of cement is done with the help of rotary packers with multiple spouts (6–12 in number) and electronic weighing arrangements. The weight of each bag conforms to the specifications laid down in different national standards. In general, the bagged cement is marketed in 50, 40, 25, and 10-kg sacks, made of either three-ply craft paper or polymers such as high-density polyethylene or polypropylene or their composites. The packers deployed are of 3000–5000 bags per hour capacity. The bagged cement consignments are well protected in transit in shrink-wrapped pallets. What is critically important in the bagging of cement is the weight control of the consignments. In most countries, if more than 2% of packed bags contain less than the specified weight, the consignments are unacceptable. Any margin of designed overweight is of course a matter of management discretion. In recent times, big bags with 0.5–2.0 tons capacity have seen common use in bagging cement.

6.7 Summary

Clinker, being a multiphase material, shows significant variation in its properties due to changes in its chemical composition, phase assemblage, and microstructural characteristics. The phase composition variations are caused by the presence of foreign elements in the lattice structure of C_3S (alite), C_2S (belite), C_3A, and C_4AF. The clinker properties are ultimately governed by the relative proportions of the above four major phases, the polymorphic forms of alite and belite, alkali-modified crystal symmetry of C_3A, and the solid solution state of the C_4AF phase.

The microstructural feature defined by the size and shape of alite and belite grains, characteristics of C_3A and C_4AF as the interstitial phases, clustering tendency of the silicate phases, mode of occurrence of free lime and free magnesia (periclase), clinker nodule porosity, clinker granulometry, etc. govern the grinding characteristics of the clinker. It has been observed that clinkers with fine granulometry, ill-formed microstructure, low porosity, high belite content, large crystal size, and high sulfate content show harder grinding properties.

Clinker is converted into cement by employing different milling systems, such as ball mills, vertical roller mills, high-pressure grinding rolls (also known as hydraulic roll presses), and horizontal roller mills. The ball mills are the oldest and most prevalent ones, although they are highly energy-inefficient. The energy considerations and the large throughput capacities required for large plants have driven the development of vertical roller mills and roller presses. These new systems are progressively surpassing the popularity of ball mills. Horizontal roller mills, or horomills as they are called, are the latest development in the field of cement grinding, and seem to combine the advantages of both ball mills and vertical roller mills. One present limitation of horomills is their low capacity of grinding, which is being enhanced by designers and manufacturers.

Selection, system design, operation, and energy conservation of the grinding process are interlinked and are also influenced by the characteristics of materials to be ground. Further, special organic chemicals called grinding aids are used in the grinding process to increase productivity and save energy.

References

1. A. K. CHATTERJEE, X-ray diffraction, in *Handbook of Analytical Techniques in Concrete Science and Technology* (Eds V. S. Ramachandran and J. J. Beaudoin), Noyes Publications, USA (2001).
2. H. F. W TAYLOR, *Cement Chemistry*, Academic Press, London (1990).
3. A. K. CHATTERJEE, High belite cements – present status and future technological options, Part I, *Cement and Concrete Research*, Vol. 26, Issue 8 (1996).
4. R. V. HARGAVE, D. VENKATESWARAN, A. K. CHATTERJEE, B. S. RANGNEKAR, Prospects of quantifications in Portland cement clinker, *Microscope*, Vol. 30 (1982).
5. W. STOIBER, Comminution technology and energy consumption, in *Process Technology of Cement Manufacture*, VDZ Congress, Verlag Bau+ Technik GmbH. Dusseldorf, Germany (2002).
6. M. SUMNER AND T. SMITH, *Cement mill process technology and the economic benefits of quality improvers*, W.R. Grace & Co, USA (2001).
7. J. YARDI, Separator technology, *World Cement*, October (2005), 87–113.
8. J. M. MARECHAL, Roller mill advantages, *World Cement*, September (2005), 91–93.
9. GERHARD SALEWSKI, Development of Loesche vertical mills for large capacity grinding, *World Cement*, July (2005), 97–101.
10. CAROLINE HACKLANDER-WOYWADT, Grinding choices, *World Cement*, September (2005), 71–77.
11. HANS PETER KLOCKNER AND HENRICH HOSE, Utilization of new slag grinding technology, *World Cement*, September (2005), 85–88.
12. DAVID S. FORTSCH, Modern slag grinding, *World Cement*, September (2005), 53–66.
13. INGO ENGEIN, Choice of slag grinding plants, *World Cement*, September (2005), 67–77.
14. H. M. LUDWIG, Influence of process technology on the manufacture of market-oriented cements, in *Proceedings of 2002 VDZ congress on Process Technology for Cement Manufacturing*, Verlag Bau+Technik GmbH, Dusseldorf, Germany (2003).

CHAPTER SEVEN

Composition and properties of Portland cements

7.1 Preamble

Prior to the development and widescale application of Portland cement, different forms of binders were in use, viz., "Roman cement," produced by mixing slaked lime with volcanic ash, or "natural cement," produced by burning a naturally occurring mixture of lime and clay. John Smeaton, a British engineer, who was entrusted to study the repeated structural failures of the Eddystone Lighthouse in Cornwall, is often credited with the discovery sometime in 1756 that cement made from limestone containing considerable proportions of clay was hydraulic in nature. This discovery was made use of in rebuilding the lighthouse in 1759, which stood for 126 years before it had to be replaced. However, little was known about the effect of these constituents on the final product, and consequently the properties of the final product were not consistent. The binders produced varied widely in quality and performance, depending on the initial raw materials used. Over the period 1756 to 1820, further advances in the scientific understanding of cements were made by Vicat and Lesage in France, and Parker and Frost in England, particularly with regard to producing a standard product with known and reproducible properties. It appears that in 1822 James Frost patented and produced "British cement" and in 1824 Joseph Aspdin, a bricklayer and mason from Leeds, took out a patent on a hydraulic cement he called Portland cement, because its colour resembled the limestone quarried at the Isle of Portland. Aspdin's method involved the careful proportioning of limestone and clay, subsequent crushing and burning to form a clinker, which was then ground to finished cement with known and reproducible properties. Aspdin established a Portland cement factory in Wakefield in England that produced Portland cement used in the construction of the Thames River Tunnel in 1828. But the cement produced then was still in the range of hydraulic limes as the raw materials were only calcined and not sintered. It was not until 1847 that I. C. Johnson showed that it was necessary to burn beyond calcination and to use the right proportions of the calcareous material and clay. Subsequently, I. D. White and Sons set up a factory in Kent, leading to the greatest period of expansion in cement manufacture, both in the UK and in Germany and Belgium. Portland cement was used to build the London sewage system in 1859–1867. The manufacture of Portland cement flourished thereafter in England and by 1890 there was a flourishing export business to the USA. The production base in the USA was set up by David Saylor at Coplay, Pennsylvania in 1871 and by 1900 domestic production there rose to 1.7 million tons, galloping to over 15 million tons per annum by 1915. In India, the production of Portland cement started in 1914.

During this 150-year history, fundamental aspects of manufacturing Portland cement and its basic chemistry have not undergone any radical change. Plenty of sustainable innovations, however, have increased the scale of operation, engineered the process more efficiently and productively, characterized and tested the products rapidly and precisely and have helped us to understand the fundamentals of chemistry behind its manufacture and application. This chapter is devoted to unfolding and understanding the wide range of composition and properties of binders belonging to the family of Portland cement.

7.2 Basic grades and varieties

Portland cement is a family of products and its grades and varieties owe their origin to the following manufacturing steps:

- manipulation of phase assemblage
- addition of special compounds
- control of the fineness of grinding
- clinker substitution by supplementary cementitious materials and fillers

The basic phase composition of plain or normal Portland cement is derived from the clinker that is ground along with gypsum to produce it, as described in the previous chapter. The resultant phase assemblage of normal Portland cement (NPC) consists of $C_3S + C_2S + C_3A + C_4AF$ + free CaO + free MgO + alkali sulfates (R_2SO_4) + $CaSO_4.nH_2O$. The first four are the principal phases, constituting 85–90%, while the other ones are present individually in small proportions totaling up to 10–15%. By manipulating the relative proportions of the four principal phases and simultaneously adjusting the product fineness, NPC is graded into different strength classes in accordance with national standard specifications. The strength classes are defined by the 28-day compressive strength of the cement mortar as tested by the procedure laid down in the relevant standards. Further to strength-based grading, NPC forms the base to formulate the following varieties of cement by the same approach of adjusting the phase proportions and product fineness:

- rapid-hardening Portland cement (RHPC or RHC)
- sulfate-resisting Portland cement (SRPC or SRC)
- low-heat Portland cement (LHPC or LHC)
- white Portland cement (WPC)

In addition, a special-purpose cement is manufactured by manipulating the phase composition and clinker microstructure, which is known as oil well cement (OWC) and which, as the name suggests, is not a general-purpose construction cement.

Another group of cements are produced by adding special compounds to the basic cement and they include:

- expansive cement
- regulated set cement
- hydrophobic cement
- colored cement

Although the above cements are derived from the basic Portland composition, they are treated as special-purpose cements and their functional issues are discussed later in this chapter.

Cements that are produced by mixing different supplementary cementitious materials (SCM) and fillers singly or in combination at the time of grinding are called "blended Portland cements." The most common varieties in this group are Portland slag cement (PSC), Portland pozzolana cement (PPC), Portland limestone cement (PLC), and masonry cement. While PSC, PPC, and PLC are general-purpose construction cements, masonry cement is a non-structural binder and is used for plastering and rendering needs.

ASTM classification of Portland cements

Hydraulic cements for general construction purposes are specified in the USA in three standards: ASTM C 150, ASTM C 595, and ASTM C 1157. There are eight types of Portland cements, specified in ASTM C 150:

1. type I: normal
2. type IA: normal, air-entraining
3. type II: moderate sulfate resistance
4. type IIA: moderate sulfate resistance, air-entraining
5. type III: high early strength
6. type IIIA: high early strength, air-entraining
7. type IV: low heat of hydration
8. type V: high sulfate resistance

The above cements are distinguished by properties such as mechanical strength, sulfate resistance, and heat of hydration, which are achieved by manipulating the chemical and phase composition of the products at the time of manufacture. The first three types have air-entraining varieties, which are produced by intergrinding small quantities of air-entraining chemicals. The air entraining varieties are important for making concrete resistant to freezing and thawing. There is also an alternative way to produce such concrete by using air-entraining admixtures in the concrete mixes, instead of using air-entraining varieties of cement.

ASTM C 595 specifies the following five types of blended cements:

i. type IS: Portland blast-furnace slag cement
ii. type IP or P: Portland pozzolan cement
iii. type I(PM): Pozzolan-modified Portland cement
iv. type S: slag cement
v. type I (SM): slag-modified Portland cement

The modified varieties of blended cements are those in which the slag or pozzolan quantity is less than their respective threshold requirements, i.e., 15% and 25% for IP and IS types, respectively. In this standard specification, as in ASTM C 150, there is provision for further categorizing the blended cements in terms of their sulfate resistance and air-entraining properties. An interesting development in specifying cements is the adoption of "performance specification" in place of "prescriptive specification," which means that cements can be classified in terms of their physical properties and environmental exposure conditions without any restrictions on prescribing the ingredients and other chemical characteristics. ASTM 1157 is the prevailing performance specification and has provisions for six types:

a. type GU: general use
b. type HE: high early strength
c. type MS: moderate sulfate resistance
d. type HS: high sulfate resistance
e. type MH: moderate heat of hydration
f. type LH: low heat of hydration

For application purposes both the prescriptive and performance standards can be used as they have their prescribed conditions of field applications. More details are available in (1).

Types of Portland cements in the european standard

In the EN standard EN 197-1:2000 the cement types are defined by two basic criteria: proportion of components and strength levels. There is a matrix of five compositional classes of cements (CEM I, II, III, IV, and V) and three strength categories (32.5, 42.5, and 52.5 MPa) (Table 7.1). CEM I is Portland cement consisting of greater than 95% clinker; CEM II is a blended Portland cement in which the designated types of additives can be added in the range of 6–35%; CEM III is a blast furnace slag cement containing 36–95% slag; CEM IV is a pozzolan cement containing 11–55% pozzolanic material, and CEM V is a composite cement containing 36–80% slag, pozzolan, or fly ash. The strength classification defines three levels: 32.5 having 28-day compressive strength of 32.5–52.5 MPa, 42.5 having 42.5–62.5 MPa, and 52.5 having strength greater

Table 7.1 Excerpts from EN 197-1:2000 for common cements

Main type	Name	Notation	Clinker (%)	Blending component (%)	Minor additive (%)
CEM I	Portland cement	CEM I	95–100	–	0–5
CEM II	Portland slag cement	CEM II/A-S	80–94	6–20	0–5
	Portland slag cement	CEM II/B-S	65–79	21–35	0–5
	Portland silica fume cement	CEM II/A-D	90–94	6–10	0–5
	Portland pozzolan cement	CEM II/A-P or CEM II/A-Q	80–94	6–20	0–5
	Portland pozzolan cement	CEM II/B-P Or CEM II/B-Q	65–79	21–35	0–5
	Portland fly ash cement	CEM II/A-V Or	80–94	6–20	0–5
		CEM II/B-V	65–79	21–35	0–5
		CEM II/A-W Or	80–94	6–20	0–5
		CEM II/A-W	65–79	21–35	0–5
	Portland burnt shale cement	CEM II/A-T	80–94	6–20	0–5
		CEM II/B-T	65–79	21–35	0–5
	Portland limestone cement	CEM II/A-L	80–94	6–20	0–5
		CEM II/B-L	65–79	21–35	0–5
		CEM II/A-LL	80–94	6–20	0–5
		CEM II/B-LL	65–79	21–35	0–5
	Portland composite cement	CEM II/A-M	80–94	6–20	0–5
		CEM II/B-M	65–79	21–35	0–5
CEM III	Blast furnace cement	CEM III/A	35–64	36–65	0–5
		CEM III/B	20–34	66–80	0–5
		CEM III/C	5–19	81–95	0–5
CEM IV	Pozzolanic cement	CEM IV/A	65–89	11–35	0–5
		CEM IV/B	45–64	36–55	0–5
CEM V	Composite cement	CEM V/A	40–64	36–60	0–5
		CEM V/B	20–38	62–80	0–5

than 52.5 MPa. The compositional classes have subclasses, depending on the level of clinker content, such as "A" (high), "B" (medium), and "C" (low). There is also a suffix to the strength parameter indicating high early strength shown by "R" and normal early strength, shown by "N," based on two-day compressive strength. In all, 27 types of common cements are defined in EN 197-1.

The blending components mentioned in the above table are explained later in the chapter. Portland composite cement, pozzolanic cement, and composite cement permit more than one constituent, while all other types permit only one main blending component. There are several others special cements for which the specifications have been ratified, including high early strength blast furnace cements (EN 197-4), masonry cement

(EN 413-1), and hydraulic road binders (ENV 13282). A few other special cements are in the pipeline for ratification. One may refer to the original standards for more details.

Cement types specified in the Indian standards

Normal Portland cement is traditionally called "Ordinary Portland Cement (OPC)" in India and the standard specification IS 269:2015 has provisions for five categories:

a. OPC 33 grade
b. OPC 43 grade
c. OPC 53 grade
d. OPC 43S grade (sleeper grade)
e. OPC 53S grade (sleeper grade)

The above grading is based on the customary practice of using the compressive strength of cement mortars at 28-day. Only the first two grades have the upper ceiling of mortar strength specified at 48 and 58 MPa respectively. In addition to the above five grades of OPC, the following types of unblended Portland cements are specified in India:

- rapid-hardening cement (IS 8041)
- sulfate-resisting cement (IS 12230)
- low-heat cement (IS 12600)
- white cement (IS 8042)
- hydrophobic cement (IS 8043)

The blended Portland cements are produced in large quantities in India and the following types are specified:

- Portland slag cement (IS 455)
- Portland pozzolana cement (part 1: fly ash-based) (IS 1489)
- Portland pozzolana cement (part 2: calcined clay-based) (IS 1489)
- composite cement (IS 16415)
- masonry cement (IS 3466)
- super-sulfated cement (IS 6909)

The composite cement standard has been adopted in 2015 and permits the use of both slag and fly ash but Portland slag cement and Portland pozzolana cement are produced with only one blending component, although in the specification of calcined clay-based pozzolana cement there is provision of mixing fly ash with calcined clay. Super-sulfated cement and masonry cement are multi-component mixes and are meant for special applications, as explained elsewhere in the chapter.

Standard specifications for cements in china

China, being the largest producer of cement in the world, has operating standards for a large number of common and special cements. The principle types of common cements have been specified by differentiating strength levels, heat of hydration, sulfate resistance, and addition of blending components. The strength levels for grading the cements have conformity with the European standards and the strength testing method is consistent with ISO procedure. The OPC variety, designated as P-O 325, has clinker content of 85–95% and the addition of slag, fly ash, or pozzolan is limited to 15%. Composite Portland cement, designated as P-C 325, however, can have lower clinker content (50–80%) and a correspondingly higher proportion of blending components.

In the context of the above discourse it may be pertinent to mention that the classification of common Portland cements in different countries have significant commonality in their basic criteria. For a comprehensive understanding of the types and varieties of cements specified in different countries one may refer to the compendium of world standards (2).

7.3 Characteristics of Portland cements

The four principal phases that constitute the bulk of Portland cement are normally present in the following proportions:

- C_3S: 55 ± 5%
- C_2S: 25 ± 5%
- C_3A: 8 ± 2 %
- C_4AF: 10 ± 2%
- $CaSO_4$: < 3.5%

Correspondingly, the broad chemical composition of normal Portland cement lies in the following range:

- SiO_2: 20–23%
- Al_2O_3: 4–8%
- Fe_2O_3: 3–5%
- CaO: 63–65%
- MgO: 2–3%
- SO_3: 2.0–3.5%
- Total alkalis as Na_2O: 0.4–1.2%
- Free CaO: 0.5–1.5%

Some of the intrinsic properties of cement, such as its water requirement for normal consistency, setting time, strength development, and soundness, and also durability properties such as resistance to chlorides, sulfates, and alkalis are governed by the above chemical and phase composition. The color of cement, however, is more dependent on raw materials and the production process; although, in the absence of specific coloring oxides such as manganese, chromium, vanadium, etc. in the raw materials, the light-gray to dark-gray color of the cement is attributed to the formation of C_4AF or its variants. The color of cement has no relation to the quality or performance of the cement and varies from factory to factory.

As soon as water is added to cement, setting and hardening begins due to chemical reactions between the cement phases including $CaSO_4$ and water. The reaction is exothermic and the process is known as "hydration." The reactivity of each of the phases is different and the ultimate trend of reactivity of the cement with time is not an additive function of reactivity of each phase but interactive in nature. The reactivity of the individual cement phases in terms of degree of hydration with time is shown in Figure 7.1, which shows that the reactivity decreases in the following order:

$$C_3A > C_3S > C_4AF > C_2S$$

However, if the strength gain with time is considered for the individual phases, the order is different, as demonstrated in Figure 7.2.

The setting process is the consequence of a change from a concentrated suspension of flocculated particles to a visco-elastic skeletal solid that can be monitored by rheological measurements. The continued development of the solid skeleton is called "hardening." Since the reactions are exothermic, an oversimplified way of following the progress of cement hydration reactions is by interpreting the heat evolution curves. Since C_3S is the most important phase in Portland cement, a typical pattern of its heat output on hydration is shown in Figure 7.3 for understanding the reaction (3).

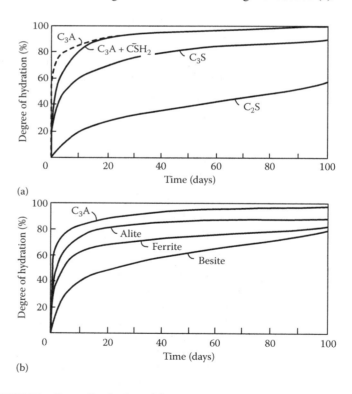

FIGURE 7.1 Rate of hydration of the pure cement compounds: (a) pastes of pure phases and (b) phases in normal cement paste.

COMPOSITION AND PROPERTIES OF PORTLAND CEMENTS 221

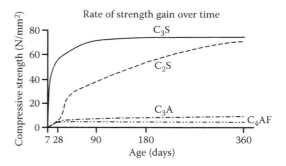

FIGURE 7.2 Strength development of the cement phases with time.

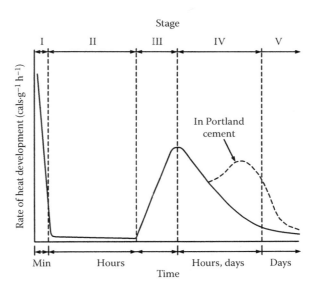

FIGURE 7.3 Stages of C_3S hydration based on heat evolution. (From J. Bensted, Hydration of Portland cement, in *Advances in Cement Technology: Chemistry, Manufacture and Testing* (Ed. S. N. Ghosh), Tech Books International, New Delhi, 2002.)

Five stages are shown in the diagram, which correspond to initial hydration (0–15 min) due to calcium ions passing into solution; induction period or lag phase (15 min–4 h), in which a slow rise of calcium ion concentration continues to take place; acceleration and setting (4–8 h) caused by crystallization of calcium hydroxide and deposition of calcium silicate hydrates in the water-filled space; deceleration and hardening (8–24 h) due to decrease in porosity of the system and delayed transportation of ionic species on the solid-liquid interface; and curing (1–28 days), when a totally diffusion-controlled process continues in an environment of decreased porosity and delayed transportation of ionic species. Thereafter the rate of hydration decreases continuously and even after a long time some amount of the compound remains unreacted in

Table 7.2 Trend of phase composition-property relationship in Portland cements

Categories	Properties	C_3S (%)	C_2S (%)	C_3A (%)	C_4AF (%)	$CaSO_4 \cdot nH_2O$ (%)
A	General purpose, high early strength, and high heat of hydration	55	20	10	10	5
B	Moderate sulfate resistance and moderate heat of hydration	50	25	7	13	5
C	Normal sulfate resistance	40	43	3	10	4
D	Low heat of hydration	28	50	3	15	4

the system. Coming to cement, it should be remembered that the effective reactivity of two cements of similar bulk chemical composition and phase composition may differ significantly due to crystal size and degree of crystallinity, concentration of crystal imperfections at phase boundaries, and polymorphic forms of the phases. Additional differences in effective cement reactivity result from differences in their particle size distribution, the distribution of calcium sulfate phases in the cement, and their solubility. Keeping aside for the present the above finer differentiating parameters, the broad relationship between the phase composition and application properties of Portland cements is illustrated in Table 7.2.

Sometimes cement shows expansion after hardening. It is then said to be "unsound." The presence of hard, unburned, uncombined lime or a relatively high amount of magnesia is normally associated with unsoundness. These compounds hydrate slowly and their reaction may start after the cement paste has hardened. The expansion that takes place as a result of delayed hydration of these oxides can be very destructive to structures.

Normal Portland cements of different strength grades are suitable for all uses where the special properties of other types are not essentially required. The uses of such cements in concrete include reinforced concrete buildings, pavements, bridges, precast concrete elements, water-retaining structures, and so on. For certain specialized applications such as high-rise buildings, prestressed concrete for highway bridge girders, railway sleepers, transmission line poles, waste disposal pipes, industrial building frames and roofing elements, etc., high-strength concrete is required, for which NPC of higher strength grades are used. The characteristics and application of Portland cements with special properties are described below.

Rapid-hardening Portland cement

RHPC is characterized by higher minimum Blaine's fineness (325 m²/kg in the Indian standard) and higher proportions of calcium silicates. It does not set any faster than OPC/NPC but, after setting, hardens and develops strength more rapidly. For proper quality assurance, it is important to specify one-day and seven-day strengths for this type of cement. It is desirable that the one-day strength of the cement is equivalent to the three-day strength of OPC/NPC and the seven-day strength should be higher than that of OPC/NPC. The cement is used in concrete, when a structure is required to carry loads earlier than what would be permitted with the use of OPC/NPC. Specific applications may include airfields, emergency defense structures, precast products, etc.

Sulfate-resisting Portland cement

This cement is more resistant to water-soluble sulfates. The problem of sulfate attack of cementitious products arises because of the volume expansion associated with the formation of ettringite in the hardened cement paste, as a result of which there could be disruption of the matrix. One should note that in the curing process of the cement, the volume expansion due to ettringite formation is of short duration and can be accommodated within the plastic paste. Later, the more stable monosulfate form ($C_3A.CaSO_4.12H_2O$) appears due to secondary transformation of ettringite. Hence, the formation of ettringite during the curing phase is non-disruptive.

In NPC/OPC, aggressive sulfate ions in solution react with the hydration products of C_3A in the hardened cement paste, i.e., the hydrated calcium aluminate crystals will produce metastable ettringite ($C_3A.3CaSO_4.32H_2O$). In addition, it is known that alkali metal water-soluble sulfates react with free calcium hydroxide to form more gypsum, which is then available to react further with the hydration products of C_3A to form ettringite. These reactions make NPC/OPC vulnerable to sulfate attack due to the expansion of ettringite.

The property of resistance to aggressive sulfates exhibited by SRPC is achieved due to its low C_3A content, limiting the total of ($2C_3A + C_4AF$) to 25% and maintaining the C_3S level below 40%. In this composition, the alumina is consumed more in forming the C_4AF phase than C_3A, thereby reducing the quantity of hydrated calcium aluminate available in the system. Simultaneously, the reduced C_3S level helps in reducing the formation of $Ca(OH)_2$. In the moderately sulfate-resisting cement the C_3A is less than 8%, while in highly sulfate-resisting cement the C_3A content is not more than 5%, along with low ferrite content in the cement. If the soil or ground water contains chloride along with sulfate it is desirable to maintain the C_3A level in the cement between 5% and 8% so that the chloride ions are arrested in the chloro-aluminate hydrate phase ($C_3A.CaCl_2.10H_2O$), thereby restricting the diffusion of chloride ions into concrete.

SRPC can withstand the attack of soluble sulfates in ground water (~ 300 ppm) or soil (~ 0.2%). It is useful in severely exposed areas such as copings, flues, retaining walls, etc. However, SRPC, like OPC/NPC, is not resistant to acids, nor is it immune to the effects of some other dissolved salts, such as magnesium compounds. The application of SRPC is particularly desirable in foundations, piles, basements, sewage water treatment plants, coastal structures, and underground construction.

Low-heat Portland cement

This cement, having low C_3A and C_3S contents, has been developed for use in large masses where the rapid evolution of heat would produce high temperatures, resulting in stresses that may lead to cracking (e.g., concrete dams and large-mass foundations, bridge abutments, retaining

Table 7.3 Mineral oxides used as pigments in coloured cements

Shade	Pigment
Brown	Manganese oxide (Mn_2O_3)
Black	Ferro-ferric oxide (Fe_3O_4); carbon black
Green	Chromium oxide (Cr_2O_3)
Red	Ferric oxide (Fe_2O_3)
Blue	Cobalt oxide (Co_2O_3)
Yellow	Synthetic material

walls, etc.). However, strength development in LHPC is slower than that in OPC and does not offer better resistance to water-soluble sulfates. In a non-aggressive environment LHPC shows better long-term properties.

White and colored Portland cement

The C_4AF phase imparts gray colour to cement and white cement, therefore, has a restricted presence of Fe_2O_3 below 0.4%. For a higher whiteness index, some manufacturing units maintain their own raw materials specifications, limiting all the coloring oxides like Cr_2O_3, NiO, V_2O_5, etc. A special quenching system forms a critical part of the manufacturing process. Oil or gas instead of coal is used as the fuel. Needless to mention that white Portland cement is used for decorative cementitious products including concrete.

Colored cements are white Portland cement with 5–10% pigment added in the form of metal earth oxides (Table 7.3). If the pigment addition exceeds 10%, the cement strength decreases. Pigments have to be finely ground as they reduce the effective surface area available for hydration, bringing down the cohesive nature of the product.

Hydrophobic Portland cement

Hydrophobic cement is basically a normal Portland cement interground with hydrophobic agents like naphthenic, oleic, stearic to palmitic acids, or even their calcium salts at concentrations, up to about 0.3%. Hydrophobic cement is manufactured to reduce its hygroscopicity to prevent hydration during storage in unavoidably humid conditions. The cement grains get coated by a film of the hydrophobic agent at the manufacturing stage to increase shelf life. This protective film is removed by abrasion when the cement is mixed with the aggregate, after which the normal hydration process begins.

Oil well cements

Oil well cements are meant for special applications in the exploration for and production of oil and gas. The essential criterion is that a slurry of cement and water with certain additives should remain capable of being pumped down to the required position before stiffening under the extreme conditions of temperature and pressure that can exist in an oil well. Worldwide practice is to follow the American Petroleum Institute's standard specifications, designated as API IOA. The same standard has been adopted in many other countries, including India. While there are eight classes of well cement for use in wells of different depths, the most

Table 7.4 Chemical requirements of API classes G and H

Parameters	Moderately sulfate-resistant type (%)	Highly sulfate-resistant type (%)
MgO, max	6.0	6.0
SO_3, max	3.0	3.0
LOI, max	3.0	3.0
IR, max	0.75	0.75
C_3S, max	58	65
C_3S, min	48	48
C_3A, max	8	3
$C_4AF + 2*C_3A$, max	–	24
Total alkalis as Na_2O, max	0.75	0.75

Table 7.5 Physical and performance requirements of API cement classes G and H

	G	H
Mix water, percent of the weight of cement	44	38
Free fluid content, max (ml)	3.5	3.5
Compressive strength 8 h curing (38°C) Atm. Pr. (Mpa)	2.1	2.1
8 h curing (60°C) Atm. Pr. (Mpa)	10.3	10.3
Pressure-temperature thickening time test	90–120 minutes	

frequently used are classes G and H. The chemical and physical requirements of classes G and H of OWC are summarized in Tables 7.4 and 7.5.

7.4 Phase-modified Portland cements

Some varieties of Portland cements are made by adding certain separately synthesized compounds to the basic cement. The most important in this category are "expansive cement" and "ultra-rapid-hardening" or "regulated-set" cement. The characteristics of these two varieties are given below.

Expansive cement   Expansive cement is cement that expands slightly during the first few days of hydration to compensate for the commonly occurring shrinkage in normal Portland cements. There are three ways of producing expansive cements from the basic Portland cement:

- K-type: intergrinding Portland cement clinker with expansive clinker containing the calcium sulfoaluminate phase ($C_3A_3CaSO_4$) and calcium sulfate as gypsum or a mixture of anhydrite and gypsum.
- M-type: blending Portland cement with high-alumina cement or aluminous slag.
- S-type: using high-C_3A clinker.

The expansive cement is used for making shrinkage-compensating concrete, chemically prestressed concrete, expansive grouting mortars, etc. Shrinkage-compensating concrete is especially suited for constructing airfields, runways, water tanks, and similar kinds of structure. The chemically stressed concrete is used for making structural elements requiring low prestressing, such as pipes, shells, folded plates, etc. Expansive grouts are extensively used for heavy machinery foundations. Controlled expansion is also a useful application for demolishing concrete structures or rocks by cracking without dust and noise pollution.

Ultra-rapid-hardening cement

This category of cement, also known as regulated set cement or jet cement, is manufactured from clinker composed of 55–60% C_3S, 5–10% C_2S, 20–25% $C_{11}A_7CaF_2$, a fluorine containing analog of $C_{12}A_7$ compound, and 4–8% C_4AF. Clinker of the above composition is ground with 8–12% anhydrite and small but varied amounts of hemihydrate, sodium sulfate, and calcium carbonate. Depending on the actual composition, the resultant cement sets in 6–60 min and develops strength up to 12 MPa in 2 h. The cement has special applications in repairing highways, bridges, railways, etc. It can also be used in making thin dividing walls, sealing inspection covers and manholes, door and window frames, and similar precast elements.

7.5 Blended Portland cements

Blended cements are produced from Portland cement or Portland clinker plus gypsum, blended or interground with pozzolana, slag, fly ash, and other supplementary cementitious materials. The progressive shift from normal Portland cement to blended Portland cements and the growing emphasis on their use in construction have been reviewed and discussed by the author in (4). The proven merits of using blended cements are:

- low heat of hydration
- improved water tightness
- enhanced resistance to sulfate and chloride attacks
- increased resistance to alkali-silica reactions
- durability in tropical environments
- resource sustainability through recycling of industrial byproducts

COMPOSITION AND PROPERTIES OF PORTLAND CEMENTS

Important characteristics and current trends of applications of the commonly produced varieties are presented later in this chapter. The addition of limestone powder as filler and the characteristics of Portland limestone cement have been discussed in Chapter 10.

Blending materials There are eight blending materials specified in En 197-1 for common cements, and they are listed in Table 7.6.

In the USA, the granulated blast furnace slag is defined in ASTM C 989 and the test methods are specified in C1073. The granulated blast furnace slag is classified into three grades on the basis of a "slag reactivity index" at 28 days (viz., 80, 95, and 115 grades). According to ASTM C 595, pozzolans are required to meet the stipulation of fineness, pozzolanic activity, and expansion as shown below:

- fineness on wet sieving on 45 μm: maximum residue of 20.0%
- pozzolanic activity index with 35% replacement in control Portland cement cured at 38°C at 28 days: a minimum of 75% of control cement strength

Table 7.6 Quality specifications of blending materials in EN 197

Type and source	Standard designation	Quality parameters	Specified values
Granulated blast furnace slag from the iron and steel industry	S	Glass C+S+MgO (C+MgO)/S	≥ 66.7% ≥ 66.7% > 1
Silica fume from the ferrosilicon industry	D	Amorphous S Loss on ignition Specific surface	≥ 85% ≤ 4.0 ≥ 15.0 m²/g
Natural pozzolan	P	Reactive S	≥ 25%
Natural calcined pozzolan	Q	Reactive S	≥ 25%
Siliceous fly ash from thermal power plants	V	Loss on ignition Reactive C Free C Reactive S	≤ 5.0% ≤ 10.0% ≤ 2.5% (if it passes the expansion test or ≤ 1.0%) ≥ 25%
Calcareous fly ash from thermal power plants	W	Loss on ignition Reactive C Reactive S	≤ 5.0% ≥ 10.0% to ≤ 15.0% (≥ 15.0 permitted subject to testing) ≥ 25.0%
Burnt shale from natural sources	T	28-day strength (for ground shale alone) Expansion (70:30 cement:shale mix)	≥ 25 MPa ≤ 10.0 mm
Limestone from natural sources	L/LL	CaCO₃ clay (by methylene blue analysis) Total organic carbon (TOC)	≥ 75.0% ≤ 1.2 g/100 g ≤ 0.50% for "L" ≤ 0.20% for "LL"

- Mortar bar expansion at 91 days with six cements containing pozzolana from 2.5% to 15.0% at 2.5% interval for alkali reactivity: maximum of 0.05%.

ASTM C 618 specifies class C and class F fly ashes for use in concrete. Further, pozzolans belonging to class N are also used and they can be either raw or calcined natural materials such as clay, shale, volcanic ash, diatomaceous earth, pumice, tuffs, and others.

In India, the granulated slag and fly ash used in the manufacture of blended cements are specified in IS 12089 and IS 3812 (part 1) respectively. For minor additions to OPC (NPC) up to 5.0%, a host of materials are permitted as "performance improvers," which have also been specified in IS 269. Apart from slag and fly ash, the performance improvers include silica fume, limestone, metakaolin, copper slag, steel slag, lead-zinc slag, and spent cracking catalyst.

The influence of the blending material on the strength and durability behavior of mortars and concrete depends on the reactivity of the blending material. The reactive materials influence the properties of mortar and concrete both by void filling effects and strength contribution, while the finely ground inert materials affect the early strength, although they may have some positive influence due to void filling.

Portland slag cement

Germany is regarded as a pioneer in the use of slag in cement and created the Eisen Portland Cement Association in 1901. But the real growth of PSC in European countries including the UK and Russia took place in the 1960s. In India, PSC was standardized for the first time in 1962. The specified range of incorporation of slag in cement is quite wide in most producing countries and in India it extends from 25% to 70%. Broadly, the chemical composition of the slag varies in the following range:

- CaO: 29–36%
- SiO_2: 30–34%
- Al_2O_3: 18–24%
- MgO: 4–12%
- S as sulfide: 0.7–0.9%
- $(CaO+MgO+Al_2O_3)/SiO_2$: 1.82–2.20

In some European countries, the alumina content in slag is relatively low (5–15%) and lime content is high (30–50%). Glass content in the granulated slag is 85–95% in most cases, which makes it substantially reactive in the blends. But as explained in a previous chapter, the slag, being hard to grind, is not a very suitable material for the intergrinding process of cement manufacture. Separate grinding and blending is a more effective option. PSC is a structural cement that can be exposed to all environments and, compared to OPC/NPC, it shows very low chloride permeability and is therefore more suitable for aggressive conditions of application.

Portland fly ash cement

The pozzolanic nature of fly ash was reported almost 75 ago, and the first significant use of fly ash as a supplementary cementitious material was in the construction of the Hungry Horse dam in Montana, USA, in the late 1940s. There has been immense proliferation of its use in cement and concrete in the last 25 years. China and India, being the two largest fly ash generating countries in the world, have focused on increased use of fly ash in building materials and more particularly in making Portland fly ash cement. This cement is a variant of Portland pozzolana cement in which the pozzolan content may vary from 15%–40%, and it is produced mostly by the intergrinding process. Some countries, including India, have restricted the upper limit of fly ash addition to 35% or even lower in binary blended Portland cement. The process of separate grinding and blending for the manufacture of Portland fly ash cement is permitted widely almost in all standards, if it is viable and essential in any specific situation. Although the reactivity of fly ash is often variable within the source of generation and between the sources, its addition to Portland cement results in the occurrence of a classic pozzolanic reaction between calcium hydroxide formed due to cement hydration and reactive silica present in the fly ash:

$$Ca(OH)_2 + SiO_2 \rightarrow \text{C-S-H}.$$

The additional C-S-H thus formed fills in porosity, making concrete or mortar less permeable to water-mediated aggressive species. It is important to bear in mind that fly ash contains substantial amounts of alumina, which can participate in the hydration reaction. Hence, in the cement–fly ash system there has to be an optimum quantity of sulfate, in addition to the amount required for controlling the setting of C_3A, so that the alumina from fly ash can participate in the formation of sulfoaluminate or aluminate hydrate phases. It is also relevant to note that siliceous fly ash (class F) is essentially pozzolanic, while the calcareous variety (class C) is normally hydraulic and cementitious. Hence, their behavior in cement mortar and concrete is not identical. Further, class C fly ash may contain hard burnt coarse lime crystals, which may result in delayed hydration and expansion. It is, therefore, essential to test the characteristics of fly ash prior to its use.

Masonry cement

Masonry cement consists of OPC/NPC with certain additions at the manufacturing stage to increase workability. The formulation consists of clinker, gypsum, pozzolana, fillers like ground limestone, and air-entraining agents. Sometimes water-retaining mineral fibers are also used in the formulation. Masonry mixes for normal plastering often consist of only OPC/NPC and sand in different ratios but such mixes are often too strong and harsh for rendering, brickwork, and blockwork mortars and the addition of lime to produce a more workable mortar often involves unpleasant working conditions. One must bear in mind that masonry cements should not normally be blended with further

Supersulfated cement

admixtures on site other than sand. It is not suitable for concrete work and for any structural applications.

Supersulfated cement comprises a mixture of finely ground granulated blast furnace slag and calcium sulfate together with Portland cement, Portland cement clinker, or another source of lime. According to the Indian Standard Specification, the slag component should not be less than 75% by weight of the total quality. Supersulfated cement contains no free lime and no C_3A and therefore confers a high chemical resistance to aggressive conditions and more particularly to water-soluble sulfates. It has been found in practice that dense concretes made with SSC and a w/c ratio of 0.45 or less have given an acceptable life in contact with weak acidic solutions (pH 3.5 upwards). The cement is finely ground (> 400 m²/kg Blaine surface) and has an intrinsically low heat of hydration. The cement, however, suffers from strength retrogression at temperatures above 40°C and, as a result, its manufacture and use are rather limited.

7.6 Characterization of cements and practical implications of properties

Testing and characterization

The characterization of hydraulic cements in general and Portland cement in particular starts with the determination of its oxide composition, measurement of "loss on ignition (LOI)," insoluble residue (IR), and SO_3 content, followed by compound composition, which is also known as "phase analysis." A conventional chemical laboratory is equipped with a drying oven, a muffle furnace, an X-ray fluorescence spectrometer (XRF) with a pressed-pellet or fused-bead preparation facility, although other sophisticated instruments like the atomic absorption spectrometer (AAS), inductively coupled plasma spectrometer (ICP), ultra violet – infrared spectrometer (UV-IR), flame photometer, etc. are made use of in certain laboratories. While the above instruments are meant for chemical analysis including major and minor oxides and even trace elements, for phase determination and quantification other types of instruments like the X-ray diffractometer (XRD), scanning electron microscope SEM, and optical polarized microscope are employed. For specific investigations involving thermal analysis, specialized instruments such as the differential thermal analyzer (DTA), thermogravimetric analyzer (TGA), and differential scanning calorimeter (DSC) serve as the important tools. For detailed crystal structure studies, sometimes a Fourier-transform infrared spectroscope (FT-IR) is used. In addition, there are several other research tools that are employed in cement research.

Approximately 85–95% of cement particles are smaller than 45 µm, with an average particle size of 15 µm. Apart from particle size, cement fineness is characterized by "specific surface area (SSA)." A quick way of determining the particle size is to carry out wet sieving of cement on a 45-μm sieve. A more sophisticated and precise method of determining the particle size distribution is to employ a laser granulometer. It is possible to recalculate the specific surface area of the ground cement from the particle size distribution data. But the routine practice in cement plant laboratories is to measure SSA by the Blaine air permeability method provided almost universally in all standards.

The "consistency" of cement paste refers to the relative mobility of a freshly mixed cement paste or its ability to flow. It provides the requirement of water for preparing cement pats and mortars. Normal consistency is determined in test pats of cement pastes with the help of a Vicat plunger. In some countries, test mortars are prepared with a specified fixed water:cement ratio or after determining the flow on a flow table within a specified range.

The setting time test of cement pastes is carried out to determine the time that elapses from the moment water is added to dry cement powder until the paste ceases to be fluid and plastic (called the "initial set") and the time required for the paste to acquire a certain degree of hardness (called the "final set"). The Vicat apparatus is used universally, following the standard procedure.

The compressive strength of cement is determined in 1:3 or 1:2 mortars prepared with the test cement and the specified standard sand, by using a uniaxial compressive strength testing machine. "Soundness" of cement refers to the ability of a hardened cement paste to retain its volume. Lack of soundness can be caused by excessive amounts of hard-burnt free lime or magnesia. The free lime-related expansion is tested by the Le Chatelier method in boiling water, while the magnesia-related expansion is determined with the help of an autoclave, following the standard procedures.

In most of the standard specifications the measurement of heat of hydration is not mandatory but, for all practical purposes, this parameter is important. It is determined with the help of conduction calorimeters and the procedures accompany the instruments used. The standard procedure for the measurement of the heat of hydration has been detailed in ASTM C 186. In the calorimetric measurement, the first heat sensing happens after a few minutes, generally 7 min, after mixing the paste. Hence, only the downslope of the first peak, corresponding to hydrolysis, is observed, followed by other peaks and valleys, characterizing the hydraulic reactivity of the test cement.

Density and related parameters

Density is not strictly a quality parameter but its principal use is in mixture proportioning. The particle density of Portland cement ranges from 3.10 to 3.25, averaging 3.15 g/cm^3. Blended cements like PSC and PPC have densities ranging from 2.90–3.15, averaging 3.05 g/cm^3. Relative density or specific gravity is a dimensionless number obtained by dividing the cement density by the density of water at 4°C. The bulk density is defined as the mass of cement particles plus air between the particles per unit volume. Portland cement on fluffing may have a mass of 830 kg/m^3, whereas on consolidation it may be 1650 kg/m^3.

Compositional aspects

The compressive strength in hydrating Portland cement pastes is developed mainly due to hydration of silicate phases. The contribution of the aluminate and ferrite phases to the development of strength is small (Figure 7.3). This small strength development of the non-silicate phases is observed only in the presence of set regulators like gypsum, as the rate of hydration of these phases and particularly of C_3A is fast. For the four broad types of Portland cement given in Table 7.2, the patterns of strength development in mortars and concrete with time are illustrated in Figure 7.4, which demonstrate that the development patterns initially follow the increasing C_3S proportions up to about 90 days of hydration and thereafter the trends are significantly reversed. Hence, the Portland cements having high early strength may not necessarily yield very durable concrete. The other critical compositional parameters and their effects are shown in Table 7.7.

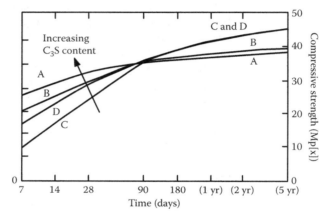

FIGURE 7.4 Trends of long-term strength of different types of Portland cement mortars.

Table 7.7 Chemical parameters and their effects on the properties of Portland cements

Parameters	Formulation/ abbreviation	Property
Loss of ignition	LOI	Pre-hydration, carbonation, and adulteration
Insoluble residue	IR	Essentially controls the quality of gypsum
Lime saturation factor	$C - 0.7SO_3$ / $2.85 + 1.18A + 0.65F$	Largely governs the ratio of C_3S to C_2S and minimization of free lime
Magnesia (periclase)	MgO	Slow hydration and possible disruptive expansion in concrete at late age, if magnesia content is high and/or the grain size of periclase crystals is large
Total alkalis	$Na_2O + 0.65\ K_2O$	Occurrence of concrete degradation due to alkali-silica reaction
Chloride content	Cl^-	Agent for reinforcement corrosion, particularly in prestressed and long-span reinforced concrete

Effect of fineness

The specific surface area of most varieties of cement, as measured by the Blaine's air permeability method, lies in the range of 280–500 m²/kg. This parameter greatly affects the setting time and the strength of the hardened cement paste because the chemical activity of the solid is directly proportional to its surface area, which increases with the increased fineness of the particles. As hydration proceeds from the outer to inner core of the cement particle, the smaller the particle, the greater the probability that nearly the whole core will be converted to C-S-H gel and calcium hydroxide crystals. For coarser particles, a considerable portion of the inner core will not be available to participate in the hydration reactions. Consequently, cement with higher fineness will develop more C-S-H gel per unit weight than a coarser cement of the same composition. This accounts for the more rapid hardening and greater early strength of a finer cement compared to that of a coarser cement. Hence where high early strength characteristics are required, cements are ground finer; for example, the specific surface of cement particles in rapid-hardening Portland cement is around 325 m²/kg compared to that for ordinary Portland cement of around 225 m²/g. Although cement particles can be produced at sizes even lower than this, too fine a cement particle size tends to give considerable shrinkage on setting, and the practically encountered values represent a compromise to optimize overall properties.

Setting behavior

Setting is a hydration-related microstructural parameter representing the onset of percolation of a network of agglomerated solid particles. Normal setting of Portland cement paste is controlled by the hydration of C_3S. Apart from the occurrence of initial and final set, the cement paste may display "false set," which is defined as a rapid increase in paste rigidity without excessive heat evolution, which during testing can be overcome by immediate further mixing to regain plasticity. False set results from rapid crystallization of gypsum from the paste liquid phase supersaturated with sulfates by excessive dissolution of calcium sulfate hemihydrate or soluble anhydrite. These phases, as explained earlier, may form during the grinding operation. One other type of setting behavior, called "flash set," occurs due to strong heat evolution causing severe stiffening, which cannot be dispelled by remixing. It happens due to an excessive and faster reaction of C_3A with water, when the soluble sulfate in the liquid is inadequate during the initial stage of hydration. "Pack set" or "silo set," as explained earlier, happens during storage but it can influence the subsequent hydration reactions. Pack set is related to the degree of gypsum dehydration during cement grinding and to the silo temperature and humidity. Gypsum in the warmer part of the silo may dehydrate, releasing water vapor that can superficially hydrate the aluminate and free lime in clinker in the colder part of the silo, affecting the flow property of cement.

Heat evolution over time

Evolution of the heat of hydration of C_3S has been shown in Figure 7.3. Since the composition of Portland cement is more heterogeneous with many compounds, the rate of heat output from Portland cement involves all the constituents and includes the heat output from the reactions of hydration products with one another (Table 7.8).

From the table it is evident that the heat output is in the order $C_3A > C_4AF > C_3S > C_2S$. The evolution of heat apparently continues for years but with varying intensity, keeping the above order of heat output more or less the same. When a normal Portland cement is hydrated, apart from the main phases, the heat evolution is further influenced by its fineness, water-cement ration, and temperature of curing. The heat of hydration values measured for different types of Portland cement are given in Table 7.9 (1).

The above results are based on tests conducted on a limited number of samples but they show a definite trend of correspondence to the basic composition of the cements. The heat of hydration data of some types of British Portland cements also confirms the same observation (Table 7.10) (3).

Irrespective of the absolute values obtained, the heat output curve of Portland cement at constant temperature gives the pattern as shown in Figure 7.5. The first peak A occurs on mixing and is of high intensity but short duration, lasting a few minutes, as in the case of C_3S hydration (see Figure 7.3). Peak A is thought to derive from the exothermic rehydration

Table 7.8 Heat evolved (kJ/kg) from the cement phases with time

Phases	3 Days	7 Days	28 Days	90 Days	360 Days	6.5 Years
C_3S	243	222	377	436	490	490
C_2S	50	42	105	176	226	222
C_3A	888	1559	1378	1303	1169	1374
C_4AF	289	494	494	410	410	465

Table 7.9 Heat of hydration for the ASTM type I to type V cements (kJ/kg)

Range of values	Type I cement	Type II cement	Type III cement	Type IV cement	Type V cement
7 days	349	344	370	233	310
28 days	400	398	406	274	–

Table 7.10 Heat of hydration data for some British cements (kJ/kg)

Cement type	3 Days	7 Days	28 Days
Ordinary Portland	260	320	360
Rapid-hardening Portland	280	335	375
Sulfate-resisting Portland	230	265	310
Low-heat Portland	195	225	290
Portland pozzolanic cement with 40% fly ash	190	225	270
Portland blast furnace slag cement with 40% granulated slag	195	230	270

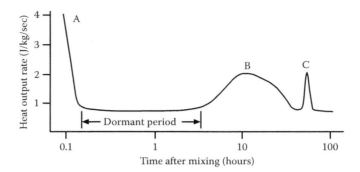

FIGURE 7.5 Typical rate of heat output from Portland cement during hydration.

of calcium sulfate hemihydrate, hydration of free calcium oxide, the heat of wetting, and the heat of solution and initial reactions of the aluminate compounds. Thereafter, the rate of heat output declines corresponding to the dormant period of cement hydration lasting up to about 4 h. Subsequently, at a time corresponding to the initial set, the rate of heat output starts to increase, reaching the broad peak B sometime after the final set. The second peak thus occurs between 6 to 12 h. This stage of heat evolution corresponds to the hydration of C_3S. In cements with a substantial C_3A content (> 12%) a sharp peak is often observed towards the end of the heat output curve. This is associated with the secondary transformation of the unstable ettringite to the stable monosulfate form ($C_3A.CaSO_4.12H_2O$). With lower C_3A content in cement there can still be a kink observed on the downward curve of the second peak. The third peak occurs between 12–90 h.

The initial release of heat is enough to raise concrete temperature in a mixture by about 2°C. With acceleration of hydration the temperature of concrete rises rapidly to about 20–40°C. In mass concrete, more than a meter thick, where the conditions are almost adiabatic, the concrete temperature rises well above the ambient and it is important to dissipate the heat rapidly in order to avoid damage in hardened concrete later. Under such circumstances, the use of cement with lower C_3S and C_3A or addition of supplementary cementitious materials in replacement of clinker is the remedial option.

7.7 Overview of the hydration reactions

When cement is mixed with water to form a paste, hydration reactions begin, resulting in the formation of gel and crystalline products capable of binding inert particles (such as aggregate) into a coherent mass. Setting has already been explained as the stiffening of the originally plastic paste of cement and water, while hardening follows setting and is the result of further hydration reactions advancing gradually into the interior of the cement particle. The strength development of cement depends on the amount of gel formed and the degree of crystallization. The hydration reactions take place through a series of stages, as already explained. Some of the physical transformations that occur in these stages are as follows:

- A dormant period immediately after mixing, where the plasticity of the paste of cement and water remains relatively constant, and any loss of plasticity can be recovered on remixing.
- The initial set, when the mix starts to stiffen at a much faster rate (normally commencing between 2–4 h after mixing at normal temperatures). At this time, the mix has little or no strength.
- The final set, when the mix begins to harden and gain strength. This occurs a few hours after the initial set and proceeds rapidly for a period of up to three days.
- The stage of steadily decreasing rate of hydration and strength gain. The hydration reactions continue to take place until either a lack of reactants or a lack of space to deposit the hydration products causes the reactions to cease.

Cement hydration reactions

The hydration reactions of cement are very complex due to the presence of several compounds. Each compound reacts in its own fashion

Cement model reactions:

- Silicate reactions

$$C_3S + 5.3H \rightarrow C_{1.7}SH_4 + 1.3CH$$
$$1 \quad 1.34 \quad\quad\quad 1.521 \quad 0.61$$

$$C_2S + 4.3H \rightarrow C_{1.7}SH_4 + 0.3CH$$
$$1 \quad 1.49 \quad\quad\quad 2.077 \quad 0.191$$

- Aluminate and ferrite reactions

$$C_3A + 6H \rightarrow C_3AH_6$$
$$1 \quad 1.21 \quad\quad 1.69$$

$$C_3A + 3C\bar{S}H_2 + 26H \rightarrow C_6A\bar{S}_3H_{32} \text{ (Aft)}$$
$$0.4 \quad 1 \quad\quad 2.1 \quad\quad\quad 3.3$$

$$2C_3A + C_6A\bar{S}_3H_{32} + 4H \rightarrow 3C_4A\bar{S}H_{12} \text{ (Afm)}$$
$$0.2424 \quad 1 \quad\quad\quad 0.098 \quad\quad 1.278$$

$$C_4AF + 3C\bar{S}H_2 + 30H \rightarrow C_6A\bar{S}_3H_{32} + CH + FH_3$$
$$0.575 \quad 1 \quad\quad 2.426 \quad\quad\quad 3.3 \quad 0.15 \quad 0.31$$

$$2C_4AF + C_6A\bar{S}_3H_{32} + 12H \rightarrow 3C_4A\bar{S}H_{12} + 2CH + 2FM_3$$
$$0.348 \quad 1 \quad\quad\quad 0.294 \quad\quad 1.278 \quad 0.09 \quad 0.19$$

$$C_4AF + 10H \rightarrow C_3AH_6 + CH + FM_3$$
$$1 \quad 1.41 \quad\quad 1.17 \quad 0.26 \quad 0.545$$

FIGURE 7.6 Cement hydration reactions at a glance. (*Note:* Numbers below reactions indicate volume stoichiometry.)

concurrently with others without causing any significant interference with them. The hydration reactions of major phases are shown in Figure 7.6.

Initially the hydrates are formed from the corresponding anhydrous phases that passed into the solution. The hydrates have lower solubility than their anhydrous forms and begin to crystallize from the solution when it becomes saturated with respect to those anhydrous phases. Hydration of C_3A occurs rapidly to form C_3AH_6 crystals, which form a film over the silicate particles, inhibiting their hydration. Gypsum added as a set regulator during grinding causes retardation of the C_3A hydration and helps to form ettringite, instead of the aluminate hydrate. The ettringite is insoluble and therefore the concentration of aluminates in solution is prevented. Ettringite then crystallizes out of solution causing a volume expansion, which is not disruptive as secondary transformation of ettringite into a more stable monosulfate hydrate form takes place almost immediately. Over the same time frame C_4AF also hydrates and forms sulfoaluminate and auminoferrite hydrates. The silicate phases hydrate at a slower rate than the non-silicate phases but the hydration of the silicates is responsible for the strength characteristics of the hardened cement. Both the silicate phases yield an amorphous mass of tricalcium silicate hydrate (a gel) and crystals of calcium hydroxide, $Ca(OH)_2$, also called "portlandite." The gel in the form of C-S-H fibers is responsible for the strength development in cement and $Ca(OH)_2$ imparts alkalinity to the hydrated cement paste, which protects the steel reinforcement in concrete from corrosion.

Microstructure of hardened cement paste

The microstructure of hardened cement paste is porous and the tentative volumes occupied by the hydrated phases and pores are as follows:

- C-S-H gel and micropores: 50%
- $Ca(OH)_2$ crystals: 12%
- Monosulfate hydrate and other hydrates grouped as AFm: 13%
- Capillary pores: 20%
- Unhydrated phases: 5%

The hydration product development over time is shown in Figure 7.7 and the volume proportions of the hydrated OPC paste components are schematically presented in Figure 7.8 (5).

Structurally, the C-S-H phase is an assemblage of amorphous and quasicrystalline hydrates. The composition is quite variable as large amounts of sulfates, alumina, and iron oxide present in the system often find their way into the gel by way of substitution. Broadly, the Ca/Si ratio in the gel varies from 1.7–2.1 and the Ca/Al ratio from 20–30. The particle size is about 0.1 µm, as determined with the help of electron microscopy.

The calcium hydroxide phase forms relatively massive crystals—two or three orders of magnitude larger than the C-S-H particles. The crystals grow within the water-filled capillary pores up to 0.1 mm size in the cement paste.

Ettringite commonly forms prismatic crystals of hexagonal cross-section with an aspect ratio of more than 10 in many instances. The particle size may range from 1–10 µm. Ettringite may not have the exact stoichiometry indicated by $C_3A.3CaSO_4.32\ H_2O$ in the cement paste

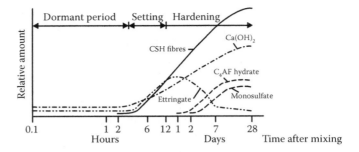

FIGURE 7.7 Hydration products in Portland cement pastes.

and, hence, the more general designation "Aft" is used for this phase. Similarly, the monosulfate phase does not normally correspond to its stoichiometry ($C_3A.CaSO_4.12H_2O$) and, hence, the designation "AFm" is preferred. The monosulfate phase forms thin platelets or rosettes of 0.1–1.0-μm dimension.

The pore sizes in the hydrated cement paste can be divided into three categories: micropores (0.5–2.5 nm), mesopores (2.5–10 nm), and macropores (10–10,000 nm). Micropores are present as an integral part of C-S-H gel. Mesopores of 2.5–10 nm can also be associated with the gel formation. Larger mesopores and macropores are formed by evaporation of water-filled spaces.

Effect of temperature on hydration and curing

Hydration of Portland cement is sensitive to temperature. Generally, increased temperature accelerates the early rate of hydration, although the rate of hydration and strength development at later ages are adversely affected. Portland cement cannot be used in sub-zero temperatures without adequate measures of raising the temperature of mixing water and the ambience of placement and curing. The effects of variable curing temperatures are, in fact, more complex. Whilst high temperatures during the early curing of concrete increase the rate of strength gain, at later stages of the curing process higher strengths are obtained at lower temperatures. It is thought that, although the C-S-H gel will be formed more rapidly at increased temperatures, its structure is coarser and less uniform than that formed at low temperatures. Since the strength of the hardened cement paste is largely a surface effect, coarser structures have reduced specific surface area compared to finer structures, and therefore reduced strength. Although the optimum temperature for maximum long-term strength gain will vary for different cements, on average it is reportedly not less than 13°C. Steam curing of concrete under pressure (autoclaving) allows the use of temperatures of greater than 100°C without causing drying out. However, autoclave treatment may have a detrimental effect on strength by producing C-S-H gel of less uniform structure, as described above. In addition, autoclaving may produce different hydration products; for example, the formation of dicalcium silicate hydrates (C_2SH), which have lower strength and increased porosity, may happen, compared to the calcium silicate hydrates ($C_3S_2H_3$) normally produced.

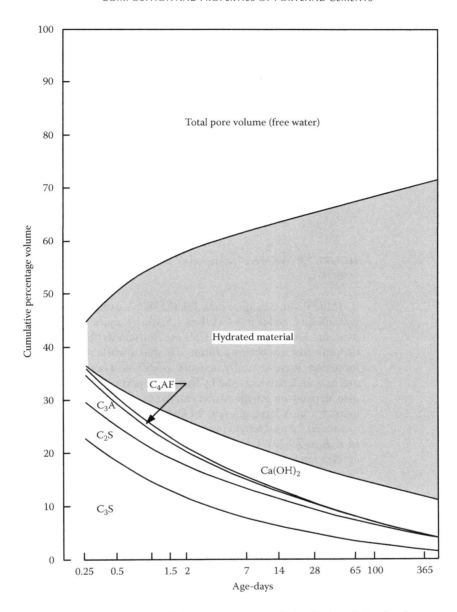

FIGURE 7.8 Percentage volumes of the components of the hydrated Portland cement paste. (Reproduced from J. A. Dalziel and W. A. Gutteridge, The influence of pulverized-fuel ash upon the hydration characteristics and certain physical properties of a Portland cement paste, Technical Report 560, Cement & Concrete Association, UK ,(986. With permission of The Concrete Society, UK.)

Hydration chemistry of slag cement

The basicity ratios, mentioned earlier, are generally used to determine the hydraulicity of blast furnace slag. Another parameter, called the F-value, is considered more applicable to define the quality of slag:

$$F = (CaO + CaS + 0.5\ MgO + Al_2O_3)/(SiO_2 + MnO)$$

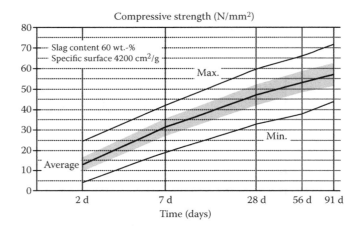

FIGURE 7.9 Range of variation of compressive strength of Portland slag cement.

If the F-value is more than 1.9, the slag is expected to be of very good hydraulicity, and if it is less than 1.5, the hydraulicity of slag is regarded as poor. These oxides and CaS are chosen, as the components in the denominator are network formers in glass, while the components in the numerator are essentially network modifiers. These equations, however, are only indicative, as the hydraulic properties of Portland slag cement also depend on its glass content, type of activation, fineness, clinker quality, etc. A large quantity of slag cement from different sources was evaluated for its strength properties and the range of values are presented in Figure 7.9 (6). On the whole, CaO, MgO, and alkali oxides within certain ranges have shown positive effects on the strength development; the alumina content up to about 15% also has a positive influence but the MnO-content has some detrimental effect on strength development.

The granulated slag does not set *per se* in water, but in the presence of activators like lime, alkalis and Portland cement will show hydraulic properties. The activators appear to function by the removal of passive surface films that act as a barrier to proper hydration. The slag cement hydrates to form C-S-H similar to that found in normal Portland cement but possibly of lower C/S ratio. Ettringite and monosulfate phases are also formed. What is not normally found in the hydration of normal Portland cement but in the hydration of slag cement is the Stratling's compound C_2ASH_8. Calcium hydroxide is generally absent as it is consumed by the reactive slag component.

Hydration chemistry of fly ash cement

Fly ash is a pozzolanic material, which means that, like all other artificial and natural pozzolanas, it combines with lime at ordinary temperatures in aqueous media to form cementing hydrating phases. All pozzolanic cements hydrate approximately as for normal Portland cement but since the calcium hydroxide liberated reacts with the pozzolana to form more C-S-H of stronger binding power, the ultimate strengths can be greater than those of normal Portland cement.

In Figure 7.10, the fraction of glass in fly ash is shown as a function of the age of hydration (7). It appears that the glassy phase starts reacting within one day of hydration. The percentage volumes of Portland fly ash cement paste hydrated over one year are shown in Figure 7.11 (5). In the long term, fly ash with good proportions of the glassy phase can contribute more to strength than an equivalent weight of Portland cement, but only if the curing conditions are such that the reactions can continue.

Use of superplasticizers in cement hydration

The complete hydration of Portland cement requires approximately 30% water. Any water added beyond this level will leave a corresponding level of capillary porosity. Since in these systems the volume fraction of water may be up to twice its weight fraction, the pore volume due to excess water can be very significant. Hence, minimization of the excess water through the use of water-reducing additives is an important issue in concrete technology. Chemical additives that can function as concrete water-reducing admixtures are designated as "water reducers" and have been classified by ASTM into two broad categories according to their effectiveness (ASTM C 1017):

- Water reducers (WR type A) that can effect water reduction of 5% or more.
- High-range water reducers (HRWR type F) that can effect water reduction of 12% or more.

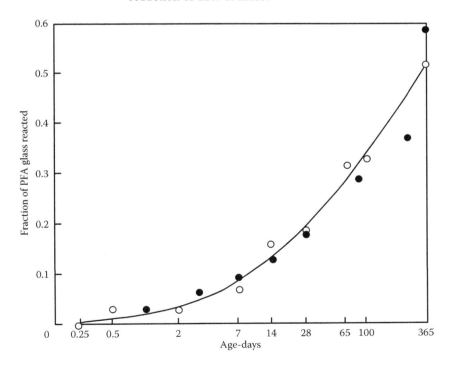

FIGURE 7.10 Reaction rate of the glassy phase in fly ash in hydrated cement. (Reproduced from J. A. Dalziel and W. A. Gutteridge, The influence of pulverized-fuel ash upon the hydration characteristics and certain physical properties of a Portland cement paste, Technical Report 560, Cement & Concrete Association, UK, 1986. With permission of The Concrete Society, UK.)

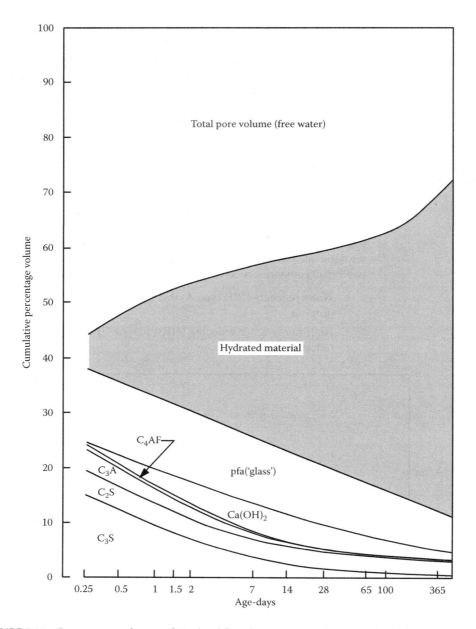

FIGURE 7.11 Percentage volumes of Portland fly ash cement paste. (Reproduced from J. A. Dalziel and W. A. Gutteridge, The influence of pulverized-fuel ash upon the hydration characteristics and certain physical properties of a Portland cement paste, Technical Report 560, Cement & Concrete Association, UK, 1986. With permission of The Concrete Society, UK.)

Since some HRWR can exhibit much higher water reduction, typically up to 30%, these are more commonly referred to as "superplasticizers." Water-reducing admixtures are further distinguished by their concurrent effect on the setting of cement, viz., accelerating (type E) and retarding (type G). Superplasticizers are organic water-soluble polymers. Use of water reducers began with "lignosulfonate"—a waste stream of the paper industry—but later, for nearly half-a-century, the superplastizer group comprised two types of polymers: sulfonated naphthalene formaldehyde and sulfonated melamine formaldehyde. Roughly a decade ago, however, the introduction of acrylic-type polymers has considerably broadened the group, which now includes dozens of different carboxylated polymers. It is important to note that the application of superplastcizers is quite complex due to variations in cement composition, the use of blended cements with different supplementary cementitious materials, and the likely presence of many other chemical admixtures in concrete. Hence, national and international standards stipulate performance requirements for superplasticizers. The stipulated tests, however, only afford a screening procedure for the selection of superplasticizers. The actual effectiveness is found out through field trials.

The key function of superplasticizers is to achieve and control the workability of fresh concrete, even at a very low water:cement ratio, without adversely affecting other features of the cementitious system, such as setting time, bleeding, air entrainment, or air void stability. The adsorption of superplasticizer molecules onto cement particles is a crucial process in the function of these admixtures. The adsorption of superplasticizers on different phases of cement is different and consequently the effects also vary. C_3A is responsible for the major part of the adsorption process, while adsorption by the silicate phases is less and occurs at a slower rate (Figure 7.12) (8). There are many instances when the cement-superplasticizer combinations exhibit poor initial workability, or premature loss of workability. Such situations are commonly referred to as cement-superplasticizer "incompatibility" problems. Figure 7.13 illustrates schematically the complexity of the Portland cement-superplasticizer system (9). The inclusion of other cementitious materials further increases the chemical complexity of the paste.

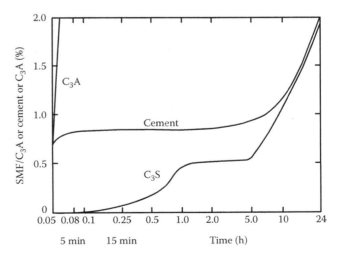

FIGURE 7.12 Adsorption of sulfonated naphthalene formaldehyde on cement and cement phases. (From N. Spiratos, M. Page, N. Mailvaganam, V. M. Malhotra, and C. Jolicoeur, *Superplasticizers for Concrete*, Marquis, Quebec, Canada, 2006.)

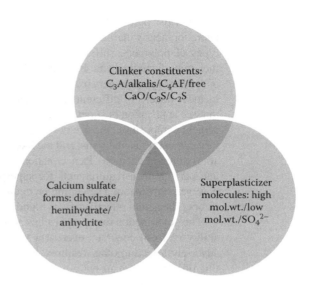

FIGURE 7.13 Inter-relation of clinker–calcium sulphate–superplasticizer. (From N. Spiratos, M. Page, N. Mailvaganam, V. M. Malhotra, and C. Jolicoeur, *Superplasticizers for Concrete*, Marquis, Quebec, Canada, 2006.)

The following chemical interactions help to understand the incompatibility issues and the ways to overcome them:

a. With sulfonated superplasticizers, the most crucial is the C_3A/$CaSO_4$/superplasticizer interaction due to competition of sulfate ions for reactions with the aluminate phase. Polyacrylate superplasicizers may not interfere in the C_3A/SO_4 reactions to the same extent.

b. Incorporation of alkalis in the C_3A during clinker formation changes the normal cubic structure of the phase into a more reactive orthorhombic form, thus impacting on reactions with sulfate.

c. The rapid solubilization of alkali sulfates increases the sulfate level in the solution early in the hydration process, altering the C_3A/SO_4 balance and the early formation of ettringite and gypsum.

d. Alkali sulfates promote coagulation of mineral particles in a slurry state, while low-alkali cements appear less compatible with the sulfonated superplasicizers. Hence, the optimum balance of alkali content is critical.

e. Increasing use of silica fume, fly ash, blast furnace slag, and fillers in Portland cement has a considerable impact on the performance of superplasticizers, although the pattern of compatibility cannot be predicted. But, for example, silica fume, due to its very high specific surface area, adsorbs significant quantitites of superplasticizer molecules, thus requiring higher dosages for a given workability. A similar effect can be envisaged for other finely ground supplementary cementitious materials, or even for very finely ground Portland cement itself.

f. Polyacrylate-based superplasticizers appear less sensitive towards variations in the composition of cement and supplementary cementitious materials than the sulfonated admixtures, but at the same time they are much less robust towards dosage fluctuations and also towards their interaction with air-entraining admixtures.

7.8 Cement for durable concrete

In almost every country, damage and distress are widely observed in recently built reinforced concrete structures. Hence, sustainable and durable construction has become crucial for human society. The emergence of cements that show progressive and steady development of long-term strength, and also blended Portland cements, is quite significant in this context. The justification for making blended cements of low clinker factor is not limited to only environmental impacts; it also satisfies the durability requirements of concrete. It is understood that the denser the microstructure of cement hydration products, the lower the shrinkage and heat of hydration. Hence, cement exhibiting a slower rate of hydration is preferable from a durability perspective. Among the cements in use, Portland slag cement with slag content ranging from 36–65% has the lowest climate change impact, followed by Portland pozzolana cement containing 15–35% fly ash. Developments are still underway for multi-component blended cements with clinker factors of 50% or less.

However, when the wide range of potential concreting materials is taken into account, the traditional specification parameters do not seem to be adequate for ensuring durability. This will be more easily understood from Figures 7.14 through 7.16. The chloride diffusion in the

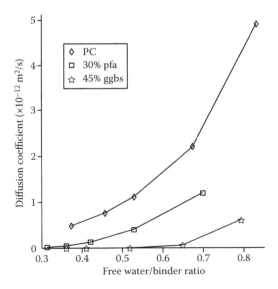

FIGURE 7.14 Chloride diffusion versus water/binder ratio of Portland cement with and without SCMs.

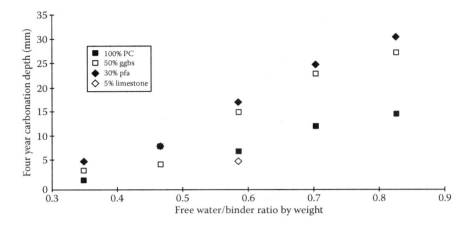

FIGURE 7.15 Four-year carbonation of Portland cement with and without SCMs.

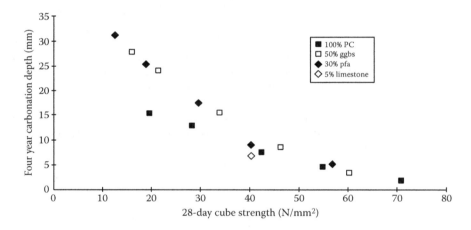

FIGURE 7.16 Relationship between carbonation and concrete strength.

blended cements, for example, is significantly improved as compared to the normal Portland cement (see Figure 7.14) but at the same time the blended cements display high rates of carbonation under identical testing conditions (Figure 7.15). It is also important to note that the carbonation rate is more related to the strength of the concrete than the cement type *per se* (see Figure 7.16). Thus, the durability performance specification is emerging as the crucial need for the ultimate use of cement and supplementary cementitious materials.

Apart from the concept of durability design specification of cement-based products, considerations of environmental impact are becoming as important as the performance of the product itself over its service life. In this context, it is important to take note of the relevance of the life cycle assessment (LCA) of construction materials. LCA for cement is essentially a holistic methodology to assess the environmental aspects of cement over its life cycle. The life cycle can be defined at three levels:

- Gate to gate: impacts are assessed while the material is handled within the plant boundary.
- Cradle to gate: the life cycle is considered from mining until cement is dispatched from the factory; in this mode, the impacts of "use phase" are left out.
- Cradle to grave: the life cycle is considered from extraction of raw materials until cement is disposed of as waste at the end of its life or use phase.

For the cradle-to-gate LCA for normal and blended Portland cements, the process starts from the mining of limestone and covers transportation, comminution and homogenization of raw materials, pyroprocessing for clinker making, transportation of fuel and its preparation for firing, transportation of gypsum and additives for cement making, finish grinding, and packing and shipment. The calculation of CO_2-e is based on the collective contributions of CO_2, CH_4, NO_2, and synthetic gases evolved during each activity, taking into account the energy content of the fuel used and the global warming potential (GWP) of the gases emitted. The conventional functional units are used as reference at each stage and for the cement production process it is related to a ton of material processed. The CO_2-e estimates for cement vary from approximately 0.73 kg to 0.79 kg, depending on various prevailing factors in a given situation. Use of the supplementary cementitious materials in cement grinding brings down the CO_2 emissions by 15–20%. It is also significantly beneficial to substitute raw materials and fuel by industrial wastes with lower embodied energy.

The environmental impact of the Indian clinker production process has been compared with other countries in a study conducted by the Confederation of Indian Industries in association with the Cement Manufacturers' Association in the country. The results indicate higher acidification and eutrophication impacts of the Indian clinker manufacturing process base, whereas the photochemical oxidation and ozone layer depletion impacts are lower, compared to other countries (Table 7.11) (10). This is only to illustrate the advantages that can be derived from LCA studies towards the future road map.

Table 7.11 Illustration of environmental impacts of the indian clinker production process

Impact category	Unit	India	Europe	USA	Rest of the world
Acidification	kg SO_2-e	2.00	1.73	1.76	1.96
Eutrophication	kg PO_4-e	0.32	0.22	0.24	0.23
Global warming	kg CO_2-e	947	937	942	964
Photochemical oxidation	kg C_2H_4-e	0.033	0.060	0.059	0.070
Ozone layer depletion	gCFC-11-e	0.00594	0.0256	0.0256	0.024

7.9 Summary

The journey of Portland cement started almost 260 years back, when British engineer John Smeaton was entrusted to renovate Eddystone Lighthouse in Cornwall in the UK. It was almost 85 years later, in 1824, that Portland cement was patented by British mason Joseph Aspdin. It took a few more decades to establish the product, to improve the manufacturing technology, and to engineer the plant and machinery for producing the cement at a reasonable scale of operation. The first seven decades of the twentieth century saw massive expansion of the Portland cement industry, with numerous engineering innovations and a deeper understanding of cement chemistry. Later, changes were also seen not only in the scale of operation and energy-efficient technology but also in huge diversification of the basic product. It was understood that normal Portland cement performs well when four principal phases, i.e., C_3S, C_2S, C_3A, and C_4AF appear in the composition in a given proportion. It was also understood that the family of Portland cement can be expanded by adopting the following specific steps at the time of manufacture: changing the phase proportions with or without adjusting the product fineness, adding special compounds to the basic composition, and blending supplementary cementitious materials including fillers at the time of grinding.

By virtue of these manufacturing strategies, normal Portland cement added to its family rapid-hardening cement, sulfate-resisting cement, low-heat cement, and white cement, in which only the relative proportions of the phases changed with or without the adjustment of fineness. Another class of products such as expansive cement, regulated set cement, and colored cement could be produced by adding special compounds to the basic cement. The third group of cements, called blended cements, was manufactured by substituting clinker partly with granulated blast furnace slag, fly ash, natural pozzolans, limestone powder, etc. The characteristics and application of these cements have been elaborated in this chapter with special reference to European, ASTM, and Indian standards.

The production of cement is accompanied with characterization and testing of properties. Important properties such as consistency, setting time, compressive strength, soundness, and their relationship with chemical and phase composition of cements have been dealt with in this chapter. Hydraulic cements perform because of their reaction with water. Since the anhydrous cements have different types of phase assemblage, their reaction with water also shows different trends. The hydrate phases that are formed and the microstructure that is developed in the hydrated cement paste have been described, along with the effects of time and temperature. The use of water-reducing chemicals is an important step in concrete making and it is essential to ensure compatibility of the water-reducing chemicals with cement. Although the chemistry is complex, the preliminary concepts have been explained. Finally, the application of cement is linked with durability of concrete and environmental impacts due to their use in construction. The life cycle assessment of cement is crucial in this context. All of these concepts have been introduced in this chapter.

References

1. P. D. TENNIS AND S. H. KOSMATKA, Cement characteristics, in *Innovations in Portland Cement Manufacturing* (Eds J. I. Bhatty, F. M. Miller, and S. H. Kosmatka), Portland Cement Association, Skokie, Illinois, USA (2004).
2. Cembureau, Cement Standards of the World, Brussels,1992.
3. J. BENSTED, Hydration of Portland cement, in *Advances in Cement Technology: Chemistry, Manufacture and Testing* (Ed. S. N. Ghosh), Tech Books International, New Delhi (2002).
4. A. K. CHATTERJEE, Special cements, in *Structure and Performance of Cements* (Eds J. Bensted and P. Barnes), Structure and Performance of Cements, Spon Press, London (2002).
5. J. A. DALZIEL AND W. A. GUTTERIDGE, The influence of pulverized-fuel ash upon the hydration characteristics and certain physical properties of a Portland cement paste, Technical Report 560, Cement & Concrete Association, UK (1986).
6. E. LANG, Blastfurnace cements, in *Structure and Performance of Cements* (Eds J. Bensted and P. Barnes), Spon Press, London (2002).
7. K. MOHAN AND H. F. TAYLOR, Effects of fly ash incorporation in cement and concrete, University Park, Materials Research Society, PA, USA (1981).
8. V. S. RAMCHANDRAn, Influence of superplasticizers on the hydration of cement, 3rd International Congress on Polymers in Concrete, Japan (1981).
9. N. SPIRATOS, M. PAGE, N. MAILVAGANAM, V. M. MALHOTRA, AND C. JOLICOEUR, *Superplasticizers for Concrete*, Marquis, Quebec, Canada (2006).
10. P. V. KIRAN ANANTH, Life cycle assessment for cement sector, Confederation of Indian Industries, Sohrabji Godrej Green Business Center, Hyderabad (2017).

CHAPTER EIGHT

Advances in plant-based quality control practice

8.1 Preamble

In modern energy-efficient and high-productivity cement plants the quality control (QC) function has turned out to be extremely crucial. A functionally specialized organizational set-up is created in all such plants with commensurate laboratory facilities to execute and oversee the following activities:

- sampling and analysis of all raw materials and fuels;
- tracking the run-of-mine limestone quality;
- raw mix proportioning and control;
- pre-blending operations;
- process control in individual unit operations like crushing, raw grinding, homogenization, burning, clinker making, cement grinding, packing and dispatch;
- chemical and physical testing of cement with a view to ensuring compliance with national and company standards;
- quality assurance and certification of cement dispatch consignments;
- investigation and solution of process and product quality problems;
- water quality monitoring;
- pollution control management;
- production and inventory reporting.

It is important to bear in mind that in large plants, the market competitiveness, complexity of limestone deposits and their exploitation, scale of operation, and high level of instrumentation and computer-aided process controls make the QC function less manual on one hand but more professional on the other. One should not forget that in some plants there may be problems particular to the location due to raw material, fuel, or

8.2 Sampling guidelines

Basic principles

The basic purpose of sampling is to collect a manageable mass of material that is representative of the total quantity of material from which it was collected. This manageable mass of material, called a sample, is subject to certain preparation procedures, which render it suitable for either physical testing or chemical analysis. The types of tests or analysis to be performed are dependent on what characteristics are required to be measured to ascertain a sample's quality.

The methods by which samples are collected, the frequency of collection, and the accuracy of the samples collected, along with the nature of the material being sampled, ultimately define their representativeness. A totally homogeneous material will require the collection of only a single sample, whereas a lumpy heterogeneous material will require the collection of many small samples, or increments, which when combined will represent the total mass, or lot, with an acceptable degree of accuracy. The increments should be collected from the parts of the lot, with the number required to be collected being dependent on the variability of the material constituting the lot. It is of fundamental importance that all particles in the lot have the same probability of being included in the final sample. For the theory and practice of sampling one may refer to (1).

Establishing a sampling scheme

The general procedure for establishing a sampling scheme consists of the following steps:

i. Define the quality parameters to be determined and the type of sample required
ii. Define the lot
iii. Define the precision required
iv. Determine the variability of the quality parameters and establish the number of sampling units (m) required and the minimum number of increments (n)
v. Decide whether to use time-basis or mass-basis sampling and define the sampling intervals in, say, minutes for time-based sampling or kilograms for mass-based sampling
vi. Ascertain the nominal top size of the material for the purpose of determining the minimum average increment masses
vii. Determine the method of combining the increments into gross samples
viii. Determine the method of sample preparation

Table 8.1 Minimum number of sampling units in a lot

Mass of lot (thousand metric tons)	Minimum number of sampling units (m)
< 5	1
5–20	2
20–45	3
45–80	4
> 80	5

Lots should be divided into a convenient number of sampling units (m), which shall not be less than the number given in Table 8.1. This number may be increased so that the sampling units coincide with a convenient mass or time.

The number of increments for a desired precision in a single lot is estimated from the following equation:

$$n = \frac{4V_I}{mP_L^2 - 4V_{PT}} \qquad (8.1)$$

Where n is the number of increments to be taken from a sampling unit, m is the number of sampling units in the lot, V_I is the primary increment variance, V_{PT} is the preparation and testing variance, P_L is the precision of sampling, sample preparation and testing for the lot at 95% confidence level, expressed as % absolute.

The above equation is obtained by transposing the equation given below:

$$P_L = \pm 2\sqrt{\frac{\frac{V_I}{n} + V_{PT}}{m}} \qquad (8.2)$$

Equation 8.2 is regarded as an estimate of the precision of the experimental results obtained on the same material. While V_I depends on the type and top size of the material, the degree of pre-treatment and mixing, and the absolute value of the parameter to be determined, in practice V_{PT} should be less than $0.02 \times V_I$ subject to a minimum value of 0.05.

If, from Equation 8.1, n is found to be impractically large, then the number of sampling units should be increased. The reference increment mass and minimum mass of gross samples for a series of nominal top sizes are given in Table 8.2.

It should be borne in mind that the minimum mass of a gross sample is a very important factor for precision but it alone does not guarantee the precision, which is also dependent on the number of increments taken to compound the sample and their variability. In ASTM C50, the minimum number of increments for particle size ranging from −6 mm to

Table 8.2 Reference increment mass and minimum gross sample mass

Nominal top size (mm)	Reference increment mass (kg)	Minimum gross sample mass (kg)
300	100	15,000
200	25	5400
150	15	2600
125	10	1700
90	5	750
63	3	300
45	2	125
31.5	1	55
22.4	0.75	32
16.0	0.50	20
11.2	0.25	13
8.0	0.15	6
5.6	0.10	3
4.0	0.10	1.5
2.8	0.10	0.65
2.0	—	0.25
1.0	—	0.10

+6–18 mm is given as 10, while the minimum weight for each increment is recommended to be 2.5, 5.0, and 7.5 kg corresponding to particle size of −6 mm, −6 mm +18 mm, and +18 mm, respectively. For randomized sampling, the practice recommended is as follows:

$$N_2 = N_1 (\text{Specific lot size (tons)}/1000 \text{ tons})^{0.5}$$

Where N_1 = the minimum increments required per 1000-ton lot, and N_2 = increments required for specified lot size.

8.3 Sampling stations in cement plants

Several sampling stations are necessary in the cement plants for offline regular controls. These are illustrated in Table 8.3.

Sampling at the stations mentioned in Table 8.3 may not necessarily be manual since various automatic samplers are available in the market. It is important to select the correct samplers for different applications. Crosscut samplers are the ones that traverse the entire stream of materials. Pneumatic retractable tube samplers can be used in transport lines. It may be borne in mind that vertical transfer points usually offer the right opportunities for extracting representative samples. One should also know and discard the undesirable sampling points in a plant, such as air slides.

Sampling procedures for hydraulic cement are specified in ASTM C 183 and C 917. For coal sampling, procedures are given in ASTM

Table 8.3 General sampling practices

Process/operation	Source/location of sample	Tentative frequency	Purpose/scope
Quarrying	Advance drilling cores Blast hole cuttings	Every 2 m Composite of each blast	Major and minor oxides analysis
Bought-out raw materials	Each consignment	Composite samples	Major and minor oxides analysis
Preblending	Stacker-reclaimer system	Periodic grab or crosscut belt sample related to stacking and reclaiming time and capacity	Homogeneity assessment based on CaO or LSF
Raw grinding	Separator duct	One- or two-hour composite samples	Determination of modulus values and sieve analysis
Kiln feed	Preheater	Four- or eight- hour grab samples	Major oxides/modulus values to ascertain homogeneity
Hot meal to kiln	Kiln inlet	Four- or eight- hour grab samples	Alkalis, chloride, sulfate, and calcination degree
Coal grinding	Transport line	Hourly or bihourly grab samples	Residues on designated sieves and ash content
Clinker	Cooler exit	Two- or four- hourly grab samples and also daily composites	Sulfate and free lime in grab samples; complete chemical analysis of daily composite
Cement grinding	Separator duct	Bihourly samples and daily composite samples	Blaine surface area and sulfate content, gypsum dehydration; daily composite samples for complete chemical analysis and physical testing
Cement packing and despatch	Packers/loaders	Daily composite samples	Complete chemical analysis and physical testing

D 2234, D 4915, and D 2013. The standard practice for sampling, sample preparation, packaging, and marking of lime and limestone products is given in ASTM C50. Practices D3665, E105, and E122 can be used to minimize unintentional bias when obtaining a representative sample.

General statistical considerations

From the QC point of view reduction of variation in kiln feed composition to a minimum ensures steady kiln operation. The whole concept of homogenization is illustrated in Figure 8.1, which shows that in practice CaO fluctuations can be dampened from 1.5 to 0.2% standard deviation or less with proper step-wise control from mining to blending under normal routine operating conditions.

To quantify homogenizing efficiency (η), the extent of chemical fluctuation can be measured in terms of standard deviations of input (α_1) to and output (α_0) from the homogenizing elements. At least 30 to 50 samples may be required to arrive at the true standard deviations, taking care of the error of analysis. In the cement production process,

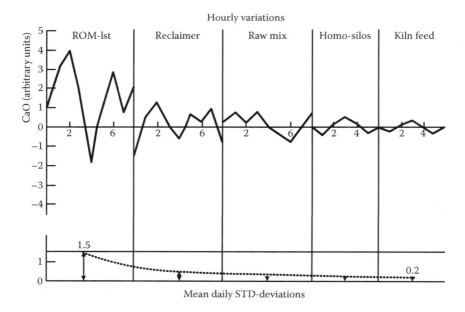

FIGURE 8.1 Dampening chemical fluctuations using homogenizing elements.

a homogenizing element is selected on the basis of input standard deviation and targeted output standard deviation.

8.4 Computer-aided run-of-mine limestone quality control

The concept of application of computer software in assessing limestone deposits and exploiting the quarries has been introduced in Chapter 1. In the present chapter the specific quality control steps involved have been elaborated. The four basic steps are:

  a. geostatistical processing of exploration data
  b. limestone deposit block model preparation
  c. determining the mining option
  d. determining the blend options

Once the mining operation and plant production start, the daily run-of-mine quality is monitored against the pre-determined targets.

Geostatistics provides many useful tools that include "variograms," "block variance," "estimation variance," "kriging," and "conditional simulation" of the deposit. The variogram tells the correlation between samples and also provides information such as the range of influence of a sample and the anisotropy of mineralization (the difference in different directions). Block variance tells the variability of the associated "volumes." In other words, it provides volume-variance relationship. Estimation variance indicates the quality of a given reserve estimate, i.e., how good or bad the estimate is. Kriging is an estimator for the grade of a block from a linear combination of the available samples in or near the block, in an unbiased manner with minimum estimation variance. Essentially, it is accomplished by assigning appropriate weightage to each of the surrounding samples. Finally, the conditional simulation refers to the simulation of the deposit by "conditioning" it or by going through all the known data points in the real deposit. It is used more as a mining tool than as an estimation tool.

Once the geostatistical processing of the geological and exploration data is done, the deposit block modeling is carried out. The deposit in question is divided into a large number of small equal blocks, whose geometry conforms to the geological settings of the deposit and the targeted operational plan. At this stage, the kriging technique is used to assign both grades and estimation error to the blocks. Specific requirements of modulus values, mineral composition, homogeneity parameters, etc., are computed for the blocks (2). A typical configuration of a deposit block model is given in Figure 8.2 as an illustration. From the deposit block model, various mining options are derived for the grade and tonnage targets, keeping the boundary and life of the mine in view. The blend options are found out in conformity with the mining options and the availability of corrective materials for raw mix proportioning. In the normal running of a plant a coordinated approach is essential in monitoring the daily mining program. This is done on the basis of blast hole drilling and online analysis of the drill hole samples, keeping the operational targets in view.

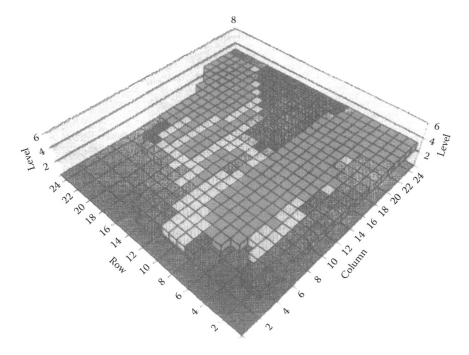

FIGURE 8.2 Schematic diagram of a deposit block model.

8.5 Preblending operation

An important adjunct to crushed limestone stockpiling and preblending, also known as bed-blending, is the sampling station. Bulk material sampling at large tonnages presents some difficulties. While it is important that at least 1% of the crushed limestone being fed to the stockpile is properly sampled, better statistical reliability will be obtained at higher sampling levels and vice versa. For example, it may be preferable to collect about 10 tons of sample per hour from a 1000-tph crusher and with suitable reduction collect about 3 to 4 tons per hour for processing in the station. One common practice is to use a moving bucket sampler cutting across the discharge from a belt conveyor. The primary sampler bucket width should at least be 2.5 to 3 times the maximum crusher product size and at least 20 to 50% longer than the maximum width of material discharge at the bucket to catch all the material cut across. After the primary stage of sampling (1:100 or 1:200), further sampling stages may involve crushing and secondary sampling (1:8), grinding in a closed-circuit ball mill and tertiary sampling (1:25). The sample conveyed to the laboratory may be in ratio of 1:20. A simplified schematic version of a sampling station is shown in Figure 8.3 (3).

As an alternative to the above traditional scheme of crushed stone sampling, a more popular practice today is to use the technique of prompt gamma neutron activation analysis (PGNAA), an online method, in which the material passes without any further size reduction through the neutron bombardment of radioactive californium 252 (see Figure 8.4) (3). The PGNAA analyzer is installed on a belt conveyor structure going to the pre-blending stockpile after the crusher. It consists of two units—a shield block assembly and an electronics enclosure. The shield block assembly houses the detectors and neutron sources, provides shielding and supports

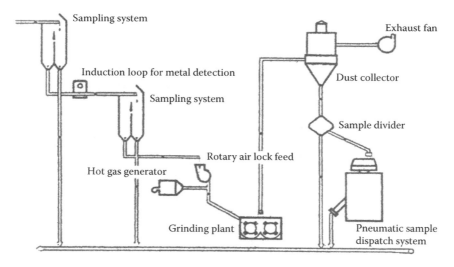

FIGURE 8.3 Schematic layout of a sampling station of crushed materials. (From H. Bergemann, Means of quality control and process optimization in cement manufacturing, *Ciments, Betons, Platres, Chaux*, No. 771, February 1988.)

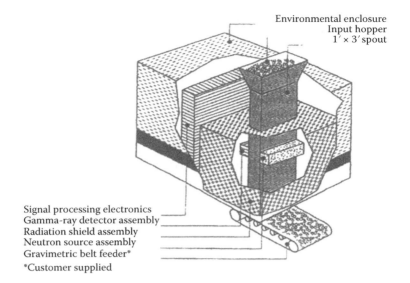

FIGURE 8.4 A cutaway view of PGNAA. (From H. Bergemann, Means of quality control and process optimization in cement manufacturing, *Ciments, Betons, Platres, Chaux*, No. 771, February 1988.)

the conveyor belt. The electronics enclosure contains the components for signal processing, computerized data analysis, a modem for remote diagnostics and connection with the operator console, which can be located at a distance of more than a kilometer. The instrument analyzes the material every minute for all major oxides, alkalis, sulfate, and chloride.

For all practical purposes, it is convenient to obtain the values of the major oxides with acceptable levels of standard deviation after approximately 10–15 minutes of integration work, as illustrated below:

- $SiO_2 \pm 0.14\%$
- $Al_2O_3 \pm 0.06\%$
- $Fe_2O_3 \pm 0.02\%$
- $CaO \pm 0.10\%$
- $MgO \pm 0.13\%$

From the interpretation of data, the process computer determines deviations from the present values and computes the material quantities that will be required of the individual components for stockpiling.

A typical pre-blending flow sheet with PGNAA of an operating plant is shown in Figure 8.5 (4). The bulk analyzer is housed in a multistory tower between the crusher and the blending bed on the stacker-reclaimer system. The crushed material passes through a feed bin that compensates for fluctuations in the quantity of material from the crusher, and into the analyzer chute. Throughput is controlled by a variable-speed conveyor belt below the analyzer. Fine correction of the raw mix is effected by adding the required corrective materials before the mill. On the whole, the performance of PGNAA for bulk analysis of crushed stone in the

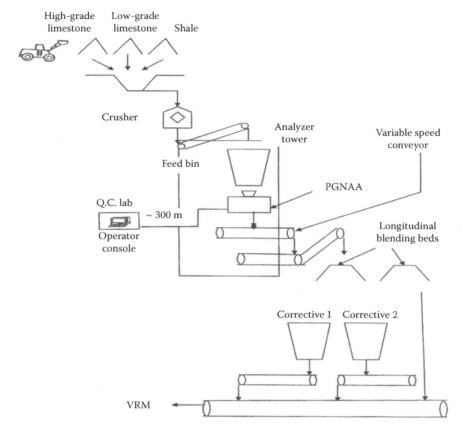

FIGURE 8.5 A typical flow chart of a pre-blending system with PGNAA. (Reproduced from M. Tschudin, *In-line control of the blending bed composition using PGNAA in Ramos Avizpe (Mexico)*, Proceedings of VDZ Congress '93 on Process Technology for Cement Manufacture, Subject 2, Bauverlag GmbH, Weisbaden and Berlin, 1995. With permission.)

cement plants has been satisfactory, as evidenced ultimately by the uniformity of kiln feed in most situations.

8.6 Raw mix control

Perhaps the most important QC aspect of a modern cement plant is the raw mix proportioning right up to its feed to the kiln. A schematic block diagram of a computer-controlled proportioning system is presented in Figure 8.6. The main sampling for mix control is done automatically. After the raw mill the representative samples are transported in capsules through a pneumatic dispatch system to an integrated sample preparation station, where all operations proceed automatically, from the accurate dosing and grinding of the sample, to cleaning and tablet pressing. X-ray fluorescence spectrometry has become the standard analytical technique in cement laboratories worldwide.

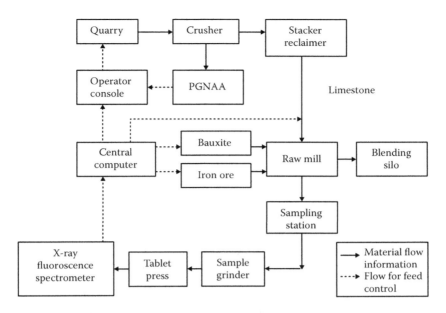

FIGURE 8.6 A block diagram of an automatic raw mix proportioning system.

X-ray fluorescence spectrometry (XRF)

In XRF, X-rays are produced by a source, which is an X-ray tube, to irradiate the sample. The elements present in the sample emit fluorescent X-rays with discrete energies that are characteristic for these elements. By detecting and measuring the energies of radiation it is possible to determine and quantify the elements present. The essential components of all spectrometers, therefore, are a source, an arrangement for irradiation of the sample, and a detection system for the emitted fluorescent X-rays. The XRF spectrometers are generally divided into two main groups: energy-dispersive (EDXRF) and wavelength-dispersive (WDXRF). The difference between the two types lies in the detection systems. The EDXRF instruments are able to measure the different energies coming directly from the samples and determine the elements present, whereas the WDXRF instruments use an analyzing crystal to first disperse the energies and then determine the elements present based on their characteristic wavelengths. The elements that can be analyzed and their detection limits depend on the spectrometer system used. The elemental range for EDXRF extends from Na to U, and for WDXRF the range is from Be to U. Generally speaking, the elements with high atomic numbers have a better detection limit than the lighter elements. The concentration range of individual elements may be highly variable from ppm level to 100%. The precision and reproducibility of XRF analysis is very high. Accuracy depends on the use of good standard reference materials (SRMs) for calibration purposes. It is, however, possible to carry out standardless analysis under certain circumstances by adopting the appropriate procedure.

In cement plants, typically, sequential and simultaneous WDXRF systems are employed. Simultaneous instruments measure all relevant elements at the same time and, hence, they are of fixed configuration for

Table 8.4 Analysis results of a raw meal sample

Oxides	Percentage determined by gravimetry	Percentage determined by XRF	Difference in percentage	Relative standard deviation
SiO_2	11.66	11.86	0.20	1.7
Al_2O_3	2.11	2.14	0.03	1.4
Fe_2O_3	1.18	1.17	0.01	0.8
CaO	45.61	45.36	0.25	0.50
MgO	1.12	1.15	0.03	2.7
K_2O	0.44	0.43	0.01	2.3
SO_3	0.29	0.30	0.01	3.4

routine use. Sequential instruments provide more flexibility and may be useful for non-routine applications.

The sample quantity is small and should be representative. The test sample is prepared either by making pressed pellets or fused beads. There could be some reduction in the analytical accuracy due to the "matrix effect" of the sample particles in the pressed tablets. Grinding the sample to less than 30 µm may help overcome this problem. With more difficult samples, one may have to take recourse to fusion with the help of fluxes like lithium metaborate or lithium tetraborate or their mixture in an induction furnace. Fused beads are essential for accurate analysis of raw materials, while pressed powder pellets serve the purpose of analysis of raw mix, clinker, cement, kiln dust, etc. In other words, pressed pellets can be used when the concentration ranges are limited for a given type of material. There are two possible methods of preparing the fusion beads:

a. using pre-calcined powders
b. using raw materials directly without pre-calcination

The method (a) gives more accurate results via a measured loss on ignition value. Generally, the fusion bead technique allows various types of samples to be analyzed with one calibration program.

The results of a typical XRF analysis of a raw meal sample from a cement production line are given in Table 8.4, which also indicates the achievable standard deviations for different oxides.

Raw meal analysis can be performed by an online analyzer locally in the production line or it can be carried out in the central laboratory of the plant. The online system comprises a complete sample preparation system including a tablet pressing facility. The analysis is carried out by an EDXRF spectrometer. As there is no time-consuming sample conveyance to a central laboratory, the online system achieves a high analysis frequency of up to six samples per hour. The higher frequency permits better raw meal control.

Monitoring of raw mix homogeneity

The homogenization process is carried out in silos either in the batch or continuous homogenizing silos, as explained earlier. In batch blending the time required for homogenization is important and can be estimated by obtaining periodic samples and by analyzing the chemical constituents to find out the input/output ratio against the blending time, the operation lasting for a long period of time. Several silos of adequate capacity, depending on the maximum mill feed rate, are needed to carry out batch blending and discharge to the kiln feed storage silos after necessary blending time. Unlike the batch blending system, continuous flow (CF) raw meal blending silos are common in plants having upstream bed blending facility. The total system is more cost effective, although performance depends on the design and operation.

Hot meal quality monitoring

The hot meal sample is taken from the hot meal chute between the bottom cyclone and the kiln inlet. The hot meal is rapidly cooled in a special sampler without allowing it to be in contact with the ambient air. The hot meal sample is prepared and analyzed in the central laboratory. The sample can be transported either manually or by the pneumatic conveying system. The analysis results are crucial for monitoring the volatiles cycle, for taking timely action to prevent high concentration of alkalis and chlorides, and also for adjusting the kiln bypass, if installed.

8.7 Kiln operation monitoring

Kiln operation is controlled primarily by adjustments to kiln feed, fuel rate, and induced drought (ID) fan speed. Irrespective of the degree of automation, most kilns are liable to be in upset conditions due to coating loss, ring build-up, material flushing, etc. The control effort is geared towards minimizing such operational instability. The principal control variables are:

a. burning zone temperature monitored by a pyrometer, or indirectly by the kiln drive power or NOx level measurement (about 1450–1500°C)

b. feed-end gas temperature by online measurement (about 950–1000°C)

c. feed-end oxygen level by online measurement (about 2.0%)

d. kiln speed (generally 2–2.5 rpm in preheater kilns and 3.5–4 rpm in precalciner kilns)

e. ID fan speed (generally operated to the limit)

f. CO measurement at the kiln inlet (generally less than 200 ppm at 1–2% O_2)

Useful information on kiln operation can be obtained from bihourly clinker samples analyzed for free lime, liter weight, and SO_3 and periodic samples of the underflow from the bottom stage of the preheater (analyzed for LOI, SO_3, Cl, and R_2O). It may be borne in mind that the volatile cycles could be minimized by avoiding hard mixes and maintaining the sulphate:alkali molar ratio between 0.8–1.2 and limiting Cl and SO_3 in hot meal entering the kiln to 1% and 3%, respectively.

The kiln control systems generally carry certain alarms, particularly for mechanical failures. From a safety angle, CO values above 0.4% should cause an alarm, while for higher values there should be tripping arrangements for the fuel supply and the electrostatic precipitator.

In the total kiln system, the clinker coolers, which are predominantly grate coolers in modern plants, also require careful monitoring for the overall stability of the pyroprocessing section. Generally, the clinker coolers are monitored by the secondary air temperature, tertiary air temperature, and clinker discharge temperature.

8.8 Cement grinding process

The overall performance of a milling circuit is best understood by its specific power consumptions. This is numerically presented in kWh/t, i.e., by the ratio of the mill motor power and the corresponding production at a given Blaine fineness. The particle size of the finished cement is another important parameter and it is determined by the following techniques:

a. Specific surface area in m^2/kg, measured by Blaine's air permeability method
b. Residue on 45- or 30-μm sieve obtained by wet sieving
c. Particle size distribution (PSD) determined by a low-angle laser light scattering method

Since the 1940s the Blaine surface area measurement has been used to determine cement quality. Despite its long usage, the limitations of the test are widely recognized. The test is slow and suffers from poor reproducibility. It is often observed that two cements with different proportions of fines can give the same Blaine value. Notwithstanding these shortcomings, this parameter still continues to form the most widely used standard specification for cement fineness in India and elsewhere. The primary reason for this is the fairly linear relationship between the Blaine's surface area and compressive strength of cements, at least in the range of 250–500 m^2/kg at all ages up to a year or so.

Sieving has also been widely used in the cement industry, although there are inherent difficulties to sieving particles of below 30 μm in size.

However, more advanced analytical techniques are rapidly gaining acceptance both in the laboratory and in the plant. The preferred method is now the laser diffraction instrument, mentioned above, in which cement particles are introduced into a laser beam as a dry powder or a fluid suspension. The particles scatter light according to their size and the method is based on the principle that the diffraction angle is inversely proportional to particle size. The angles are measured and converted into the particle size distribution. Dry analysis is preferred for cement as no sample pre-treatment is involved. Dry powder feeder units

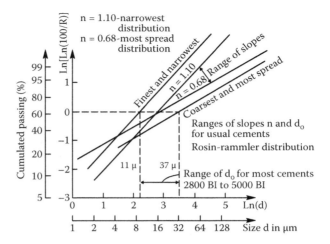

FIGURE 8.7 RRSB plot of cement particle size distribution.

usually rely on pressure to force a sample through the laser beam into a collecting device. They accelerate the material close to the speed of sound in order to shear or break up agglomerations.

Rosin-Rammler-Sperling-Bennet (RRSB) plot of particle size spread

Since the above two methods of characterizing the cement fineness do not clearly show the integrated overall distribution of particle size of ground cement, attempts have, therefore, been made to completely characterize the particle size distribution by a graphical presentation with some characteristic values, which is popularly called RRSB plot (see Figure 8.7). In this plot there are two constants, d_0 and n, which are hardly identical for two different cements in terms of their particle size distribution. The RRSB plots of 150 cements have shown the following ranges:

Blaine specific area: 240–570 m²/kg

Slope (n): 0.69–1.30

Constant (d_0): 11.0–42.0 μm

It is important to note that, apart from the above characteristic particle size distribution of the product, the cement grinding process cannot be effectively monitored without determining the phase composition, microstructure, quality, and quantity of supplementary cementitious materials, and extent of gypsum dehydration in cement. These aspects are discussed below.

8.9 X-ray diffractometry for phase analysis

Unlike XRF, which analyzes the elements present in cement, X-ray diffraction analysis (XRD) determines how the elements are combined into compounds or phases. Bragg-Brentano diffractometers are almost universally used for such purposes, and produce patterns with diffracted intensity and resolution. It is generally observed that it is possible to identify crystalline compounds that are present in quantities at more than 1%. Further, there are techniques available to provide semi-quantitative and quantitative analysis of the crystalline constituents. The application of XRD to cement-related materials had been reviewed by the author in an earlier publication (5), which can be referred to for recapitulation of the fundamentals of XRD.

There are several factors that interfere with the accurate analysis of cement by XRD. Some of the interfering factors are chemical and some are physical. The chemical ones include solid solubility in the compounds, polymorphism of the phases, degree of crystallinity, etc. The physical factors are the variable absorption of X-rays by the phases, preferred orientation of the particles, and so on. It is important to understand that even if the diffractometer is properly aligned, the slits, monochromator, and filter of the instrument are suitably selected, and the measurement time is rightly chosen, the above chemical and physical parameters can substantially alter the peak positions and their width and intensity, making the phase quantification imprecise and irreproducible. To overcome these problems in applying XRD to cement phases more precisely, the Rietveld refinement technique has been adopted (6), which is a complex crystallographic technique to refine the X-ray diffraction profile of powder samples. The technique solves the peak superposition problem and corrects the physical effects, mentioned above, to a great extent. In the last decade, the Rietveld refinement-based quantification system has been installed in several plants all over the world for the following purposes:

- kiln control or pyroprocessing
- mill control or comminution
- finished cement quality

The phases that are mostly quantified through the Rietveld-based XRD and their relevance to the process or quality-related problems are summarized in Table 8.5.

The above process and product evaluation are carried out in plants with the help of an X-ray diffractometer provided with completely

Table 8.5 Relevance of phases for product/process control with XRD

Compounds	Source material	Process/property relevance
Free CaO	Clinker	Degree of burning
C_3S/C_2S	Cement and clinker	Product strength
C_3A	Cement and clinker	Product setting time
C_4AF	Cement and clinker	Product color
$CaSO_4.nH_2O$	Cement	Dehydration during milling
$CaCO_3$	Blended cement	Limestone addition to cement
SiO_2	Blended cement	Pozzolana addition to cement
C_2AS	Blended cement	Slag addition to cement

automated Rietveld software and precise working files for each phase to be quantified. Topas BBQ is the most commonly used software, although other options are available.

8.10 Online quality control in cement plants

Whilst the need for analyzing raw materials, kiln feed, clinker, and cement can be fulfilled by a conventional offline X-ray fluorescence spectrometer, additional real-time analytical information is frequently solicited by the QC personnel for more comprehensive process control. The latter may include such parameters as free lime and phase composition of clinker, free quartz in raw mix, dehydration of gypsum in cement, etc., which require an X-ray diffractometer, as explained above. Thus, in high-capacity modern plants it is desirable to have both an X-ray fluorescence spectrometer and an X-ray diffractometer for chemical and phase analysis, respectively. In addition, as mentioned earlier, for particle size analysis one needs a laser granulometer. Such facilities, when housed offline in the central laboratory, fail to provide real-time data for process control. Hence, over the years the concept of linking the instruments to the process through automation has gained importance. Use of bulk analyzers eased the approach quite significantly.

The main objective of automation is uniform quality, more stable kiln operation, and longer availability of the kiln and mill systems. For example, it has been observed in a plant that the installation of the laboratory automation system resulted in the reduction of the standard deviation of clinker LSF from 1.03 to 0.57 (Figure 8.8).

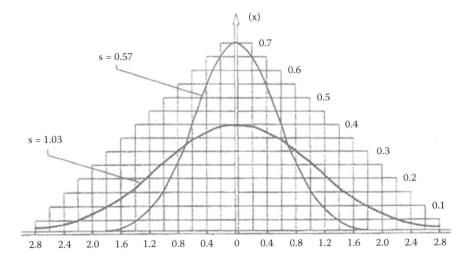

FIGURE 8.8 Clinker composition variation before and after automated quality control.

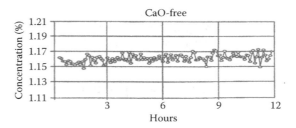

FIGURE 8.9 Long-term stability (12 hours) of free lime in clinker burning.

Online systems for chemical analysis

It appears that the first online XRF analyzer for raw meal control was installed in cement plants in the early 1990s by the joint efforts of Outokumpu and FLS Automation. More or less during the same period, the Applied Research laboratories of Switzerland introduced the online "Total Cement Analyzer" facility by integrating XRF monochromators, XRF goniometers, and an XRD system into a single platform (7). This facility was capable of conducting full chemical analysis of all concerned materials as well as the analysis of free lime in clinker, limestone, and slag additions to cement. The long-term stability of free lime in clinkers obtained by the use of the above integrated system is illustrated in Figure 8.9.

Another example may be given of two types of modular online XRF analyzers developed by IMA Engineering: Quarcon, an XRF bulk material analyzer for conveyor belt installations, and IMACON, an analyzer for fine materials. With these analyzer systems, the chemical compositions of all vital process streams can be analyzed and all important process stages can be put under fast real-time online control. A typical IMACON analyzer includes a measuring head assembly, which is installed inside an air-conditioned measuring head and sampling unit, and the probe electronics set, which contains the analyzer electronics and the local control panel. The IMACON analyzer is connected to the process management computer station, which acts as the user interface. The process management station is equipped with the necessary calculation, reporting, and calibration software, as well other requirements. The analytical results showed, for example, that the accuracy of calibration of the analyzer was 0.15 for CaO in raw meal and the precision standard deviation from repeat measurements and standard deviation from replication tests for the same oxide were 0.08 and 0.15, respectively (8).

Online kiln monitoring for clinker phases

In order to overcome the shortcomings of offline phase analysis for process control, online X-ray diffractometry has made its appearance. This has provided a basis for a new level of control where the controlled parameter is the clinker mineralogy including percentages of alite, belite, ferrite, aluminate, free lime, alkali sulfate, magnesia, etc. Furthermore, by monitoring the various polymorphic forms and crystal symmetry of clinker phases, kiln operation can be adjusted through precalciner parameters, kiln speed, burners, coolers, etc. in order to ensure production at the optimum cost and quality.

The system offered by FCT ACTech COSMA operates on the basis of continuous sampling and sample preparation (9). The ground sample passed through COSMA is approximately 70 kg of powder per day, continuously returned to the process by pneumatic transport. The instrument uses a patented full area detector and it also employs Rietveld refinement of X-ray diffraction patterns. The unit automatically updates the analysis of the process stream every minute. With frequent analysis comes the trend setting of the results and the ability to control processes online. By using the online XRD analyzer in combination with PGNAA, it is possible to monitor the clinker phases and to apply feedback control to adjust the raw material set points in real time (Figure 8.10). In addition, the unit measures free lime in real time (Figure 8.11). The result can be transmitted to the plant's automatic kiln control system (Fuzzy, Lucie, Linkman, etc.) to accurately control the burning zone conditions. While monitoring the kiln, it is possible to detect high free lime for a number of reasons. In one situation, the kiln feed may be incompletely burnt due to the kiln being cold, leaving high free lime. In another situation, a hot kiln can also generate high free lime if the LSF was high. Such diverse situations can be detected and corrected with the help of online phase analysis.

The online analytical system can also be based on the near infrared radiation spectra of materials, instead of neutron activation and X-ray fluorescence. No radioactive sources are necessary for this analyzer. The ABB SpectraFlow CM100 raw materials analyzer powered by SOLBAS

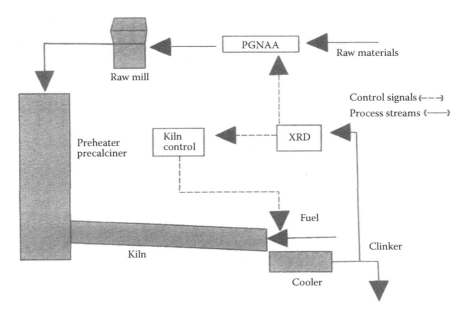

FIGURE 8.10 Total kiln control based on XRD and PGNAA. (Reproduced from P. Storer and C. Manias, *World Cement*, February, 69–78, 2001. With permission of World Cement, UK.)

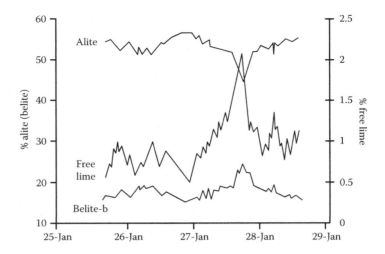

FIGURE 8.11 Real–time trend of free lime with alite and belite. (Reproduced from P. Storer and C. Manias, *World Cement*, February, 69–78, 2001. With permission of World Cement, UK.)

technology is an example of the use of this technology. At Kipas Cement in southern Turkey two Spectralflow Online Analyzers are installed with a control software adjusting the set points of the weigh-feeders since 2014.

Online cement quality control

One of the important cement quality control measures based on phase analysis pertains to measuring the sulfate content in the finished cement by quantifying the gypsum ($CaSO_4.2H_2O$), the hemi-hydrate ($CaSO_4.0.5\ H_2O$), and the anhydrite ($CaSO_4$) phases, which helps to determine the degree of dehydration occurring in the mill. The other QC measures include limestone addition by measuring calcite, slag, and fly ash additions on the basis of their pre-determined indicator phases or their glass content. By using an online XRD analyzer in a cement milling circuit any variations in the quality of gypsum can be compensated automatically. Similarly, variations in the quality of the limestone or slag or any other supplementary cementitious material can be taken care of in the process in real time. For example, if the silica content of the limestone increases, the feeder can be sped up, provided that the LOI limits are not exceeded. The circuit for gypsum, limestone, and clinker control is shown in Figure 8.12 (10).

Needless to say, the continuous real-time monitoring of the mineralogy of clinker and cement during the burning and grinding processes opens up the possibility of more precise process control.

Reference materials for calibration

Chemical or phase analysis, be it offline or online, requires standard reference materials (SRM) without exception, and may occasionally need secondary specially prepared reference materials. There are international and national organizations for sourcing such materials. The National Institute for Standards and Technology (NIST) in the USA is the most widely known sourcing organization. The National Council for Cement and Building Materials in India is a nationally approved source for a

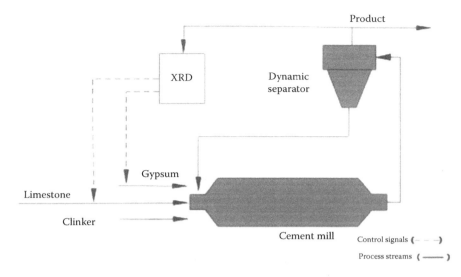

FIGURE 8.12 In-process control of gypsum and limestone addition with the help of an XRD analyzer. (Reproduced from D. Rapson and P. Stor er, Improving control and quality, *World Cement*, March, 67–71, 2006. With permission of World Cement, UK.)

large number of reference materials. A catalogue titled "Geochemical Reference Material Compositions" by Potts, Tindle, and Webb lists the reference material, its composition, and the supplier to facilitate the appropriate selection of materials. All SRMs are provided with certified sheets of analytical values. It is, however, important to bear in mind that all SRMs should also be checked and verified, whenever necessary, by using some alternative analytical procedures.

SRMs are particularly needed in the present contexts for the following objectives:

- performance assessment of the analytical instruments
- evaluation of analytical accuracy
- developing calibration curves for analysis
- estimating peak shifts in XRD
- determining the fineness of cement by Blaine's air permeability method

Performance assessment of XRF and other analytical instruments for cement and raw materials is an essential step in QC. The performance requirements, for example, for analyzing hydraulic cements are described in the standard method for wet chemical analysis ASTM C 114. An illustration of the assessment of an XRF spectrometer with reference to the above standard follows (11).

The NIST Portland cement SRMs were run on an instrument that was under performance assessment for two consecutive days. A shorter analysis time (500 s) was selected as it would have high impact on precision and accuracy. The SRM sample weight was 4.0 g. The accuracy and precision of measurements for the major oxides are shown in

Table 8.6 Accuracy and precision obtained in trial performance of an XRF spectrometer

Oxide	Accuracy (%)	Precision (%)
SiO_2	0.01 (0.20)	0.07 (0.16)
Al_2O_3	0.06 (0.20)	0.05 (0.20)
Fe_2O_3	0.05 (0.10)	0.02 (0.10)
CaO	0.01 (0.30)	0.07 (0.20)
MgO	0.10 (0.20)	0.03 (0.16)
SO_3	0.06 (0.10)	0.06 (0.10)
K_2O	0.009 (0.05)	0.008 (0.03)
Cl	0.002	0.0002 (0.0003)

Table 8.6, which also shows the corresponding permissible values in parenthesis. The maximum difference between the measured mean values and the certified values is the accuracy and the maximum difference of the duplicate results is the precision. The table, thus, displays the acceptable performance of the spectrometer.

It is relevant to note that in all cement manufacturing units it is desirable to have secondary reference materials in the central laboratory for preparing the calibration curves for limestone, clinker, and cement. The range of variations of these materials is not the same in each plant; hence, it is better to prepare secondary reference materials suitable to the specific locations rather than depending on SRMs. The secondary reference materials are prepared from the local mine and plant substances covering the ranges likely to be encountered in analysis. The individual samples of materials are analyzed by two or more analysts by using three different analytical methods, one of which is the standard wet method of analysis. The mean values, if they are statistically acceptable, are used for calibration purposes.

It is important to note that a standardless analysis program cannot be totally discarded in plant laboratories, particularly when there is proliferation in the use of industrial wastes. Hence, the analytical facilities should be capable of carrying out such analytical programs. There is an increasing preference for ICP for this purpose.

8.11 Flue gas analysis

Exhaust gas generated through combustion processes is called flue gas or stack gas. Its composition depends on the type of fuel and the combustion conditions, including air volume ratio. Many flue gas components are polluting in nature and need to be eliminated or minimized by special cleaning procedures before the gas is released to the atmosphere.

The exhaust gas as generated is called raw gas; after removal or minimization of pollutants it is called clean gas. The main flue gas components that are relevant for cement plants are:

Nitrogen (N_2): the main constituent of air (79% vol). This enters into the combustion system with the combustion air and is again released into the atmosphere as a carrier of wasted heat. Minor quantities take part in the formation of nitrogen oxides along with the nitrogen released from the fuel.

Carbon dioxide (CO_2): produced in all combustion processes and a contributor to the green house effect through its ability to filter heat radiation.

Carbon monoxide (CO): formed essentially due to incomplete combustion of fossil fuels. In enclosed spaces it is dangerous, as at concentrations of only 700 ppm in breathed air it will be fatal within a few hours. The threshold value in any work place is only 50 ppm.

Water vapor (H_2O): hydrogen contained in fuel reacts with oxygen and forms H_2O. At higher temperatures it may stay in the flue gas; at lower temperatures it may condensate.

Oxygen (O_2): the portion of oxygen that has not been consumed in the combustion process remains in the flue gas and provides a measure of combustion efficiency.

Nitrogen oxides (NO and NO_2 together as NOx): in combustion processes nitrogen from the fuel forms "fuel–NO" and at high temperatures forms the combustion air "thermal–NO." This NO will react with oxygen in the stack and form harmful NO_2. Technologies such as selective catalytic reaction (SCR), staged air supply, etc, to reduce the formation of NOx have been discussed in another Chapter 9.

Sulfur dioxide (SO_2): formed by the oxidation of sulfur primarily present in fuel. The threshold value in a work place is 5 ppm. It may lead to formation of sulfurous (H_2SO_3) and sulfuric (H_2SO_4) acids, both of which have damaging properties. Scrubbing technologies are used to clean flue gases from sulfur oxides. In cement plants SO_2 is transferred to the product instead of being released to the atmosphere.

Solids (dust, soot): solid pollutants in flue gases originate from the incombustible components of solid or liquid fuels.

In flue gas analysis, "mass" concentration and "part" concentration are common and used in parallel. The mass unit is in mg or µg, while the part concentration is expressed in ppm. Because of the variation of gas volume with temperature and pressure changes, it is necessary to indicate the temperature and pressure values existing at the time of measurement. Alternatively, the measured concentration values may be converted into corresponding values at the standard normal conditions.

A gas has its standard normal volume at a pressure of 1013 ml air (hpa) and a temperature of 273 K (corresponding to 0°C). The conversion formula is as follows:

$$C_2 = C_1 \times \frac{T_1 \times p_2}{T_2 \times p_1} \tag{8.3}$$

where C_1 is the measured concentration, C_2 is the converted concentration, T_1 is the gas temperature during measurement (273 + actual in °C), T_2 is the normal standard temperature (273 K), p_1 is the gas pressure during measurement in hpa, and p_2 is the normal standard pressure (1013 hpa).

Extractive sampling and measurement techniques

Gas analysis is mostly carried out using the extractive sampling method. A representative portion of the gas is extracted from the process stream with the help of a sampling probe and is then conditioned before it is fed into the analyzer (see Figure 8.13).

Sampling probes are available in different designs and materials for temperatures up to 1200°C and above, if necessary. A coarse filter is fixed at the head of the probe and certain probes also hold pressure and temperature sensors in the head region. The sample gas cooler is located between the sample probe and the analyzer. Flue and process gases always carry a certain amount of moisture that may exist as water vapor at high temperatures (above the dew point) or as liquid at low temperatures (below the dew point). Sample gas coolers are used to cool the sample gas and thus to dry it at a constant level, e.g., at 4°C. The cooling procedure results in the generation of condensate, which is discharged by a pump. The sample lines (made of metal or plastic) are heated, if necessary, to keep the gas above the dew point. A big advantage of extractive sampling is that more than one analyzer may be operated with one probe; similarly, one analyzer may be connected to several sampling points, with a sample line switching unit.

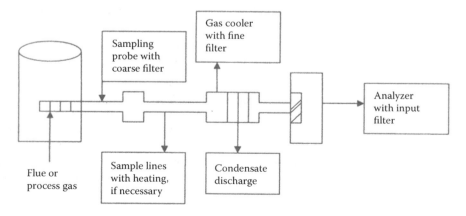

FIGURE 8.13 Extractive gas sampling system.

A variety of measurement techniques are utilized to determine the concentration of different gases in a gas mixture. Some illustrations are given in Table 8.7

From the table it is observed that for the measurement of oxygen one of the three techniques is used: paramagnetic, potentiometric, or solid-state electrolytic. Oxygen is the only gas with a considerably high value of paramagnetism, due to which oxygen molecules are attracted by a magnetic field. This effect is used to determine the oxygen concentration in gases. The design could be different from analyzer to analyzer.

Another common oxygen measurement method uses a *solid-state electrolyte* sensor made of zirconia ceramics coated with layers (electrodes) of porous platinum as catalysts on two surfaces. At temperatures above 500°C oxygen molecules ionize in contact with the platinum electrodes with greater mobility and move through the molecular holes of the zirconia cell. As the two electrodes are in contact with gases of different oxygen concentration (sample gas and reference air), oxygen ions will move through the zirconia cell from one electrode to the other and generate a differential voltage across the electrodes. From this differential, by comparison with the reference gas, the oxygen content of the sample gas can be determined. Since this kind of zirconia probe is capable of measuring oxygen under very harsh conditions, it is very often used in direct process flow measurements. It may be relevant to mention that the oxygen concentration measured with an online zirconia probe is generally lower than measurements obtained with the extractive analyzers. The difference is caused by the differences of moisture content and temperature between the online gas sample and the extracted gas sample.

Oxygen as well as CO, SO_2, and NOx can be determined by *ion-specific potentiometry*. The sensors are filled with an electrolytic solution and multiple ion-specific electrodes are placed in the solution with a gas permeable membrane. Detailed design and operating principles differ, depending on the gas component.

Infrared radiation is absorbed by gases such as CO, CO_2, SO_2, or NO at a wavelength that is specific for the gas. As IR radiation passes through a measuring cell filled with the gas to be analyzed, an increase

Table 8.7 Measurement techniques for different gases

Techniques	O_2	CO	CO_2	NOx	SO_2	H_2S	Hydrocarbon
Non-dispersive IR		✓	✓	✓	✓		✓
Non-dispersive UV				✓	✓		
Paramagnetism	✓						
Chemiluminiscence				✓			
Flame ionization							✓
Potentiometry	✓	✓		✓	✓	✓	
Solid-state electrolyte	✓						

in the gas concentration causes a corresponding increase in IR absorption and a decrease in the radiation intensity received by the IR detector. In this context two different approaches for IR radiation are in use: dispersive and non-dispersive. In the dispersive mode, the IR radiation is passed through a prism or grating before entering the measuring cell. Only two wavelengths of the entire spectrum are used—one that the component in question absorbs and another which is not absorbed. The ratio of absorption at these two wavelengths is a measure of the gas concentration. In the non-dispersive mode, a broad band of radiation from a lamp enters the measuring cell. When a component that absorbs IR radiation is present in the gas sample, the radiation intensity at the specific region of the spectrum gets reduced, which is proportional to the concentration of the gas component. Generally, a solid-state IR detector is used for measurement.

Certain gas components such as SO_2 or NO absorb *ultraviolet radiation* as well. For these gas components, the UV method competes with the IR method.

Chemoluminescence detectors are used for low concentrations of NO and NO_2. Such a detector is made up of an ozone generator, an enrichment and reaction chamber, and a photomultiplier component. NO reacts with ozone resulting in the emission of characteristic light radiation or chemoluminescence. The intensity of the emitted radiation is proportional to the NO concentration.

In the *flame ionization* method, the sample gas containing hydrocarbons is introduced into a flame (usually hydrogen/air) in which the hydrocarbons are ionized. The positively charged ions are collected by a negatively charged collecting electrode, generating a current proportional to the hydrocarbon concentration.

In addition to the above, there are other types of sensors used in gas analysis that are based on thermal conductivity, optical properties, etc.

Special considerations for gas sampling systems in cement plants

As already mentioned, special considerations include the sampling probe and accessories to condition the sample gas so that its physical properties are acceptable to the analyzer. The system is designed and arranged such that a correct representative sample of the gas to be determined is extracted and transported to the analyzer without composition change and at the correct temperature and flow rate. The sample gas should be free of dust particles and excessive water vapor (especially so in the case of infrared analyzers, because of cross-sensitivity problems).

For trouble-free, continuous, and reliable operation of the analyzer, the following conditions should also be ensured:

a. absence of leaks of gas sampling lines;

b. satisfactory cleaning of the sample gas lines; and

c. constant removal of condensates from the gas inlet lines.

The analyzer should be located as close to the sampling point as possible to reduce delay in indication. The analyzer should be protected from radiant heat and direct heat from the sun, so that the operating temperature of the analyzer does not exceed the permissible value. Sample lines should be of correct cross-section: large-diameter pipes and lines will increase the "dead volume" upstream and result in longer dead time of the indication, whereas smaller cross-sections may have the danger of clogging.

Specially constructed probes are used for extracting the sample gas from closed vessels or wide pipes/ducts. The probe must be installed in such a way that the sampling point is away from the wall of the duct/pipe/vessel, where local effects may make the gas sample unrepresentative.

A heated ceramic external filter is used in the probe. Heating is done by an electric heating sleeve. The object of the heated ceramic filter is to ensure that filtering is carried out before condensation occurs, since condensation at or before filtering will cause clogging of the sampling system.

Water-cooled, twin probes are used for extracting samples of gas at high temperatures, such as at the inlet of a rotary kiln. Here, one probe is used for extracting sample gas while, at the same time, another probe is cleaned by air- purging. This can be done at fixed intervals, either manually or by means of an automatic scavenging system. This will ensure an uninterrupted supply of sample gas to the analyzer.

Attention should be paid to the correct installation of the gas sampling lines with proper gradients. Condensate traps or water seal-type catch pots should be provided at low-points. Sharp bends, restrictions, and leakages should be avoided. An electric gas cooler is installed after the probe to cool the sample gas, to eliminate excess water vapor, and to maintain the dew point at a constant value of $+ 2°C$. Otherwise, the water vapor would condense along the lines, causing several problems such as blockage of the sampling lines and unstable flow of gas, damage to the analyzer cells, intermittent passage of water droplets through the analyzer cell, and absorption of soluble gases and corrosion of the components of the sampling system. A pump is used to suck the sample gas into the analyzer, in case the pressure in the sampling line is low (less than 20 mbar). The output of the analyzer should be freely let into atmosphere so that the analyzer does not operate at an overpressure of more than 20 mm WG, otherwise the readings of the analyzer may be affected.

Correct and regular maintenance is very essential to ensure the trouble-free and consistent operation of gas analyzers. Online analyzers provide continuous monitoring of the process and enable more meaningful and faster corrective actions in contrast to manual laboratory analysis. The latter method has the disadvantages of discontinuity in the analysis and delay before the results are available for any corrective action, besides involving more personnel and consequent human errors. But online automatic analyzers seem to have the serious problem of requiring constant and careful attention. This is partly due to the very hostile ambient conditions in which analyzers have to function. The more important reason is, perhaps, a lack of understanding as well as a lack of information and spares. It is common experience that for obvious reasons such as dust, contamination, and physical damage, the sampling system, particularly the sampling probe, needs maximum attention and care, especially when sampling is taken at the inlet of a rotary kiln.

Gas analysis in the cement production process

The main objectives of gas analysis in cement production are:

a. to obtain process information in order to control the combustion and calcination processes in the kiln and preheaters, respectively;

b. to actuate interlocks and warning systems for eliminating or reducing explosion hazards; and

c. to control pollution.

In order to achieve these objectives, flue gas from coal-fired kilns is analyzed to determine the presence and quantity (% volume) of oxygen (O_2), carbon monoxide (CO), nitrogen oxides (NO_x), and sulphur oxides (SO_x). The justification for analyzing the above components is that O_2 in the flue gas is the best expression for the combustion in the kiln and possible precalciner firing, and CO indicates poor combustion and direct explosion risks.

Thus, oxygen and CO are the main components of interest in flue gas analysis for the purposes of process information and interlock. SO_x and NO_x are primarily determined for pollution control. These components should be measured at a location as close to the emission point, into the free air, as possible. Sometimes, NO_x is also being used to control burning zone temperature.

The combustion process is controlled effectively with the help of CO and O_2 contents in the flue gas because these constituents:

a. allow unambiguous correlation with the combustion conditions; and

b. are essentially independent of fluctuations in the elemental composition of the fuel, whereas the CO_2 level in the gas varies considerably with fuel composition.

If the flue gas analysis indicates high concentrations of CO and low oxygen level, one may conclude that all the heat in the fuel has not been fully liberated. Sometimes the flue gas may contain hydrogen (H_2) and hydrocarbons (C_mH_x). Hydrogen sulfide (H_2S) may also occur if there is sulfur in the fuel. On the other hand, if the fuel:air ratio is too low (i.e., very high "excess air") energy will be wasted, though all the fuel might have been burnt completely. This is because heat will be spent up in heating the excess air. Further, too large an "excess air" will encourage the formation of SO_2 and NO_2, leading to corrosion and pollution.

Interlock and warning systems based on gas analysis

In certain proportions, mixture of CO and O_2 is explosive. Typically, an amount of 8–12% CO is dangerous when mixed with more than 6% O_2 and, if ignited, an explosion can occur. Practice shows that the CO content increases very rapidly when combustion is incomplete, for example, with insufficient combustion air. Analysis of CO in the flue gas can, therefore, be used to warn and switch off certain parts of the process. An electrostatic precipitator (ESP) in the cement making process is a potential igniter (due to corona discharges and/or flash-over between the electrodes) and has to be switched off if CO concentration in flue gas exceeds a certain limit. If the hot gases from the kiln are used for drying purposes in the raw mill circuit, precautions should also be taken to isolate the circuit.

Measurements of CO are generally taken after the preheater or just before the electrostatic precipitator. Considering the delay in the cyclones and total analysis time, the alarm warning is set at 0.4% CO and the ESP tripping at 0.8% CO. These values give the operator in the control room, or on the burner's platform, reasonable time for corrective actions as well as avoiding unnecessary switching off of the electrostatic precipitator.

Gas sampling points and probes in a cement plant

Sampling points are located according to the main purpose of gas analysis. Thus, for example, for combustion control of the rotary kiln, the kiln inlet (where "raw meal" is fed) is the most suitable point for gas sampling. This will minimize the response time of analysis and minimize disturbances due to infiltrated air. The purpose of sampling and the types of probes used are briefly discussed below:

  a. Kiln inlet: usually, O_2 concentration is measured for optimizing the combustion process in the kiln. As mentioned above, this is the most suitable sampling point for avoiding penetration of undesired air due to leakages in the kiln inlet seal. But this measuring point is very difficult due to high gas temperature, high concentration of dust, danger of condensate formation, and mechanical damages by material charge. Again, the difficulties encountered with a wet kiln with flue gas at a temperate of roughly 150 to 200°C differ from those of a dry kiln where the exhaust gas at the sampling point may have a temperature of about 1100°C. The sampling probe is of the water-cooled type, often called a "wet probe." The water-cooled twin probe, with automatic scavenging or air-purging arrangement, appears to be a suitable type, though not totally free from difficulties. The metal shape and orientation of the sampling probe are chosen to minimize damages due to material-charge, high temperature, and contamination. Although O_2 is usually measured at the kiln inlet, the probe also facilitates the measurement of CO and NO_x.

  b. Precalciner outlet: oxygen content is measured to monitor the calcining/sintering process. The probe is of the "dry" type, that is, air is used for cooling the probe. A measuring/cleaning cycle is usually established, so that compressed air is admitted into the probe at regular intervals to remove the dust. "In-situ" (zirconium oxide) probes can also be used, with arrangement to continuously supply the "reference air" to the probe. Also, arrangement has to be made to apply calibration test gases into the probe for periodical checking of calibration.

  c. After the preheater tower: oxygen and carbon monoxide are measured with "dry"-type probes since the temperatures encountered at this location are not high. The main objective of the gas analysis is to ensure that explosive gas mixture of CO and O_2 are not formed.

d. SP inlet: CO is monitored with standard flue gas sampling probes. Since the ESP is a potential igniter, careful and close monitoring of CO at the inlet is very important. As mentioned earlier, "alarm" and "TRIP" circuits and other interlock systems are actuated by the appropriate CO concentration values at the ESP inlet. Safety being the prime consideration, the analyzer and the associated alarm/interlock circuits need very careful attention and maintenance to ensure the required reliability of the entire system.

e. In the stack (after the exhaust fan): for pollution control, the analysis of NO_x and SO_x is important. Standard sampling probes can be used because of the relatively low temperature. Continuous "online" analyzers are not generally used for measurement, since periodical analysis with portable instruments may be sufficient for pollution control purposes.

8.12 Process measurements

Needless to say, parametric measurements are essential in process control and the quality of process control depends on the accuracy, reproducibility, and reliability of measurement methods. Generally speaking, the cement manufacturing process requires measurement of:

- temperature
- flow rate
- pressure
- gas analysis for O_2, CO, NO_x, and SO_x
- power
- particle size distribution
- chemical constituents

The above measurements are also necessary for carrying out heat and mass balance of all unit operations. Since we have already dealt with the last four parameters, in this section we shall concentrate on the measurement of the first three.

Gas flow and velocity measurements

The measurement of both air/gas velocity and volumetric airflow can be accomplished using a pilot tube. The mechanism works by converting the kinetic energy of the flow into potential energy. It is used to concurrently determine the difference between static, dynamic, and total pressure of a fluid.

The device was invented by the French engineer Henri Pitot in 1732 and consists of a small cylinder positioned in the pathway of the fluid. One side of the cylinder is open in order to allow the fluid to enter. Once the fluid enters the tube, it cannot flow further because the cylinder does not have an outlet. Inside the tube, a diaphragm enables separate measurement of the static pressure and stagnation pressure. The static pressure is the pressure when the fluid enters the tube, while the stagnation pressure, also known as the total pressure, is the pressure when the fluid comes to rest. The Pitot tube is a double-walled, generally nickel-plated, metal tube with the bottom end bent at a right angle (Figure 8.14). The measurement system also requires a differential pressure transducer and a computer setup that includes the necessary hardware and software to convert the raw transducer signals into the proper engineering units. The incorporation of sensors to measure the air temperatures, barometric pressure, and relative humidity can further increase the accuracy of the velocity and flow measurements. The Pitot tube measures air or gas velocity directly by means of a pressure transducer, which generates an electrical signal that is proportional to the difference between the total pressure and still air (static pressure). The volumetric flow is then calculated by measuring the average velocity of an air stream passing through a passage of a known diameter. When measuring volumetric flow, the "passage of a known diameter" must be designed to reduce air turbulence as the air mass flows over the Pitot tube. Also, the placement of the Pitot tube in the passage will influence how accurately the measured flow tracks the actual flow through the passage. Calibrating the measurement system in a wind tunnel can further improve the accuracy of the velocity and the flow measurements.

The difference between the total pressure and the static pressure is known as the "dynamic pressure." The basic relation between the total pressure and static pressure is given by:

$$P_t = P_s + \frac{\rho V^2}{2} \tag{8.7}$$

$$\text{Or } V = \sqrt{\frac{2(P_t - P_s)}{\rho}} \tag{8.8}$$

where V is the gas or fluid velocity, P_t is the total pressure, P_s is the static pressure, and ρ is the gas density.

In order to compute the velocity, it is necessary to know the density of the gases, which depends on the composition and temperature. The kiln exhaust gas generally contains O_2, CO_2, N_2, and H_2O in certain proportions. Based on their percentages, the average density is calculated. Hence, the moisture determination becomes important.

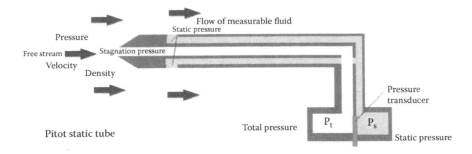

FIGURE 8.14 Basic construction of a Pitot tube.

Velocity measurement by anemometers

An anemometer is a gadget that can either be mechanical or electronic in construction. The mechanical type is made up of a wheel connected to a revolution counter and a chronometer. This is generally used at the fan inlet at ambient temperature in the velocity range of 0–25 m/s. It may get damaged if used at higher velocities.

Electronic anemometers are made up of four-bladed wheels that rotate at a speed proportional to the air velocity. These rotations are registered electronically and the impulses are connected to an analog linear signal proportional to gas velocity. This type of anemometer is sensitive to dust and cannot be used when the gas is dust-laden. However, it can be used at up to 120 m/s and at temperatures up to 300°C.

Since in a cement plant the velocity measurements by an anemometer are often difficult, the most practical approach is to use a pilot tube that can be inserted through a small hole in a duct and connected to a U-tube water gauge or any other differential pressure gauge. The gas flow rate (m³/s) is obtained from the duct area in square meters multiplied by velocity in m/s.

The flow rate can be converted to normal temperature and pressure as follows:

$$Q_{NTP} = \frac{Q \times 273}{(273 + t)} \times \frac{P_a t_m + P_s}{10,330} \tag{8.9}$$

Temperature measurements

Temperature measurements are carried out with the help of resistance thermometers, thermocouples, and pyrometers.

The resistance thermometers are based on the principle of increase in electrical resistance with temperature. The resistance variation is measured with a wheatstone bridge. The range of temperature measurement depends on the resistance of the material used. Generally, it is limited to about 600°C.

Thermocouples are composed of the wires of different compositions. If the soldered ends of the two wires are kept at different temperatures (Figure 8.15), electrical current flows through the wires, forming a continuous circuit.

The choice of wires used in the thermocouple depends on the temperature to be measured, the atmosphere of measurement, and the linearity of the electromotive force (emf). Generally, chromel-alumel thermocouples are used up to 1100°C, while for higher temperatures (1450°C), Pt–Pt/Rh thermocouples are used. Fe-Constantan thermocouples are used for low-temperature measurements. The performance of thermocouples is adversely affected by the presence of high CO in the atmosphere. A typical use of thermocouple assembly is shown in Figure 8.16.

Optical pyrometers are instruments through which the brightness of a hot object is compared visually with that of a source with a known standard brightness–temperature relationship. Infrared pyrometers work on the principle that each object emits infrared radiation corresponding to a certain temperature. These pyrometers measure the rate of energy

ADVANCES IN PLANT-BASED QUALITY CONTROL PRACTICE 283

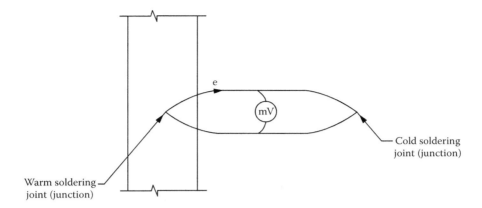

FIGURE 8.15 Schematic diagram of a thermocouple.

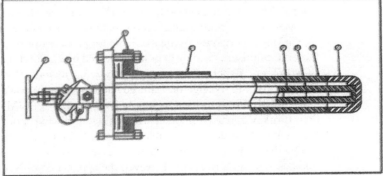

FIGURE 8.16 Sketch of a thermocouple assembly and its installation in the plant for monitoring reduction zone temperature.

emission per unit area over a relatively broad range of wave lengths. This obeys the Stefan–Boltzman equation:

$$W = E\sigma T^4 \tag{8.10}$$

where W is the energy, E is the emissivity ranging from 0 to 1, T is the degree Kelvin, $\sigma = 4.88 \times 10^{-8}$.

For high-temperature measurements a two-color pyrometer is often preferred, but this can be sensitive to dust in the gas flow.

Another temperature measurement method is the use of acoustic pyrometers. It operates on the principle that the speed of sound in a gas is dependent upon the temperature of the gas. By sending an acoustic signal from a transmitter through a volume of gas at a known distance from a sound receiver, and measuring the time taken for the sound to travel from the source to the receiver, the average temperature of the gas present along the straight-line path between the transmitter and receiver can be determined. The acoustic pyrometer in the kiln hood or cooler can optimize cooler fans by returning maximum temperature as secondary or as tertiary air to the precalciner.

8.13 Total process control system

For the last 25 years the structure of control systems for cement plants has been based on the distributed control system (DCS), in which programmable logic controllers (PLCs) are dispersed throughout the process system, each to control a certain section of the process. The entire system may be networked for communication and monitoring. The fully developed DCS comprises SCADA (supervisory control and data acquisition) software plus HMIs (human machine interfaces), which are usually PCs. The process conditions for each production section are monitored and the resulting signals transmitted by wire or optical cable to each RTU (remote terminal unit) or PLC via I/O (input/output) devices. The control signal generated from the RTU or PLC then returns through I/O to each executive device. The RTU or PLC performs PID control loop functions, data compiling and formatting, and manages alarm logic and interlocking control. The communication between HMI, master station, and controllers uses an open communication protocol such as ethernet or TCP/IP.

Various expert systems are available for process optimization. ABB's Expert Optimiser is widely popular, followed by FLSmidth Automation's ECS/Process Expert. The emerging systems include Pavillion8 MPC and Powitec's Pit Navigator. Lafarge had developed its own captive system, LUCIE. Polysius used to supply its Polexpert KCE/MCE as part of their turnkey contacts. In other words, there is a wide choice of systems, from which the selection suitable for a given plant can be made.

On the technology front, there is continued dependence on fuzzy/rule-based control. Recently, however, the application of Neural Net technology and Model–based Predictive techniques appears to be increasing. Within the model-based Predictive Control there are many different approaches adopted by the suppliers. The new technologies are being attempted more in mill control systems than in kiln optimization. The key issue in process control is still connected to the introduction of novel soft sensors; Powitec's "Predictor" and FLSA's new RGB camera software are examples of such soft sensors that are likely to play an increasing role in the application of technology. Without a good-quality soft sensor for product quality, real advancements in process control cannot be achieved.

8.14 Summary

Quality control (QC) in a modern large-capacity cement plant is a highly critical function. There is an organizational set-up in each manufacturing unit devoted to this function that covers a wide range of activities starting from the mine to the dispatch of cement. This activity is carried out with the help of a well-equipped central laboratory and dispersed online measuring and control facilities. Expertise in chemistry and chemical technology is essential in this functional area.

Representative sampling of different materials is the foundation of QC function. Hence, the basic principles of sampling, involving concepts of "lot" and "sampling units," minimum mass of gross samples, etc., are the fundamentals that one should understand in executing the QC function. Thereafter, one should take stock of the sampling stations and sampling practices in a cement plant, which covers the entire process. Each unit operation in a cement plant has its own features and specialty in quality control. The initial stages of QC function are quarry optimization and achieving the targeted run-of-mine quality, and the preblending operation as carried out in blending beds and bulk belt analyzers, leading to proper raw mix proportioning and its homogenization to yield the right quality of kiln feed. The control of the kiln operation is quite critical in order to convert the kiln feed into a clinker of the targeted quality. Then comes the criticality of the cement milling process.

There is extensive use of X-ray fluorescence spectrometry (XRF) and X-ray diffractometry (XRD) in the cement manufacturing units. Whilst XRF is the most widely used facility for finding out the elemental composition of cement, clinker and raw materials, XRD is the principal tool to characterize those materials in terms of their compound or phase composition. Both the X-ray techniques are in effect supplementary to each other and are employed in offline and online modes. The offline facilities are housed in the central laboratory, while the online equipment and auxiliaries are dispersed at different sections of the process but are interconnected to the central control rooms through their operating consoles. The quantification of clinker and cement phases and the adoption of online control techniques are progressively gaining importance in the plants for real-time analysis, monitoring, and process corrections. There is increasing preference in online XRF facilities and bulk belt analyzers working on the principles of neutron activation or even infrared radiation. The analytical online facilities are linked up with the process control arrangements for real-time corrections, whenever there are deviations of chemical and phase compositions from the respective set points.

The QC function depends on proper selection and use of standard reference materials and secondary reference materials for reliability of analytical results. This aspect should be adequately understood. Another area in QC function is flue gas analysis, as the measurement of flue gas components is as critical as the characterization of materials in the process control measures. A cement technologist has to specifically understand the role of oxygen and carbon monoxide in kiln operation. Understanding of the cement process remains incomplete without knowing the significance of process parameters like gas flow, pressure, and temperature. These measurements are essential for heat and mass balance of the unit operations in general and the kiln system in particular. Finally, overall process control is an integrated process that uses various inputs and expert systems, a brief summary of which has been provided in this chapter.

References

1. F. F. PITARD, *Pierre Gy's Sampling Theory and Sampling Practice*, CRC Press, Florida (1993).
2. P. C. SOGANI AND K. KUMAR, Technological advances, in *Modernization and Technology Upgradation in Cement Plants* (Eds S. N. Ghosh and Kamal Kumar), Akademia Books International, New Delhi (1999).
3. H. BERGEMANN, Means of quality control and process optimization in cement manufacturing, *Ciments, Betons, Platres, Chaux*, No. 771, February 1988.
4. M. TSCHUDIN, *In-line control of the blending bed composition using PGNAA in Ramos Avizpe (Mexico)*, Proceedings of VDZ Congress '93 on Process Technology for Cement Manufacture, Subject 2, Bauverlag GmbH, Weisbaden and Berlin (1995).
5. A. K. CHATTERJEE, X-ray diffractometry, in *Handbook of Analytical Techniques in Concrete Science and Technology* (Eds V. S. Ramachandran and James J. Beadoin), Noyes Publications, New Jersey, USA (2001).
6. R. A. YOUNG, *The Rietveld Method*, *The IUCR Monograph on Crystallography*, Oxford University Press, Oxford (1993).
7. R. YELLEPEDDI AND D. BONVIN, Integrated cement analysis, *World Cement,* June, 1999, 73–78.
8. J. RAATIKAINEN, The new quality control tool, *International Cement Review*, June, 1999, 79–83.
9. P. STORER AND C. MANIAS, On-stream XRD analysis, *World Cement*, February, 2001, 69–78.
10. D. RAPSON AND P. STORER, Improving control and quality, *World Cement*, March, 2006, 67–71.
11. D. WISEMANN, Precise and accurate X-ray analysis, *World Cement*, June, 2007, 43–44.

CHAPTER NINE

Environmental mitigation and pollution control technologies

9.1 Preamble

Environment and development, once thought to be separate issues, are now considered as complementary to each other so as to ensure that human progress does not exhaust the resources of future generations. Safeguarding the environment requires sustainable development that guarantees a perpetuation of resources and skills that will meet tomorrow's needs. Thus, sustainable development would demand more than pollution control and more than mere compliance with environmental regulations. It would necessitate pragmatic technological initiatives involving preservation of the biosphere, selection of cleaner technologies, and gainful recycling of wastes.

The cement industry is responsible for the discharge of a large quantum of pollutants into the atmosphere. The contribution is not limited to the real-time industrial operation of a cement plant, but also encompasses a whole range of activities covering the backward and forward linkages. The upstream stage is essentially the stage of obtaining raw materials and energy inputs, which gradually leads to depletion of natural resources—ecological disturbances like soil erosion and deforestation, noise, and emissions of gases like carbon dioxide. At the process stage, the problems are due to emissions of particulate matters and gases affecting the ambient air quality, including greenhouse gases. The downstream stage would essentially relate to the wastes that are generated during the process and their disposal, which in the absence of a viable waste management system would result in the loss of valuable land on one hand and increasing accumulation of wastes and pollution of the environment on the other. The concept of sustainable industrial development would seek a solution to all the above problems through one or more of the following means:

a. Prevention. This approach requires tackling of the environmental problems at the generation stage itself. Instead of following end-of-the-pipe pollution treatment measures, which are often quite expensive to implement, the emphasis is laid on optimum resource utilization during the production process itself so as to reduce the generation of pollutants at source.

b. Containment. Whatever has been emitted after making efforts for pollution prevention at source has to be contained through an end-of-the-pipe approach. Pollution control devices like fabric filters and electrostatic precipitators used in the cement industry are examples of such containment means.

c. Utilization and disposal. This comprises (i) utilization of wastes, the generation of which cannot be prevented, to make value-added products; and (ii) environmentally sound disposal of wastes that can neither be prevented nor utilized. Utilization of secondary fuels and raw materials in the cement industry are examples of this strategy.

This chapter deals with the status and development of technologies towards the above approaches of environmental mitigation and pollution control.

9.2 Pollutants emitted into the atmosphere during manufacture of cement

The most important polluting substances within the meaning of the directive of most of the countries in the world are:

- dust
- sulfur dioxide and other sulfur compounds
- oxides of nitrogen and other nitrogen compounds
- organic compounds, in particular hydrocarbons (except methane)
- heavy metals and their compounds
- chlorine and its compounds
- fluorine and its compounds

The above list excludes carbon dioxide emission, as it is strictly not a pollutant but a greenhouse gas. Carbon dioxide discharge from the cement industry is quite significant. For example, in producing a ton of Portland cement clinker, 0.70 to 1.0 tons of CO_2 gas is discharged. More than 50% of the emission comes from the calcination of limestone and the rest from the fuel burning. Taking into account the global production of different varieties of Portland cement, it is estimated that about 3.0 billion tons of CO_2 emission per year is attributable to the cement industry, which constitutes a substantial proportion of gross emission of all industries. The topic of CO_2 emission and climate change has been dealt with in detail in Chapter 10 of this book and, hence, the subject is not discussed further here.

Principal pollutants

The pollutants of principal concern from the manufacture of cement are:

* Dust. The generation of dust from cement manufacturing plants is recognized as a major problem and has received considerable attention, both from various regulatory authorities and from the industry itself. Dust emissions are classified into two categories, depending on their sources:

    a. suspended solid particles (SPM) in exhaust gases, originating directly from the process;
    b. fugitive dust (FD) resulting from entrainment and re-entrainment of dust by wind, air venting, or vehicle movement.

This classification of dust categories is important, since the impact of the associated emission is likely to vary significantly depending on the dust source. Fugitive and dusty air emissions may be expected to settle mainly within the confines of the plant and its immediate vicinity, whilst low-density buoyant process emissions generated in cement kilns, once dispersed from tall stacks, could conceivably contribute to long-range and trans-boundary air pollution. It should be further noted that most of the dusts are products of the various manufacturing stages that can be returned to the process; they are generally not wastes or byproducts requiring disposal. It is therefore economically desirable to design and operate plants with high dust capture efficiencies.

* Sulfur dioxide (SO_2). Sulfur may be a minor constituent of the raw materials utilized to manufacture clinker, normally being present in the form of metal sulfides and sulfates. The amount of sulfur present will vary widely according to the nature and location of the deposits used. During the preheating and calcining operation, sulfides and most sulfates are decomposed to yield sulfur dioxide (SO_2). Further, compounds of sulfur are common constituents of most fuels used in cement kilns, with sulfur levels as shown in Table 9.1. On combustion of the fuel, the sulfur compounds

Table 9.1 Typical sulfur contents of various fuels

Fuel	Sulfur content (% weight)
Coal	0.5–2.1
Lignite	0.2–0.8
Used tyres	1.3–2.2
Petroleum coke	2.0–6.0
Other alternative fuels	0.5–16.0
Fuel oil	1.0–5.0

present are oxidized to SO_2 and pass through the kiln with the hot gases, which after preheating or drying of raw materials are routed to the stack for dispersion into the atmosphere. SO_2 emissions can increase when, for example, the sulfide contained in the material to be burnt is incompletely converted into sulfate. Normally, however, only a small fraction of the SO_2 generated within the kiln (whether originating from the raw materials or from the fuel) is released to atmosphere, since it is mainly incorporated into the cement clinker by chemical combination.

* Nitrogen oxides (NOx). The formation of nitrogen oxides associated with the combustion of fuels is a well-established (if not yet fully understood) phenomenon. There are two principal mechanisms for NOx formation:

  a. Fuel NOx, whereby nitrogen-containing compounds chemically bound in the fuel react with oxygen present in air to form various oxides of nitrogen.

  b. Thermal NOx, whereby some of the nitrogen components in the combustion air react with the oxygen present to form various oxides of nitrogen.

The principal parameters influencing the formation of fuel NOx are the quantity of excess air fed to the burners and the nitrogen content of the fuel. The extent of thermal NOx formation is directly influenced by the flame temperature as well. Owing to the high temperature involved during the primary firing step, thermal NOx dominates the NOx reaction mechanisms for NOx generated in rotary kilns. When a secondary firing step is utilized, fuel NOx reaction mechanisms may dominate, owing to the lower flame temperature involved.

Pollutants generally of lesser concern

The following additional pollutants may also (depending on specific local circumstances) be associated with the manufacture of cement, although the scale of emissions is usually significantly less than that for the principal emissions listed above.

- Trace elements. In common with virtually all mineral deposits, both the raw materials and the fuels utilized in cement manufacture contain trace elements in various proportions, including possibly alkaline and transition (heavy) metals. The specific elements and the exact amounts present in these mineral deposits may vary widely from the one location to another, as indicated in Table 9.2.

  During the cement production process, at the high temperatures experienced in cement kilns, some trace elements tend to vaporize and move with the flue gas. These vapors then tend to condense (mainly on dust particles) and are principally incorporated in the clinker, although a relatively small portion leaves the kiln system with the exhaust gases. In specific cases where the concentration of trace metals in the raw material is high, an accumulation of these metals can occur in the flue gas system of the kiln. By removing

Table 9.2 Indicative concentrations (mg/kg) of trace elements in raw mix and fuel

	Raw mix	Fuels Coal	Lignite	Fuel oil
As	3–15	9–50	0.3–9	0.01–0.1
Cd	0.04–0.15	0.1–9	0.1–2.4	0.02–0.4
Hg	0.02–0.14	0.1–0.4	0.03–0.3	–
Pb	4–15	11–270	0.8–6	1–34
Tl	0.2–0.8	0.2–4	0.07–0.3	0.02–0.12
Cr	23–29	5–80	0.9–8	2–4
Ni	18–23	20–80	0.6–1.9	5–43
V	32–102	30–50	2–7	2–117
Zn	31–47	16–220	1–70	5–85

some of the dust from the burning process the level of trace elements emitted to atmosphere can be reduced.

- Carbon monoxide (CO) and hydrocarbon emissions, as a result of incomplete fuel combustion, are generally negligible due to the requirement to operate the kiln (for reasons of product quality) in an excess-oxygen condition. Emissions of CO can rise due to incomplete combustion in both the primary and secondary firing stages. To ensure a low CO emission concentration, the fuel and air must be proportioned consistently and the mixing of fuel and combustion air has to be complete. When using coarse-sized fuel in the secondary firing stage, the air:fuel ratio must be increased to a safe margin. Measurements show that the CO concentration can be up to 10 g/Nm^3 under operating conditions. High CO concentrations can also occur at transient start-up or upset conditions of the kiln system, though these are generally of short duration.

- Hydrogen sulfide is not normally present in the gases leaving the kiln and only arises on rare occasions when disturbed operation results in the occurrence of strongly reducing conditions.

- Fluorine compounds in the feed constituents are partly volatilized into gaseous acidic fluorides at the high temperatures present in the kiln. However, such acidic compounds are immediately neutralized (in the form of solid calcium fluoride) by the alkaline constituents in both the feed and clinker. Thus, 88–98% of the fluoride content of feed materials is trapped in the clinker and the remainder deposits on dust particles and is mainly removed via the dust arrestment equipment. Practice has indicated that no gaseous fluorides are emitted from cement plants under normal operating conditions.

- Chlorine compounds behave in a similar manner to those of fluorine. Any acidic chloride formed in the kiln is immediately neutralized (to form harmless sodium and potassium chlorides) and trapped in the clinker and on dust particles.

Basic concepts of pollution monitoring

Fixing the emission limits for any atmospheric pollutant is obviously contingent upon the availability of suitable measurement devices and assessment techniques. The pollution control agencies of all countries recognize this fact with provisions that the measures utilized to monitor compliance of atmospheric emissions have to be approved by the relevant authorities. In addition, all monitors must be regularly calibrated using the recommended national standard. The method adopted should give results consistent with other such methods that may be adopted elsewhere. The technological status is highlighted below:

A. Dust
- Particulate matter in exhaust streams from kilns and clinker coolers should be monitored continuously. There are no instruments that give continuous absolute measurements of particulate concentration. For statutory regulation, standard methods of short-term sampling and analysis (carried out in accordance with the recommended national standard) should be employed. However, for operational control purposes the following monitoring devices are available.
    – Corona power recorders (for indirect monitoring of electrostatic precipitator efficiency);
    – Opacity (photometric) meters (for indirect monitoring of dust burdens).
- Monitors should be zero-checked and calibrated at regular intervals. Since monitor indication is to some extent influenced by the particle properties (size, shape, colour), the calibration of such monitoring systems is no longer accurate if the operating conditions or the processed raw materials or fuels of the monitored plant change appreciably. Several calibration curves corresponding to different operating conditions must therefore be obtained. Provision for manual testing should be installed for use in the event of breakdown or suspected malfunctioning of the continuous monitors.
- For critical emission streams, preset high-level alarms should be fitted to monitors. An analysis of occasions when any emission limit is exceeded, including the duration and amount, is also useful for operational control.

- For emissions entailing smaller exhaust gas volumes, where bag filters are utilized and where continuous particulate monitors are not fitted, pressure drop and flow rate indicators (with appropriate alarms) can be fitted to warn of bag failures.

B. Other pollutants

- Oxides of nitrogen in the kiln exhaust gas can be used to influence combustion control and should be monitored continuously in new plants.
- Oxides of sulfur do not normally require monitoring as these compounds are mainly absorbed in the product and emissions are low. Only where the volatile sulfur content of raw materials or fuels is excessive and might cause significant emissions are dry gas sampling systems necessary.
- At least for new plants, trace element control necessitates initial analysis of the raw materials and fuels so as to ensure that volatile trace elements are not present in amounts likely to result in excessive emissions. For existing plants, where levels of trace elements have been established, it should only be necessary to analyze any new raw materials or fuels.

The status of and progress made in the above aspects of pollution control and monitoring for environmental mitigation in the late 1990s has been published by the present author (1,2). The evolution of pollution control techniques for the cement industry around the same period were also dealt with by some other authors (3,4), as environmental mitigation strategies for the cement industry underwent intense evolution during that period. The present discourse is based on collation and updating of the previous publications.

9.3 Generation and broad characteristics of dust

It is estimated that about 2.0–2.6 tons of raw materials, gypsum, coal, etc. are processed to produce a ton of cement. About 5–10% of these finely pulverized materials remains suspended in gas and air streams and is substantially removed before the gas or air is discharged into atmosphere. The volume of gas or air to be de-dusted varies between 6–15 m^3 per kilogram of cement produced, depending on the design of a plant (Table 9.3).

For any particular application, the most appropriate abatement technique depends on the characteristics of the emission source. The basic information includes gas flow rate, dust burden, dust particle size, humidity, and temperature.

Table 9.3 Dust generation and vent gas volume

Source	Dust generation	Vent gas/Air volume
Crusher	15–50 g/m^3	50 m^3/min-m^2 of feed hood opening
Raw mill—gravity discharge	20–80 g/m^3	4x mill volume per minute
Air-swept raw mill (roller mill)[3]	300–500 g/m^3	1.66–2.33 m^3/kg product
Coal mill—gravity discharge	20–80 g/m^3	4x mill volume per minute
Drying-grinding coal mill	100–120 g/m^3	2.0–2.7 m^3/kg product
Dry kiln	40–60 g/Nm	1.37–2.3 Nm^3/kg clinker
Semi-dry kiln	20–30 g/Nm^3	2.0–3.0 Nm^3/kg clinker
Wet kiln	30–50 g/Nm^3	2.8–3.0 Nm^3/kg clinker
Clinker cooler	10–15 g/m^3	1.2–1.8 Nm^3/kg clinker
Cement mill	200–300 g/m^3	4x mill volume per minute
Packing plant	20–40 g/m^3	35–40 m^3/min per filling spout

It is necessary to characterize the dust loads generated at each stage in the cement manufacturing process in order to ensure that compatible dust control equipment is integrated into the production plant at every emission point in order to contain the emissions of dust into the atmosphere. More specifically, the various dust loads generated should be characterized in order to:

- select dust arresting equipment of the desired capture efficiency in order to adequately control emissions;
- choose the above equipment with such reliability that plants can operate them effectively at all times;
- determine the capacity of the dust collector so as to meet the loads expected under both normal and upset conditions.

The essential data required to characterize the various dust loads generated at any point in the cement manufacturing process include the following:

- volume of exhaust gas stream under normal conditions
- temperature of the exhaust gas stream
- composition of the exhaust gas stream
- particle density
- particle resistivity
- particle size distribution (psd)

The primary sources of generation of fugitive dust are conveyors, transfer points, leakages, open stockpiles, discharge to and from hoppers, and so on. The total quantity of the particulate emission from all fugitive dust sources is quite substantial and according to a rough estimate it varies from 10–30% of the total emission from a cement plant.

It is understood that the dust arising from different processing units varies in composition starting from raw materials and ending in cement via clinker and coal. The compositional range of dust generated from coal-fired kilns is as follows: LOI (23.8–36.3%), SiO_2 (9.9–19.2%), Al_2O_3 (1.8–6.4%), Fe_2O_3 (1.8–3.5%), CaO (38.6–46.8%), MgO (0.7–2.6%), Na_2O (0.04–0.59%), K_2O (0.29–3.41), SO_3 (0.5–6.2%), Sol. Cl^- (0.05–0.39%). The recycling of kiln dust in the process is obviously dependent on the composition in general and more particularly on the alkali, sulfate, and chloride contents. The measurement of trace elements in kiln dust is important on two accounts: one, in relation to the water permeation characteristics in dumping areas, and two, in relation to the fixation of regulated metals in clinker. Generally, Mn, Zn, Cd, and Pb tend to concentrate in dust more easily than other elements studied. Leaching studies have shown that Sr, Zn, and Sb exhibit high leaching tendencies, irrespective of their initial concentrations. Most other trace elements did not respond to leaching within 72 h within the detection limits of the measuring instruments (3–5 ppm, barring As, for which it was 0.1 ppm). The particle size of dust varies widely from less than 1 μm to over 100 μm, as illustrated in Table 9.4.

The biochemical and physiological effects of dust emission were studied in the 1970s and 1908s. The National Institute for Occupational Safety and Health in the USA, based on at least two large studies involving over 2500 workers in 16 factories, came to the conclusion that the standard mortality rate of cement workers was almost consistently below the average of the population. No specific respiratory diseases or detrimental attacks had been attributed to the presence of dust in the plant ambience. However, the biochemical and botanical effects of cement process dust were found to be varied in soil and plants of different species. Upon hydration, cement dust forms a hard crust on the soil surface and renders it relatively impervious, reducing its porosity and water holding capacity. The soil pH is generally increased due to presence of lime, magnesia, and alumina. Some retardation in microbial activity

Table 9.4 Particle size of dust emitted and captured

Size (μm)	Kilns and raw mills	Cement mills
> 90	–	1.4–3.1
90–75	–	1.0–2.3
75–60	–	1.9–3.3
60–45	–	5.4–10.1
45–30	0.3–1.8	16.8–22.0
30–20	0.5–5.0	31.2–39.5
20–10	12.1–41.8	52.0–68.0
10–5	30.4–74.3	68.3–86.5
< 5	25.7–69.9	13.5–36.7

in soils with cement dust deposition was linked to increased levels of organic carbon and decreased levels of nitrogen. It was also observed that the effects of cement dust on leafy plants and crop plants were highly species-specific.

On the whole, the sheer nature of the manufacturing process, the quantity of dust generated, and the consequent impacts on the upkeep of the plant, machinery, and working environment, as well as the varied effects on plantation, had compelled the cement production units to adopt and improve the dust pollution control facilities.

Operational controls and design measure

A cement plant, new or old, is designed to generate as little dust as possible. Irrespective of the process and plant equipment, the fundamental requirement for the objective of low dust generation is its operational stability, attained through proper chemistry, consistency of raw materials and fuel, and the corresponding processing parameters. The plants should operate uninterruptedly from one end to the other as frequent start-ups and shut-downs add to the problem of dust generation. The uninterrupted operation is always backed by predictive and preventive maintenance. In addition, all modern plants are provided with

- covered storage for raw materials
- delivery of materials with minimal dust generation
- storage of clinker in silos
- conveyors designed with wind boards to avoid spillage
- enclosed, vented transfer points with minimum free fall
- provision of aspiration hoods, wherever necessary
- a cleaning mechanism for returning conveyor belts
- enclosed clinker conveyors
- wet suppression of fugitive dust with surfactants
- adequate size for the generation of dust

In practice, it has been possible to estimate the dust loads generated at each stage of the process in order to ensure that compatible dust collection equipment is integrated into the plant at every emission point.

Trends and progress in dust abatement technologies

As far as the exhaust gases from kilns, coolers, mills, and dryers are concerned, the contemporary technologies are limited to electrostatic precipitators (ESP), fabric filters (FF), and gravel bed filters (GBF). The GBFs are capable of cleaning gases with very high temperatures and they also function well at low humidity. In the cement context, their use is limited to clinker coolers. Hence, for other parts of the process the overall technological comparison is narrowed to ESPs and FFs. The use of cyclones in the cement industry is essentially restricted to pre-cleaning operations as integral components of ESPs and FFs. Table 9.5 gives the conditions of use of the three main dust collectors in different stages of the process and Table 9.6 provides the operating features of the equipment as compared to cyclones. In the following sections the design

Table 9.5 Conditions of use of the main dust collecting equipment in different stages of the process

Section	Gas temp. (°C)	Dew point (°C)	Particle size < 10 μm (%)	Dust collector
Crusher	30–45	20–25	20–30	FF
Raw mill and kiln	140–170	50–70	75–85	FF/ESP
Clinker cooler	200–250	20–30	15–25	ESP/GBF
Coal mill	60–80	35–45	60–75	FF/ESP
Cement mill	70–90	60–70	20–50	FF/ESP
Packing plant	30–45	20–25	20–50	FF

Table 9.6 Operating features of the main dust collectors as compared to cyclones

Parameters	Cyclone/multicyclone	Fabric filter	ESP	Gravel bed filter
Efficiency (%)	80–95	99.99	99.99	99.99
Pressure drop (mmWG)	150	150	25	150
Cut diameter (μm)	3–10	0.5	0.1–0.8	0.5–2.5
Capital cost	Low	High	Very high	Moderate
Operating cost	Low	Moderate	Low	Moderate
Maintenance	Nil	Periodic	Periodic	Tedious
Secondary pollution	Nil	Nil	Nil	Nil
Particle size (μm) collection	20	Submicron	Submicron	1
Operating temperature (°C)	Very high	260	150	400

and operating features of all four dust collectors have been discussed in more detail.

Cyclones and multicyclones

Cyclones and multicyclones (Figure 9.1) are the conventional dust collectors used in the cement industry. These are based on the principle of separation of particles by centrifugal force, which is developed by tangentially introducing the gas into the cyclone. Cyclones that are usually operated in parallel sets of identical units for treating large gas flows are known as multicyclones. Cyclones and multicyclones, being low-efficiency devices, are not able to meet the present stringent emission standards and are therefore generally used as pre-collectors.

The advantages and limitations of cyclone/multicyclones are as follows:

Advantages:
- easy to construct and operate
- low initial and running cost

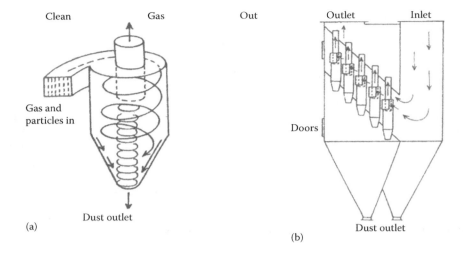

FIGURE 9.1 Schematic diagrams of a typical (a) cyclone and (b) multicyclone.

- moderate pressure drop
- low maintenance due to lack of moving parts
- capability of operation with high dust loading
- high-temperature and high-pressure applications
- useful for handling a wide range of gases and dust

Limitations:
- very low collection efficiency for fine particles
- plugging and erosion problems
- problems with light or needle-shaped material
- potential problems with abrasive dust

Fabric filters/ bag filters

A fabric filter is a unit in which the dust-bearing gas is passed unidirectionally through a fabric in such a manner that the dust particles are retained on the "dirty" gas side of the fabric while the cleaned gas passes through the fabric to the clean gas side, from where it is removed by natural and/or mechanical means. All fabric filter systems consist of the filter medium, cleaning device, dust collection hopper, isolation enclosure or housing, prime gas mover, and the necessary sensing devices and operational controls. Particle collection in a fabric filter is based on several collecting mechanisms, such as inertial impaction, interception, diffusion, and gravitational sedimentation, or a combination of these (see Figure 9.2).

The inertial impaction mechanism involves a collision of the particles with the fiber. The greatest impaction collection efficiency is obtained by either an increase in velocity or particle size. In the case of a collection mechanism involving particle interception, the particles have less inertia and can barely follow the streamlines around the obstruction.

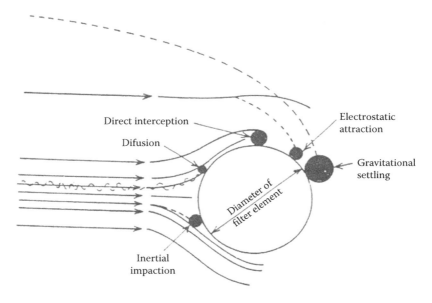

FIGURE 9.2 Filtration mechanisms in FF.

The particle is almost completely immersed in the viscous stream around the fiber, which is enough to slow it down so that it grazes the fiber and stops. The collection efficiency increases with decreasing fiber diameter or with increasing particle size. The mechanisms of impaction and interception usually account for 99.9% of the collection of particles larger than 1 µm. The diffusion mechanism works for very small particles (0.01–0.5 µm). Collection of these fine particles is a result of random motion. Gravity settling may account for an appreciable fraction of particle collected because of the long gas stream residence time. Gravity settling is greater for larger dense particles and low gas velocities.

Pressure drop in a fabric filter occurs due to inlet and exit losses, fabric and residual dust, and formation of dust cake. The resistance of the fabric is only 10% of the total resistance and hence relatively insignificant. One of the major factors considered in the design of FF is the air:cloth ratio (A/C) or filtering velocity, defined as the actual volumetric gas flow rate divided by the filtering area. The air to cloth ratio generally varies between 0.4–2.5 m^3/min-m^2. The total area required for a given cleaning duty is calculated from the air:cloth ratio.

Both natural and synthetic fibers are used as filter medium. The types of fabric and their characteristics are given in Table 9.7. These fabrics can either be felted or woven. Felted fabrics are composed of randomly oriented fibers, compressed into a mat. Woven fabrics have a definite long-range repeating pattern. A cross-section of a woven fabric shows considerable porosity in the direction of fluid motion. Felted fabrics are generally two to three times thicker than woven fabrics. More severe cleaning methods are needed for dust cake discharge from felted fabric than from woven fabrics. Felted fabrics are used for a higher air:cloth ratio. Woven fabric generally uses low-energy cleaning.

Table 9.7 Filter fabric characteristics

Fiber	Chemical name	Operating temp. (°C) Long	Short	Resistance to Abrasion	Minerals acids	Organic acids	Alkalis
Cotton	Cellulose	80	110	Good	Good	Good	Good
Wool	Protein	90	120	Good	Fair	Fair	Poor
Nylon	Polyamide	90	120	Excellent	Poor	Fair	Good
Orlon	Poly acrylonitrile	115	135	Good	Good	Good	Fair
Dacron	Polyester	135	160	Excellent	Good	Good	Good
Polypro-pylene	Olefin	90	120	Excellent	Excellent	Excellent	Good
Nomax	Polyamide	220	260	Excellent	Fair	Excellent	Good
Fiberglass	Glass	290	320	Poor-fair	Excellent	Excellent	Poor
Teflon	Poly tetrafluoro ethylene	230	260	Fair	Excellent	Excellent	Excellent

Several chemically and mechanical fabric finishing treatments of these felt fabrics are available that enhance bag life, promote ease in cleaning, provide better dimensional stability, and adjust fabric permeability. The various fabric finishing techniques include heat setting, scouring, glazing, napping, calendaring, resin treating, etc. Ultimately these techniques enhance the ability of fabrics to provide high dust collection efficiency.

Fabric filters are generally used in almost all the sections of the cement plant. These are also extensively used for control of fugitive dust emissions. Glass fiber filters have very high temperature resistance capability and are used in the kiln section.

In fabric filters the dust particles are deposited in voids and after filling the voids a cake starts to build up on the fabric surface, which does most of the filtering. When the dust layer on the fabric becomes too thick, pressure drop increases and requires cleaning of the fabric. Depending upon the characteristics of the fabric, dust, etc., three types of cleaning are used, i.e., mechanical shaking, reverse air cleaning, and high-pressure pulse jet cleaning. Low-ratio filters are generally cleaned by mechanical shaking and reverse-air cleaning methods. Mechanical shaking methods, which are used when the bag length is 2–3 m, are now obsolete as they require high maintenance (Figure 9.3).

In reverse air bag filters, the dust-laden gases are passed through the fabric, where the dust is collected on the inside of the bags since the dirty air travels from inside to outside. The clean air passes through the clean air plenum. The bag cleaning is done by re-introducing a part of the clean gas/air from the outside to the inside. This dislodges the dust from the bag inside and allows it to settle in the dust hopper, from where it is discharged by the dust conveying system. The cleaning is usually effective and can be triggered after the completion of a preset time interval or

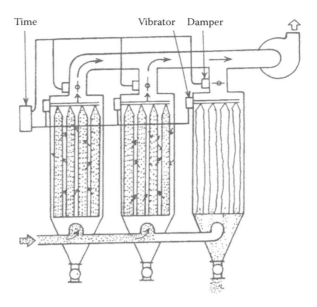

FIGURE 9.3 Fabric filter with mechanical shaking.

a predetermined pressure drop. Air requirements for cleaning are about 40–50 Nm³/1000 m³ of gas at a pressure of 1.5–2.0 kg/cm² (Figure 9.4).

In pulse jet filters the dust is deposited on the outside of the bag, while air passes through the bag. The bags are mounted on wire cages to prevent collapse. The cleaned gas is taken out through the clean air plenum. The bags are cleaned by compressed air injector, placed at the top of the bags, and the dust falls into the dust hopper and is discharged through the dust conveying systems. The compressed air required is about 2 Nm³/1000 m³ of gas at a pressure of 5–7 kg/cm². Pulse jet cleaning is mostly used in fabric filter systems in all sections except in kilns, where reverse air is used (Figure 9.5).

Condensation of moisture and temperature are the main operational problems associated with fabric filters. Humidity control is probably the most common problem if a hygroscopic dust is involved. One remedy is to make sure that the bag house is cleaned before being left out of operation. Fabric filters cannot perform properly if a gross temperature overload occurs. Fabric must be selected to withstand the temperature conditions. When the temperature of the bag house falls below the dew point, condensation takes place, causing clogging. To avoid this problem the bag house must be insulated. Notwithstanding the above problems, the FFs are preferred more and more in the cement industry primarily due to their high collection efficiency coupled with insensitivity for minor process upsets.

The preference for FFs is further triggered by the introduction of a PTFE surface filtration medium, with which pressure drops may be reduced to as low as 50–120 mm H₂O and bag life may increase to more than four years. The PTFE filter medium is able to operate at a high A/C

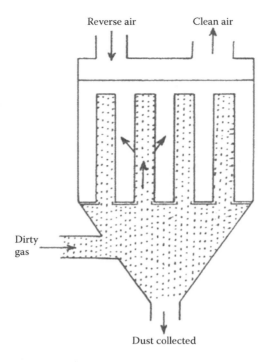

FIGURE 9.4 Fabric filter with reverse air cleaning.

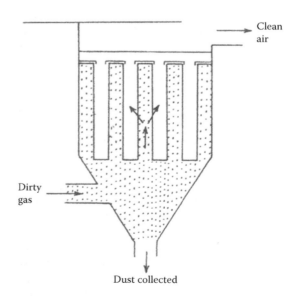

FIGURE 9.5 Fabric filter with pulse jet cleaning.

Table 9.8 Illustrative performance of PTFE pulse-jet filters

Parameters	Kiln and raw mill	Cement mill	Coal mill
Air volume	350,000 m^3/h	1700 m^3/min	2667 m^3/min
Inlet temp. (°C)	140	100–123	65–75
A/C ratio	1.26	1.10	0.91
Inlet dust load (g/m^3)	> 100	75–95	Not applicable
Emission (mg/Nm^3)	0–7.2	< 10	< 10
Max. ΔP (mm H$_2$O)	90–95	30–40	90–105

ratio, thus reducing the size of the FF. The performance of PTFE pulse-jet FFs is illustrated in Table 9.8 (5).

On the whole, the advantages and limitations of fabric filters can be summarized as follows:

Advantages
- very high collection efficiency (99.99%)
- moderate power consumption
- low capital cost on simpler application
- collection of product in dry conditions
- wide variety of fabric for various applications
- performance unaffected by minor process upset

Limitations
- gas cooling often required
- high maintenance cost
- dew point problem leading to blinding of bags
- fire and explosion hazard
- periodic bag replacement

Electrostatic precipitators

Electrostatic precipitation differs from other cleaning methods in that the force required to separate the particulates from a gas is electrostatic in nature and is applied directly to the particles themselves. This electrostatic force results from the electric field. As a consequence of force being directly utilized, ESP accomplishes gas-solid separation with less energy than any other gas cleaning systems.

An ESP consists of four major components: the enveloping structure (casing), discharge electrodes, collecting electrodes, and insulators. The casing is a gas-tight shell with hoppers to receive the collected dust, inlet and outlet gas plenums, and an inlet gas distributor. The structural members are usually outside the shell. The units are safe to touch and work at design temperature when fully energized. The most popular design of ESP is the wire-in-duct design where collecting electrodes are arranged in rows to form gas ducts within which discharge electrodes are spaced along the gas flow (Figure 9.6). The ESPs are available in both horizontal and vertical gas flow design as well as for dry and wet collection. The vertical gas flow with dry collection is used in coal mills, while the horizontal gas flow dry collection precipitator is most widely used in all other industrial operations.

Discharge electrode types are many and various, but all are required to provide points and sharp edges in order to promote the formation of corona. The diameter of discharge electrodes is generally 2.5 mm. Spiralized wire- and ribbon-type electrodes are widely used. Collecting electrodes are usually sheets of well-reinforced, light-gauge metal or tubes. The thickness of collecting electrodes is generally 1.2–1.6 mm. The collecting and discharge electrodes are cleaned by various types of cleaning, i.e., rapping mechanisms. Insulators physically support the high-voltage discharge system and isolate it from grounded components. Special materials of construction are needed for high-temperature service and to lessen the possibility of surface breakdown.

Electrostatic precipitation is based on the three basic steps:

1. electrical charging of the dust particles
2. collection of the charged particles in the electric field
3. removal of the precipitated dust from the collecting plates

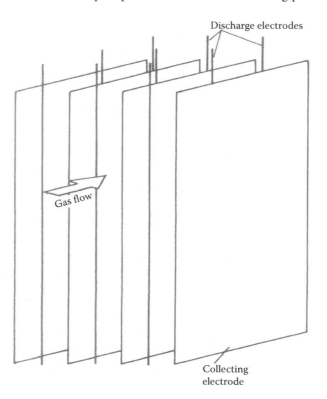

FIGURE 9.6 Schematic diagram of collecting and discharge electrodes arrangement.

The operational principle is shown schematically in Figure 9.7. When a potential gradient is established in the gas between a thin wire and the parallel plates, ionization of gas molecules takes place. The electrical breakdown or ion discharge is known as "corona discharge," which transforms the gas from insulating to conducting state. Corona discharge can either be positive or negative. Negative corona is the most suitable for industrial use as most of the gases are electron-negative in nature. In ESPs using negative corona, the discharge and collecting electrodes act as negative and positive (earthed) electrodes, respectively. After gas ionization, negative ions collide with the dust particles, transfer the charge, and cause them to migrate towards the collecting electrode. The positive ions move towards the discharge electrode.

ESP consists of discharge electrodes connected to a negative-polarity, high-voltage DC (70–100 KV) source, which together with the earthed collecting plates creates electric fields through which dust-laden gas passes, as explained above. The discharge electrodes can be wired, serrated tapes, or rigid members having protrusions with a small radius of curvature. Because of the small radius of curvature, the electric field in the immediate vicinity of the discharge electrodes becomes so concentrated that the electrical breakdown strength of the gas is exceeded. Local electrical breakdown in the form of corona discharge occurs and ionizes the gas, creating positive ions and free electrons. The positive ions are captured by the negative discharge electrodes, while the electrons that are initially propelled towards the collecting plate quickly become absorbed by the electronegative gas molecules, particularly H_2O, O_2, and SO_2 contained in the gas. The negative ions created by this electron attachment process have low mobility compared to the free electrons and they drift towards the collecting plates, forming an

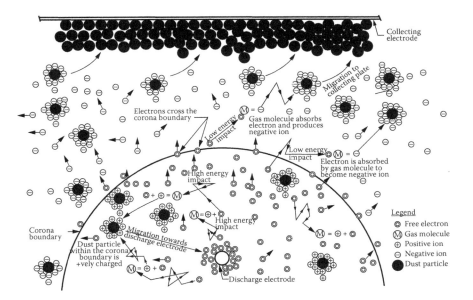

FIGURE 9.7 Operational principle of an ESP.

ionic space charge that stabilizes and determines the current flow in the precipitator.

In the flow of negative ions between the discharge electrodes and collecting plate, the suspended dust particles will acquire an electric charge by two distinct processes: field charging and diffusion charging. Under the influence of the electric field and hydrodynamic forces, the negatively charged particles move towards the grounded collecting plate with a velocity that is commonly referred to as the migration velocity. When the charged dust particles reach the collecting plates, they adhered to it while giving off their charge, forming a dust layer. The deposited dust is periodically dislodged by the rapping and is dropped in to the bottom precipitator hopper, from where it is removed continuously or periodically by suitable extraction equipment.

The efficiency of an ESP system depends on its collecting area, gas flow rate, and migration velocity. The effective migration velocity is the velocity with which dust particles move towards the collecting electrodes. It is very difficult to determine the migration velocity and it is generally computed from the charging and precipitator field strength, particle radius, gas velocity, and conductivity of the particles. The migration velocity has a relation with the particle diameter (Figure 9.8) and generally varies between 7–14 cm/s for the materials in the cement production process. An increased migration velocity is experienced with increased gas velocity but an increased re-entrainment also takes place. Therefore, proper arrangement is made to have uniform gas velocity at which migration velocity is high with no re-entrainment. Generally,

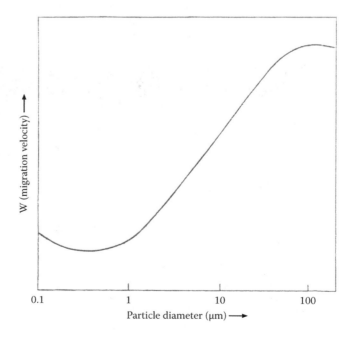

FIGURE 9.8 Relationship between migration velocity and particle diameter.

1 m/s velocity is kept at the inlet of ESP. An increase in gas temperature causes a lower migration velocity and finally a decrease in efficiency.

Resistivity is a key factor for the operation of ESP. It is defined as resistance to the current flow, i.e., reciprocal of conductivity. It is expressed in ohm.cm. Experimental determination of resistivity is difficult and is either measured at the site or at a laboratory. For proper operation of ESP, the resistivity of dust should be between the range 10^6–10^{11} ohm.cm (Figure 9.9). If it is less than 10^6 ohm-cm the particles on reaching the collecting electrode lose their charge very rapidly and are re-entrained, consequently reducing the efficiency of ESP. If the resistivity is more than 10^{11} ohm-cm, it becomes difficult for the dust particles to lose their charge to the collector and consequently they adhere strongly to the collector and will resist dislodgement. The dust layer on electrode creates impedance to the flow of current, which results in a voltage drop across the dust layer. If the voltage drop across the dust layer is greater than the voltage breakdown, then a reverse-ionization condition (back corona) develops, which is undesirable for efficient precipitation operation. Resistivity depends on absorbed moisture, temperature, and gas conditioning. An increase in moisture content or dew point reduces the resistivity of dry dust by improving the surface condition and consequently increasing the collection efficiency (Figures 9.10 and 9.11).

The temperature of the gas leaving the cyclones in a five/six-stage preheater kiln is in the range of 320–360°C, which is cooled with water in conditioning tower to bring down the temperature between 150–180°C. Water cooling at the kiln inlet is better than cooling with air because of the higher moisture content and lower specific gas volume. The resistivity

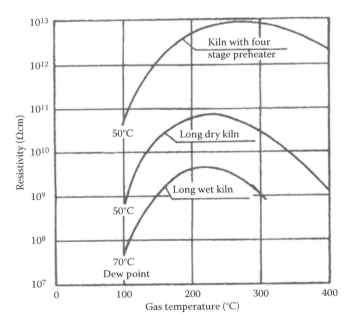

FIGURE 9.9 Resistivity of dust from kilns and coolers.

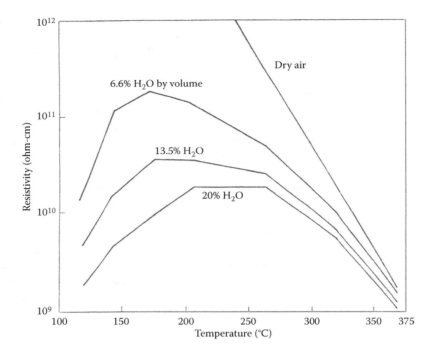

FIGURE 9.10 Moisture conditioning of kiln dust.

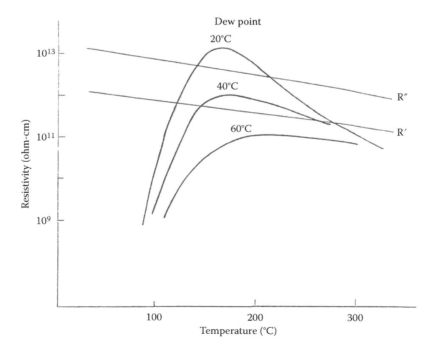

FIGURE 9.11 Temperature resistivity of kiln dust at different dew points.

at temperature 150–180°C is reduced, and hence higher voltage can be applied. The amount of water required for the cooling tower is quite high (0.15 kg/kg clinker) and may be a problem in a water-scarce region. The cooling tower also needs careful maintenance.

There are many commonly encountered problems in an ESP; some are related to dust characteristics, some are attributable to constructional features, and others are purely operational. The dust-related problems arise mainly from high resistivity. Both the chemical composition, particularly in terms of sulfur and alkali contents, and particle size distribution are important. While the composition defines the resistivity, to a large extent the particle size distribution controls the migration velocity. The low resistivity of dust does not normally lead to operational problems, while high resistivity results in back-corona and re-entrainment of dust in the gas flow.

The problems of design and construction relate to improper electrode alignment, distortion in the collecting plate, vibrating corona wire, air ingress into the hopper or gas duct, and so on. The operational problems arise from excessive dust load, process upsets, excessive dust deposition on electrodes, improper rapping, overflowing hoppers, etc.

Advances in high-power switching technology made it possible to develop a "pulse energization" system, in which high-voltage pulses are repetitively superimposed on DC-based voltage. Pulse energization makes it possible to attain more favorable electrical conditions for high-resistivity dust than what is obtained with conventional DC energization. Further, it makes it possible to change the current input without losing a well-distributed corona current. The technology has helped the introduction of "hot ESPs," where the gases coming directly from the kilns are treated at around 330°C. ESP size has also been reduced with pulse energization in hot ESPs with high dust resistivity. Other improvements in ESP technology include microprocessor-based controls for voltage supply and rapping, wide duct spacing (400–500 mm), an ESP management system, etc. Various designs of discharge and collecting electrodes have also evolved. The result of these improvements has been the achievement of high precipitation efficiency and more operational reliability. Progressively, the application of ESP has been extended to clinker coolers, in which the unstable operation of the kiln sometimes leads to excess air temperatures of up to 450°C and dust loads of 50 g/Nm^3. The cooler ESPs are generally designed with wide ducts (400 mm) and moderate gas velocity (0.7–0.9 m/s) to reduce wear due to abrasive clinker dust.

In an overall assessment, the advantages and limitations of ESP can be summarized as follows:

Advantages
- high collection efficiency (99.99%) for fine particles
- low pressure drop
- low operating costs
- ability to handle a large gas flow

- wide gas velocity range
- low treatment time (0.1–10 seconds)

Limitations
- high initial costs
- explosion hazards
- complicated operational procedure
- sensitivity to process upset

A major disadvantage of ESPs is the risk of explosion from incomplete combustion of fuel. ESPs are now provided with automatic de-energization controls, when the CO content in the kiln gas reaches a pre-set level.

Gravel bed filter

The gravel bed filter is a later introduction to the cement industry. It is a de-dusting facility with fully automatic cleaning of the filter bed from its dust load within the equipment itself. It uses a horizontally arranged filter bed consisting of almost round quartz grains of uniform size (1.5-mm diameter) with a rough surface. The quartz grains are highly wear-resistant, so there is no serious wear of the filling in the bed when the dust is removed (Figure 9.12).

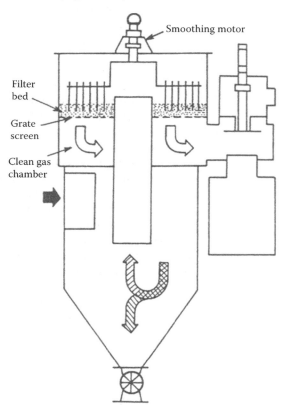

FIGURE 9.12 A schematic diagram of a gravel bed filter.

ENVIRONMENTAL MITIGATION AND POLLUTION CONTROL TECHNOLOGIES 311

The cyclone gravel bed filter was introduced in Germany in 1966 but it is now available more widely in many other countries. A gravel bed filter consists of several units of equal size connected in parallel. Each unit consists of a lower cyclone pre-separator and an upper gravel bed filter. During the cleaning cycle, which may be, for instance, 3 minutes every 90 minutes, the unit is isolated from the gas stream by means of the three-way valve, and cleaning air from an air blower—or sucked in by the negative pressure—is admitted to the gravel bed in a reverse-flow direction. During the cleaning process, the gravel bed is stirred by a rotating rake-shaped double arm. The cyclone gravel bed filter can withstand temperatures of up to 475°C.

As already stated, the gravel bed filter is mainly used in clinker coolers with high efficiency. The pressure drop across the filter, however, is rather large (about 120–200 mm H_2O), which would reflect in high operational costs. Further, initial costs are fairly high, as compared with fabric filters or electrostatic precipitators.

Reducing fugitive dust

As explained earlier, the fugitive dust mainly originates from stockpiles used as storage or for pre-blending and homogenization of raw materials, viz., limestone, coal, clay, etc., handling and transportation of raw materials from one point to another as required by the process, loading and unloading operations, and from leaks of ducts, fans, etc. The total quantity of the particulate emission from all the fugitive dust sources is quite substantial and, according to a rough estimate, it varies from 10–30% of the total emission from a cement plant. The methods employed for fugitive dust control in the cement industry are local exhaust ventilation systems and water spray systems. Proper housekeeping is also important to maintain cleanliness of the environment. The fugitive dust control methods appropriate for various sections in a cement plant are given in Table 9.9. Water spraying systems include industrial dust suppression facilities such as fog cannons, piston-drive sprinklers, and innovative nozzle designs.

Local exhaust ventilation is employed to control the movements of dust-laden gases by placing appropriate loads on the top of various fugitive dust-generating points and causing its flow through suitably designed ducts and finally to a dust collector for cleaning. The velocities to be maintained in the ducts are optimized to ensure minimum air flow rate and maximum dust pick up. The main application areas of hood in the cement plants are:

- crusher discharge points mainly for limestone and coal
- filling and discharge of silos and bins
- discharge to and from hoppers
- belt conveyor transfer points
- packing plant

In the context of dust pollution control it is important to recognize two particle size-related parameters, viz., PM_{10}, and $PM_{2.5}$, which refer to particles smaller than 10 μm and 2.5 μm respectively. Since the dust particles of such small size are particularly harmful to health, the control

Table 9.9 Fugitive dust control methods for various sections

Fugitive emission sources	Planting/ vegeta-tive cover	Paving gravel	Wind barriers	Sweeping and cleaning	Enclosing	Hood and ducting	Reducing drop height	Water spray
Crusher discharge					X	X	X	X
Screening					X	X	X	X
Conveyor transfer points					X	X	X	X
Discharge to and from hoppers					X	X	X	
Silos and bins						X	X	
Stack cleaning						X	X	
Loading and unloading			X					
Paved roads			X	X				
Unpaved roads	X	X	X					
Open storage piles	X		X					
Construction sites		X	X					
Exposed areas	X	X	X	X				

of such particulate matter assumes special importance. It is equally important to note that the high-efficiency particulate matter control systems used in cement plants are designed to meet the requirement of controlling the emission of such dust particles.

9.4 Sulfur dioxide emissions

The levels and causes of SO_2 emissions in cement plants vary widely. The pattern is influenced by the content and form of the sulfur in both the raw feed and fuel and by the complex sulfur volatile cycles that build up in the kiln system. The emissions are also affected by the type of process used. It has been discussed earlier that sulfur enters the pyroprocessing system in the form of sulfates or sulfides, such as pyrites, from raw materials and also as other sulfur compounds from fuel. At temperatures in the range of 1100–1400°C, sulfur in the kiln may form gaseous SO_2 and, unless absorbed by the raw materials, SO_2 will be emitted in the exhaust gases. In the four- or five-stage preheater kilns, a significant proportion of raw meal is calcined in the lower stages where the temperatures are in excess of about 820°C. There is intimate contact between the free lime and SO_2, when both the oxides are present, to cause the following reactions:

$$CaO + SO_2 + \tfrac{1}{2} O_2 \longrightarrow CaSO_4 \tag{9.1}$$

$$CaO + SO_2 \longrightarrow CaSO_3 \tag{9.2}$$

In excess air the calcium sulfite oxidizes to calcium sulfate:

$$CaSO_3 + \tfrac{1}{2} O_2 \longrightarrow CaSO_4 \tag{9.3}$$

The calcium sulfate passes down the kiln and, as the temperature rises to 1400°C and above, it decomposes again, following the reverse path of Equation 9.1. The SO_2 is carried back to the kiln inlet and reacts again with calcium and alkalis in the raw meal to form calcium sulfate, alkali sulfate, or even complex compounds like calcium langbeinite ($K_2SO_4.2CaSO_4$). These compounds eventually become incorporated in clinker and are carried out of the kiln. It shows that, although there could be substantial circulation of sulfur compounds within the kiln, practically all the gaseous SO_2 leaving the kiln will be arrested by the bottom preheater cyclone and returned to the kiln.

In the context of the sulfur cycle it is important to clarify that sulfides in raw meal oxidize at temperatures of 400–600°C in the upper stages of the preheater to form SO_2. Enough free lime may not be available in this part of the preheater to react with SO_2 but the following reaction may take place to form sulfites:

$$SO_2 + CaCO_3 \longrightarrow CaSO_3 + CO_2 \tag{9.4}$$

Table 9.10 Sulfur measurements in a preheater kiln

Location	Temperature (°C)	Sulfides in materials (%)	SO_2 in gases (mg/Nm^3)
Preheater feed	80	0.55	–
Preheater exit	400	–	2490
Cyclone 1	415	0.49	2375
Cyclone 2	650	0.13	1860
Cyclone 3	820	0.17	0
Cyclone 4	890	0	–
Clinker	–	0	–

The unreacted SO_2 will escape with the preheater gases and the sulfites will be carried down to follow the reaction path described earlier. Measurements taken on a four-stage preheater kiln system fed with raw meal containing relatively high sulfides are shown in Table 9.10 (6).

From the data it is evident that in this specific instance the sulfides were oxidized mainly in the first two cyclones leading to a high SO_2 level in the preheater exhaust and the zero reading at cyclone 3 showed that no fuel sulfur reached the preheater exit.

In precalciner kilns, because there is a high proportion of calcined material in the vessel, compared to the cyclones of a simple preheater kiln, there is plenty of free lime to absorb SO_2. If in the precalciner kiln system with grate cooler there is tapping of tertiary air, there will be sufficient free lime in the material entering the kiln riser pipe and in the lower preheater stages for the absorption of SO_2 leaving the kiln. In preheater-precalciner kilns provided with a bypass duct and ESP, there is less opportunity for the SO_2 in the bypass gas to be absorbed by the dust and a substantial proportion of SO_2 may escape to atmosphere. If the preheater exhaust gases are passed through the raw mill, there will obviously be considerable absorption of SO_2 in the mill. The roller mills turn out to be the most effective machine for combined drying and grinding, and consequently to utilize large quantities of kiln gases to treat raw materials containing up to 20% moisture. This mode of operation obviously allows maximum absorption of SO_2 from the kiln gases in the raw mill.

Unlike steel manufacture and power generation, the cement production process has a unique in-built desulfurization mechanism, by virtue of which 70–90% sulfur gets trapped in the clinker and 10–30% only appear in the stack.

Prevention and abatement opportunities

The best way to prevent or reduce SO_2 emissions would be to reduce the sulfur input into the system, which may not be practical in most circumstances. The alternative approaches include process adjustments, online abatement and end-of-the-pipe abatement. The maintenance of oxidizing conditions ($O_2 > 2\%$) at the back-end of the kiln and in the lower stages of a preheater is crucial to promote fast reaction between free lime and the sulfur dioxide. Similarly, for the same reason, reducing conditions should be avoided at the gas outlet of a precalciner.

The online abatement involves the injection of calcium hydroxide ($Ca(OH)_2$) to those preheater zones, where there is formation of SO_2. The injection system consists of a storage silo for the calcium hydroxide, proportioning and conveying equipment, and a pneumatic injection unit. The effectiveness of the gas-solid reaction depends on uniform distribution of lime in the gas stream. Operating results have shown that the SO_2 emission can be reduced by up to 70%, provided the proportion of lime is much in excess of the stoichiometric requirement. The process will generate calcium sulfate as a byproduct, the disposal of which will have to be taken care of in an environment–friendly manner. In view of the investment issues for the lime injection process, attempts have been made to add slaked lime to the preheater meal in pre-estimated proportions, but the emission reduction results have not been very satisfactory. The principle of the end-of-the pipe treatment of the exhaust gases for SO_2 reduction remains the same but the system layout differs. The offline abatement techniques include wet scrubbing and dry absorption of SO_2. In wet scrubbing the gases are passed through a heat exchanger and then reacted with limestone slurry in an absorber. The gases need to be reheated to achieve dispersion from the chimney. Wet scrubbing reduces the SO_2 emissions quite effectively. A dry system for absorption of SO_2 was developed by Lurgi in 1990's, which involved the use of a circulating fluidized bed of calcium hydroxide mixed with raw meal. Both the technologies have been tried out in cement plants. The installation of the scrubbing system, wet or dry, in any plant is an investment issue and calls for a proper cost-benefit study before implementation.

9.5 Nitrogen oxide emissions

Nitrogen oxides (NO and NO_2), commonly called NO_x, occur, as explained earlier, essentially due to the combustion of fossil fuels. The oxidation of nitrogen compounds in the fuel produces "fuel NO_x" and the molecular nitrogen in combustion air gives rise to "thermal NO_x." Occasionally raw materials can contain appreciable amounts of chemically bound nitrogen, which can convert to NO_x in the temperature range of 300–800°C. The NO_x volume from the raw materials is small and is not comparable to nitrogen oxides formed from fuel and air. Between the two nitrogen oxides, the proportion of NO_2 is always much smaller. The variations in NO_x formation are caused by the process parameters. For the dry process plants the commonly measured values are often as high as 800–1050 mg/Nm³, while the statutory emission limits in many countries are 500 mg/Nm³. Hence, the reduction of NO_x emissions has become almost unavoidable in most cases.

Primary methods for NO_x emission reduction

One way of reducing NO_x emission is to adopt "primary methods" of process control. The primary methods are designed to act on the following influencing factors to minimize the generation of NO_x during the combustion process:

a. flame shape and temperature
b. combustion air temperature
c. excess air in the combustible gases
d. residence time of fuel
e. nitrogen content of fuel

The most important measure is to ensure stable kiln operation. It is claimed that low-NO_x kiln burners of certain design help in achieving lower emissions. The design of low-NO_x burners varies in detail but essentially the coal and air are injected into the kiln through three or four concentric tubes with only 6–10% primary air. The results of NO_x reduction by using such burners are apparently modest.

Abatement of NO_x formation in precalciners is another primary approach in modern plants. Some "denoxing" designs of precalciners aim to partially burn the fuel in a reducing atmosphere and to complete combustion by the subsequent injection of air (Figure 9.13).

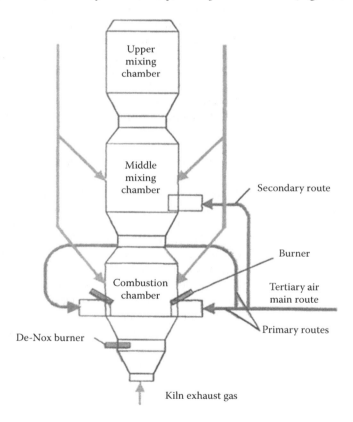

FIGURE 9.13 A precalciner with multistage combustion for reducing NOx emissions. (Reproduced from S. Yoshida, The latest developments in pyroprocessing, *World Cement*, December, 2002. With permission.)

(An extension of this concept is "multistage combustion." The concept, developed by Polysius in early 1990's, involves four stages of combustion starting from the burning zone. The kiln gases encounter a burner at the kiln inlet, which produces a reducing atmosphere and decomposes a part of the nitrogen oxides generated in the burning zone. Further, at the precalcining stage the fuel is burnt with a quantity of tertiary air producing a reducing atmosphere there. In the last combustion stage the remaining tertiary air is fed into the system as "top air" for residual combustion.)

Secondary methods for NO$_x$ emission reduction

Secondary reduction measures reconvert the NO$_x$ already formed back into elementary nitrogen either by selective non-catalytic reduction (SNCR) or selective catalytic reduction (SCR). The principle of SNCR is based on the reaction of the NO$_x$ with ammonia or ammonia water to convert it to N$_2$:

$$NH_2 + OH \longrightarrow NH_2 + H_2O \tag{9.5}$$

$$NO + NH_2 \longrightarrow N_2 + H_2O \tag{9.6}$$

The reaction occurs in a temperature window of 900–1100°C. If the temperature is too low, there will be unacceptable escape of ammonia into the atmosphere, and if it is too high, the ammonia itself converts into nitrogen oxides. The technology has been in practice in a limited number of plants and it is important to understand that the temperature window of the reaction defines that the process is suitable for preheater-precalciner kilns. Because of safety hazards in the use of ammonia, it is probably preferable to use ammonia water (7).

In the SCR technique the NO$_x$ reduction process is carried out on a noble metal-loaded catalyst surface with ammonia injection at a lower temperature of 350°C without any adverse effect on sulfur-bearing gases. Although the process is used in the power sector, it may not be very cost-effective for cement production.

It is relevant to mention here that in a cement plant utilizing sewage slurry as fuel, the kiln exhaust gases contain sulfur dioxide, ammonium compounds, heavy metals, hydrocarbons, and residual dust from the upstream electrostatic precipitator. Exhaust gases pass through an activated coke-packed bed filter installed in combination with an SNCR unit (8). The filter contains lignite coke, which has excellent absorption properties. Figure 9.14 shows a filter unit of this type in a schematic form. The filter performed as a pre-cleaner for the exhaust gases by removing most of the pollutants before the gases were subjected to the SNCR process for removal of NO$_x$. It was a plant-based trial conducted jointly by a cement plant and Polysius in Europe, which was technically successful in bringing down the emissions of the multiple pollutants to below the detection levels.

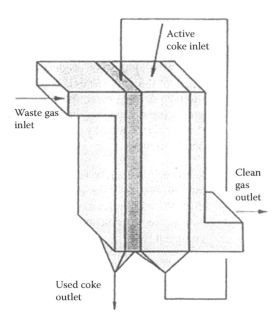

FIGURE 9.14 Schematic diagram of active coke-packed bed reactor POLIVITEC. (Reprinted from W. Rother, Exhaust gas emissions, *International Cement Review*, January, 1996.)

Ambient air quality standards

The ambient air quality (AAQ) standards do not refer to a particular process or industry. They define that the pollutant concentration for a given area in ambient air should not exceed a certain limit. The standards are set for both gaseous emissions and suspended particulate matter (SPM). The standards prescribed in India by the Central Pollution Control Board consist of 12 pollutants with different time-weighted averages. The maximum annual limits for the four major pollutants of relevance are given in Table 9.11 along with the 8-h limit for carbon monoxide.

The industrial units are also required to monitor AAQ and stack emissions within industrial premises. The basis for compliance in such cases is defined by the number of samples to be collected for stack emission and installation of AAQ monitoring stations. For example, the standard requirements that were first introduced in 1998 for the cement plant premises are given in Table 9.12. However, currently, the emphasis is on

Table 9.11 Ambient air quality standards notified by the Indian Central Pollution Control Board

Category of area	Annual standard limits ($\mu g/m^3$)				8-h limit (mg/m^3)
	PM_{10}	$PM_{2.5}$	SO_2	NO_2	CO
Industrial, residential, rural and other areas	60	40	50	40	2
Ecologically sensitive areas	60	40	20	30	2

Table 9.12 Criteria for AAQ and stack emission monitoring

Plant capacity (1×10^5 TPA)	Number of AAQ stations	Stack emission monitoring frequency (weeks)
< 1.0	–	8
1.0–3.0	2	4
3.0–6.0	3	2
> 6.0	4	1

regulating the annual arithmetic mean of minimum 104 measurements in a year at a particular site taken twice a week 24 hourly at uniform interval. The 24-hourly or 8-hourly or 1-hourly monitored values, as applicable for different pollutants, should be within the stipulated limits for 98% of the time in a year.

It is important to note that in all the countries attempts are being made to meet stricter norms of AAQ, particularly for new projects.

9.6 Noise pollution

Noise, being an unwanted sound, can turn into a pollutant when it is experienced as objectionable or harmful. Noise interferes with our activities at three levels:

- auditory: interfering with the hearing mechanism
- biological: interfering with physiological mechanisms
- behavioral: interfering with psychological/sociological behavior

The above three levels of noise interference affect human beings in different ways, as outlined in Table 9.13.

Noise level in cement plants

In cement plants, noise is generated by machinery such as crushers, grinding mills, fans, blowers, compressors, conveyors, etc. The noise levels emitted in cement plants are known to vary in general from 80–120 dBA. The sources and their general noise levels are given in Table 9.14.

Table 9.13 Effects of noise on human beings

Auditory	Non–auditory Biological	Psychological
Hearing loss	Constriction	Anxiety/nervousness
Tinnitus	Gastro-intestinal modifications	Fear
Reduced production	Skin resistance alteration	Misfeasance
Communication	Respiratory modification	Sleep stage alteration and wakening

Table 9.14 Noise generation in cement plants

Equipment/machinery	Noise levels (dBA)
Excavator	90–100
Pneumatic drill	105–115
Dozer/dumper	90–105
Crusher	90–105
Screens	90–115
Raw mill	80–100
Coal mill	100–105
Preheater fan	85–90
Primary air fan	90–95
Cooler fans	90–95
Mill drive room	95–105
Kiln drive	90–100
Conveyor drive	85–90
Compressor	100–120
DG set	100–115
Cement mill	95–105

Table 9.15 Specified limits of noise in India

Category of Area	Limits in dBA Day	Night
Industrial	75	70
Commercial	65	55
Residential	55	45
Silence zone	50	40

There are prescribed standards in different countries with respect to noise. For illustration purposes, the limits for ambient noise quality in India are given in Table 9.15.

Some international regulations on noise pollution are shown in Table 9.16.

Noise pollution reduction

Reduction measures for noise pollution involve three elements, viz., the source, the transmission path, and the receiver, i.e., the working population. Noise control at source is generally tackled by using vibration damping pads under the machine base or isolating the vibrating components from the main body of the machine. The choice of the measures to be adopted is based on the frequency spectrum of noise. The noise control in the transmission path is accomplished by erecting barriers or enclosures in between the source and the workmen. In the construction of such structures, sound-absorbing materials are used, the selection of which is site-specific and

Table 9.16 Examples of international regulations on noise level (dBA)

Exposure time	International Standards Organization	Department of Defense, USA	OSHA, USA
8 h	90	90	90
4 h	93	94	95
2 h	96	98	100
1 h	99	102	105
30 min	102	106	110
15 min	105	110	115
7.5 min	108	114	120
3.75 min	111	118	125

depends on the acoustic properties. Noise prevention measures at the receiver end involve the use of personal gear for ear protection.

9.7 Selected monitoring techniques

Monitoring of dust and gaseous emissions and other pollutants is very important in the production process. Reduction and compliance would depend on the effectiveness of the monitoring measures adopted. The monitoring measures should be online and continuous as far as possible, if the devices are reliable and can be installed and operated conveniently. Alternatively, manual monitoring systems will have to be taken recourse to, although such techniques will bring out only sporadic data.

Corona power-based method for monitoring ESP performance

A dependable and cost-effective method for continuous surveillance of the performance of electrostatic precipitators can be achieved by monitoring of the corona power (precipitator voltage multiplied by precipitator current) applied to the ESP collecting plate area. This provides a simple but meaningful indication of how well the installed precipitator capacity is utilized. The ratio of applied corona power P to precipitator collecting plate area A may be combined with the specific collecting plate area A/Q (i.e., the ratio of collecting plate area A to gas flow Q) into a useful, single key indicator for precipitator performance efficiency termed specific corona power $P/Q = P/A \times A/Q$.

Practically all new precipitator installations employ microprocessor-based transformer/rectifier controllers that have the capability of providing digital as well as analog signals for the corona current and voltage of each individual transformer/rectifier. By electronic multiplications and additions of signals, it becomes possible to totalize the corona power of all transformer/rectifier sets, and thus to generate this signal.

Corona power recorders are subject to some inherent limitations when used for monitoring the performance of precipitators. The correlation between precipitator performance (efficiency) and specific corona power is not universal and will vary from plant to plant, as well as with changing plant operating conditions. Other factors that may influence the correlation are disconnected transformer/rectifier sets, the use of intermittent energization or pulse energization, and the occurrence of back ionization in connection with operation on high-resistivity dust. It is necessary, therefore, to undertake emission measurements at varying corona power inputs, in order to establish the actual relationship between precipitator performance and corona power for a specific plant and specific plant operating conditions.

Opacity/photometric measurement

For continuous monitoring of the performance of dust collectors, opacity meters are sometimes installed in stacks. The system is based on the principle of photometric measurement of opacity of the dust-laden gas stream and correlating the measured value with the dust concentration level by calibration. The instrument measures the reduction (O) of the intensity of a beam of light passing through a known length of the gas stream

$$O = 1 - I/I_0 \qquad (9.7)$$

where O = opacity (1 = 100%), I_0 = intensity of light beam entering gas stream, and I = intensity of light beam leaving gas stream.

The light transmission I/I_0 is related to the light path length and the light attention coefficient by the logarithmic expressions. The light attenuation coefficient (Σ) is assumed to be proportional to the particulate mass concentration $C(g/m^3)$ of the gas flow:

$$\Sigma = F \times C \qquad (9.8)$$

The factor F, however, depends very much upon particle size distribution and the optical properties (refractive index) of the particles, and also depends on their density. The correlation factor F exhibits very large variations from one cement kiln installation to another, and can also exhibit large variations for test series at the same plant at different times, depending on raw mix, fuel, and kiln operation. Consequently, the opacity meters cannot be considered suitable as means for surveillance of particulate emission limits specified as mass concentrations. This type of continuous monitoring device, in particular if they are less complex and maintenance-free in design, may have some merits for use on the major gas streams (kiln exhaust) for the guidance of plant operators.

AQM monitors

Instruments are available for the continuous indicative or quantitative monitoring of dust concentrations down to 0.001 mg/m^3 with high precision and stability. The design is based on the proven technology of scattered light, and measures the scatter of visible light returning from particles passing through a chamber located in the sampling unit. The sampling unit is positioned to within two meters of the emission point being monitored and uses an integral purging clean air supply with automatic control of the light intensity, to ensure continuous operation.

Manual monitoring In the manual method a controlled gas quantity is drawn at isokinetic condition through a nozzle and probe inserted in the stack and passed through a pre-collecting cyclone and back-up filter and subsequently vented through a vacuum pump. The dust concentration is determined gravimetrically by determining the quantity of collected dust and by monitoring the gas flow drawn through the system. It is understandable that the manual method of stack sampling does not reveal the entire spectrum of dust emission, especially under the upset conditions.

In order to assess the total suspended particulate matter, CO, SO_2, and NOx in and around a cement plant, air samplers are used. The equipment for the purpose generally utilizes the principle of drawing a measured quantity of air by vacuum pump through a filter and the dust concentration determined gravimetrically.

The instrument used for the characterization of noise is known as a sound level meter. It consists of a microphone that gives an electrical signal proportional to the pressure and is indicated as decibels by the indicator. The average exposure to noise over a period of time is measured by an instrument called a noise dosimeter.

9.8 Current environmental outlook

The last one-and-a-half decades have seen increasing environmental awareness in human society in general, with the inherent expectation of better environmental quality. Growing interest and concern has led to stricter implementation of regulatory measures. In tandem, the industry itself has felt the compulsions of improving internal efficiency and cost savings. The net result has been a push towards the adoption of the Environmental Management Systems under ISO 14000 and the effective introduction of cleaner production technologies. The requirements of use with guidance are given in ISO 14001 and the general guidelines on principles, systems, and supporting techniques are described in ISO 14004. The core elements of ISO 14001 are shown in Figure 9.15. It is important to mention here that unlike compulsive compliance with pollution control norms under regulatory pressures, another strategy has emerged in many countries to undertake voluntary environmental initiatives. This approach is a participatory one between the industry and the authorities through the conclusion of voluntary agreements. In addition, other management tools like "eco-rating," life cycle assessment (LCA), etc. are also in practice.

The EMS efforts and other pulls and pushes have helped the industry to accomplish the following:

   a. quarry reclamation through conservation, stockpiling, and use of top soil and overburden; re-contouring of slopes of more than 30% to minimum erosion and run-off; planting native vegetation
   b. reduction of stack emissions of dust to 30 mg/Nm^3
   c. limitation of stack emissions of SO_2 to 100 mg/Nm^3 unless special dispensations are called for due to pyrite content in raw materials exceeding 0.25%

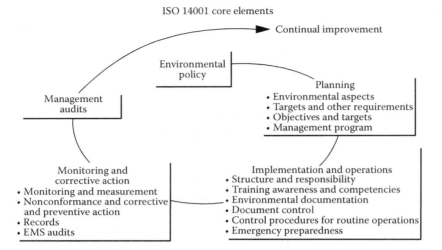

FIGURE 9.15 Core Elements of ISO 14001.

    d. limitation of stack emissions of NO_x to less than 1000 mg/Nm^3, even in mixed streams of precalciners in preheater kilns

    e. limitation of HCl emission to 10 mg and HF emission to 1 mg/Nm^3

    f. limitation of total organic carbons (TOC) emission to 10 mg/Nm^3

    g. limitation of emissions of Hg, Cd, and Tl and their compounds to 0.05 mg/Nm^3

    h. limitation of heavy metals (Sb + As + Pb + Co + Cr + Cu + Mn + Ni + V and their compounds) to 0.5 mg/Nm^3

    i. limitation of dioxin and furan emissions to 0.1ng/Nm^3 at 10% O_2

    j. maintenance of ambient air quality in terms of yearly mean for dust and gaseous emissions at below 100 μg/m^3

    k. control of noise pollution below safe limits

    l. high level of waste recycling

    m. effective recycling of cooling water with full compliance with the relevant standard

    n. commitment to reduce CO_2 emission

Thus, sustainable development has become an integral part of growth of cement production and its application all over the world.

9.9 Summary

For many years, the cement production was not considered environment-friendly mainly due to "visual" dust pollution associated with the manufacturing process. The most prudent and emphatic step taken by the cement industry was to minimize and control the dust emission. Although the cement production process has never been a chemically polluting one, it has several other emissions, besides the dust.

They include sulfur dioxide, nitrogen oxides, organic compounds, heavy metals, halogens, and carbon monoxide. The emission of carbon dioxide is not strictly regarded as a pollutant but has its environmental role as a greenhouse gas. The abatement techniques for dust emission primarily include the use of fabric or bag filters, electrostatic precipitators, and gravel bed filters, supported by cyclones as pre-collection systems. The primary techniques for reducing the generation of nitrogen oxides are based on the use of low-NO_x burners or the use of staged combustion in the modern precalciner kilns. The secondary methods suitable for the cement production process are the selective "non-catalytic" and "catalytic" reactions (SNCR and SCR). The emission of sulfur dioxide is not a major issue in the cement plants. In special situations, where the sulfur dioxide emissions are unusually high, the technology of online injection of calcium hydroxide or end-of-the-pipe scrubbing is adopted. The cement production units also cause noise pollution due to its heavy machinery with moving components. Appropriate measures are taken to keep the noise pollution below the safe limits. The cement plants need to be provided with monitoring systems for both dust and gaseous pollutants. For dust emissions the corona power-based system for monitoring the performance of ESPs is installed. Sometimes the opacity meters are used for continuous monitoring of stacks. Although the continuous monitoring systems are preferred, they have their limitations and problems. Hence, the manual methods of gas sampling and measurement of pollutants become essential.

In combating environmental problems, the cement industry has always attempted to make use of the best available technologies to prevent releases of prescribed substances, to render harmless the discharge of potentially harmful materials, and to provide the best practical option to minimize the damaging environmental impacts on operating the process. These strategies have finally led the industry to adopt ISO 14001 certification quite extensively. The Environmental Management System involves declaration of the organization's environmental policy, setting targets of improvement on a year-to-year basis, careful documentation, periodic site inspection, and third-party audits.

In conclusion, it is important to note that in terms of environmental benignancy the cement plants being installed today are superior to those built a decade ago. With all the new initiatives and emerging concepts, the plants of tomorrow will far exceed the expectations of the human society.

References

1. A. K. CHATTERJEE, Environmental technology management of the cement industry in India and neighboring countries, Engineering Foundation Conference, University of New Hampshire, USA (1994).
2. A. K. CHATTERJEE, How clean and environmentally benign the cement plants can be: a technological commentary, Proceedings of the 7th NCB International Seminar on Cement & Building Materials, Vol. 2, New Delhi (2000).

3. S. B. YADAV, Pollution control technology in cement industry, in *Energy Conservation & Environmental Control in Cement Industry* (Eds S. N. GHOSH and S. N. YADAV), Vol. 2, Part II, ABI, New Delhi (1996).
4. K. K. DAS NAG, Pollution control for dust emissions in the cement industry, *Indian Cement Review*, January (1992).
5. C. P. GANATRA AND I. K. KIM, Trends in filtration and pollution control in Asia, *World Cement*, February (1998).
6. British Cement Association, Prevention and abatement of SO_2 emissions, The Cement Environmental Yearbook, Tradeship Publications, UK (1997).
7. British Cement Association, Prevention and abatement of NO_x emissions, The Central Environmental Yearbook, Tradeship Publications, UK (1997).
8. S. YOSHIDA, The latest developments in pyro-processing, *World Cement*, December, 2002.
9. W. ROTHER, Exhaust gas emissions, *International Cement Review*, January, 1996.

CHAPTER TEN

Trends of research and development in cement manufacture and application

10.1 Preamble

Following the coinage in the mid-1990s of the term "disruptive innovation" by Harvard Business School Professor Clayton Christensen, technologies are classified into two categories: "sustaining" and "disruptive." Sustaining technologies are the ones that undergo successive improvements, often incremental in nature, pushed by cost as a driver, while disruptive technologies refer to innovations by which a product or a service takes root initially in simple applications at the bottom of the market and then relentlessly moves up the market, eventually displacing an established product or a process technology.

When one looks at the history of the Portland cement industry, which spans almost two centuries, one is bound to be convinced that this is a unique example of sustaining technology. In this time span the Portland cement industry has grown phenomenally to reach a production level of over 4 billion tons globally. Portland cement has become the most extensively used building material and a broad relationship between cement consumption and per capita income of a country has led the world to an economic conviction that the cement consumption of a country is an excellent indicator of its growth and progress. Needless to say, this growth story has certainly been backed by numerous "sustaining innovations" from academia, research institutions, and the industry itself, supplemented by adaptable developments drawn from the metallurgy, machine fabrication, and electronics industries. Since all these innovations have been incremental and sustaining in nature, the present chapter is set to identify past research attempts that could have been disruptive in nature, had they been taken to the stage of commercialization, and the probable reasons for their stalling. This chapter also tries to emphasize that in the last decade there has been an acute change in research and development (R&D) thrusts because of environmental compulsions, which have the potential of leading towards disruptive technologies if pursued innovatively. When this potential is going to be realized is an important matter of prognosis.

10.2 Sustaining technologies in the growth of the cement industry

Joseph Aspdin patented Portland cement in England in 1824, and its development was furthered by Isaac Johnson between 1838 and 1911 through essential improvements in product and process; some even consider Isaac Johnson to be the real inventor of Portland cement. While many industrial developments took place more by facts or situations that were observed to exist or happen than by scientific findings, a few outstanding scientists laid critical milestones in the cement industry, which are summarized in Table 10.1. These fundamental findings remain relevant even 200 years later.

Alongside the unraveling of cement chemistry and material science, significant engineering strides were made in the second half of the twentieth century to substantially increase the scale of operation and to introduce technologies that were less energy-consuming and less man power-intensive (Figure 10.1) (1). These advancements led to the adoption of automation, instrumentation, and computerization in the manufacturing process and to visible improvements in pollution control. In fact, the second half of the twentieth century is dotted with a large number of incremental process improvements. The more important advances in process and equipment are compiled in Tables 10.2 and 10.3.

The net effect of all these developments resulted in shaping up the industry with the following advanced engineering features:

- The scale of the clinker-making process went up to about 12000 tons/day from a system comprising a single kiln, multiple preheater strings, separate line precalciners, and grate coolers.
- Large raw grinding systems with vertical roller mills or hybrid systems consisting of hydraulic roll presses and ball mills, provided with efficient separators.

Table 10.1 Critical milestones in Portland cement chemistry

Year	Unravelling of cement science	Innovator/investigator
1887	Reporting on basic cement hydration	H. Le Chatelier
1895	Introduction of soundness test	W. Michaelis
1904	Standard specification of cement	Am. Soc. Civ. Eng
1919	Concept of w/c vs strength	D. Abrahams
1924	Phase composition of cement	R. Bogue

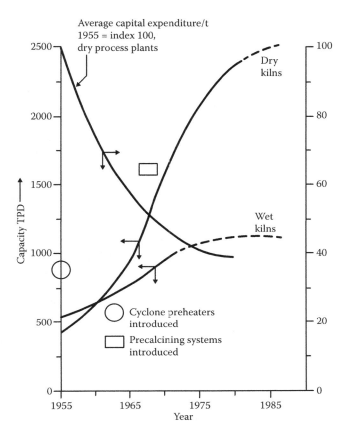

FIGURE 10.1 Changes in technology and scale of operation in the second half of the twentieth century. (From A. K. Chatterjee, Modernisation of cement plants for productivity and energy conservation, in *Cement and Concrete Science & Technology* (Ed. S. N. Ghosh), Vol. 1, Part 1, ABI Books Pvt Ltd, New Delhi, 1991.)

- Various system designs for large cement grinding capacities with ball mills, vertical roller mills, hydraulic roll presses, etc.
- The specific electrical energy consumption reached a level of 70 kWh/t of cement.
- The lowest specific thermal energy consumption touched a level of 680 kCal/kg clinker.
- Kiln burners with five channels to fire multiple fuels, including alternative waste fuels, were designed and installed.

Commercially successful modern grinding and burning systems that are operational at different plants all over the world today have already been dealt with in previous chapters. The attempt here is to focus on the futuristic developments in equipment, production processes, and cement products.

Table 10.2 Tentative technological milestones in the raw and finish grinding processes

Year	Mill systems
1891	Invention of tube mills by Davidsen
1904	Introduction of tube mills with air separators
1920–1940	Proliferation of tube mill systems; introduction of closed-circuit raw mills with bucket elevators, airswept mills, combined drying and grinding systems, etc.
1950–1960	Introduction of closed-circuit cement grinding with air separators, classifying liners, changing grinding media sizes, minipebs for fine grinding, etc.
1960s	Concept of grinding aids; improved metallurgy of liners and grinding media
1970s	Increased mill size with power transmission ratings of 6 MW and above; introduction of roller mills of improved design including vertical-ring roller mills and centrifugal suspended roller mills, etc.
1980s	Introduction of roller presses; adoption of vertical roller mills for cement grinding; high-performance separators, etc.
1990s	Use of pre-grinding concept for increasing clinker grinding capacity; introduction of horizontal roller mills

Table 10.3 Course-changing developments in pyroprocessing

Year/period	Inventions and innovations
1850s	Batch-type small-capacity shaft and bottle kilns
1877	Continuous rotary kilns patented by Crompton
1896	Practical development of rotary kilns in the USA by Hurry & Seamen
1928	Design of Lepol grate kiln by Lelep; no commercialization until 1935
1934	Suspension preheater patented in the then Czechoslovakia
1950	Commercialization of suspension preheater system by Humboldt
1963–1966	Conceiving of precalcination system; first trials in Germany in 1966
1998	Development of short two-pier rotary kilns
1993–2003	Scores of significant but incremental developments in precalciners, burners, coolers, and control systems

10.3 Status of potentially disruptive pyroprocessing technologies

It is evident that the basic rotary kiln technology for making Portland cement clinker has undoubtedly stood the test of time. But, at the same time, it has always been found that the rotary kiln technology is not an efficient process of clinker making. Hence, substantial research efforts were directed in the 1980s towards developing new pyroprocessing systems in which there would be no rotary kilns. The processes that were then researched are briefly described below.

Fluidized bed clinker making process

A historical account of the development of the fluidized bed clinker making process is available in (2), starting from the Pyzel process, followed by the Fuller Company's pioneering attempt and subsequent Japanese pilot plant trials held separately by IHI-Chichibu, Kawasaki, and Sumitomo. Later, however, Kawasaki Heavy Industries pursued on their own development to scale up the process by designing, constructing, and operating pilot plants of three progressively increasing capacities, viz., 2 t/d, 20 t/d, and 200 td. The system configuration is shown in Figure 10.2 (3). It was claimed that the system could take various coals

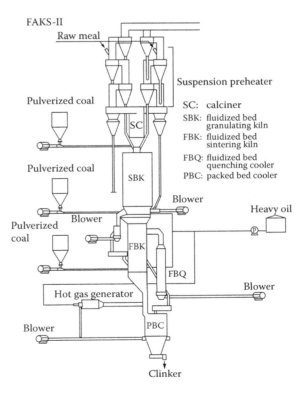

FIGURE 10.2 Schematic diagram of the fluidized bed pyroprocessing system of KHI design. (From Kawasaki Heavy Industries Ltd, Catalogue No. 3F2428 on Fluidized bed advanced cement kiln systems (FAKS), Japan, 2000.)

Table 10.4 Projected scale-up data for commercial FB plants

Parameters	200 t/d	1000 t/d	3000 t/d
Heat consumption (kCal/kg clinker)	771	713	690
Power consumption (kWh/t clinker)	43.0	41.5	37.2
Fluidized bed kiln dia. (m)	2.5	5.7	9.7
FB Q cooler dia. (m)	0.7	3.6	8.6
Tower width (m)	10.0	17.0	26.0
Tower length (m)	8.0	16.0	26.1

from low calorific values to low volatiles, could offer 10–25% better thermal efficiency, a 10–25% reduction in CO_2 emission and 40% or more reduction in NO_x emission, a 10–30% saving in construction costs, and a 70% reduction in space. Scale-up data for the commercial plants of different capacities then developed are given in Table 10.4. Apparently, the scaling-up of the process beyond 3000 t/d was found problematic due to an enlarged kiln diameter, and in terms of the specific heat consumption the system did not offer any significant improvement over the relatively more efficient rotary kiln systems. On the other hand, the capacities of rotary kilns continued to increase for the economy of scale. Thus, in a highly competitive cement market, the technology of fluidized bed clinker making appears to have lost the race. However, its relevance for special cements—including white cement—should not be overlooked.

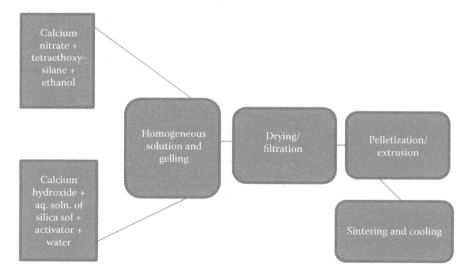

FIGURE 10.3 The sol-gel process scheme for manufacture of cements. (From C. H. Page, C. H. Thombare, R. D. Kamat, and A. K. Chatterjee, Development of sol-gel technology for cement manufacture, in *Advances in Cementitious Materials* (Ed. S. Mindess), *American Ceramic Society*, Ohio, USA, 1991.)

Sol-gel process

Two alternative versions of the sol-gel process of cement manufacture were attempted in the production of clinker (see Figure 10.3) (4). The basic technical feasibility of producing β-C_2S in the lime-silica system was established. The effect of dopants like boron oxide, sulfate, magnesia, and fluoride was observed in manipulating the crystal size and the hydraulic reactivity and not in any polymorphic transformation of the calcium silicate. However, dopants like alumina and fluoride in combination appeared to help in the formation of C_3S at temperatures much lower than its threshold temperature of stability (1280°C). In all the experiments, the molar C/S ratio ranged from 2.26 to 2.38. A white cement of very high reflectance and high compressive strength levels could be produced by this technology in the system C-A-S-fluorine at a temperature of 1200°C. The technology also appeared feasible for making colored clinker with suitable chromophores. Hardly any further work has been reported on the applicability of this process in cement making. The up-scaling possibilities and the viability of this process are worthy of re-examination.

Microwave heating

Synthesis of clinkers for gray, white, and colored cements by microwave processing was reported in (5). ACC Limited in India had a collaborative project with the Materials Research Laboratory, Pennsylvania State University, USA in this field. The experiments were conducted in a 900-W 2.45-GHz multi-mode cavity with a turntable to rotate a sintering packet in which the sample was loaded. The basic facility and working scheme are shown in Figure 10.4. A single pellet sintering was carried out.

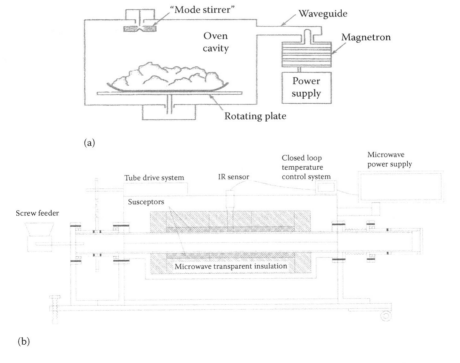

FIGURE 10.4 The microwave heating facility and its working scheme for clinker making. (a) A microwave oven components. (b) A rotary tube microwave heating system.

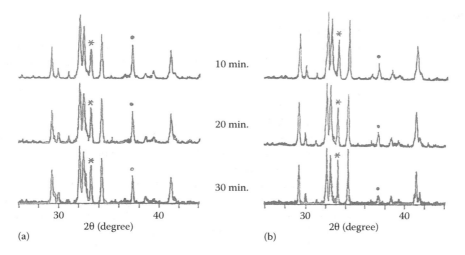

FIGURE 10.5 Comparison of X-ray diffraction patterns of clinkers produced by (a) microwave heating and (b) an electric resistance furnace.

The results showed that microwave processing enhanced the clinkering reaction to a certain extent for both the gray and white cement samples. This effect was maintained by the lower temperature of clinker formation as compared to the parallel runs of the same raw meals in a conventional fast-heating electric furnace. The clinkering temperature in microwave heating was found to be almost 100°C lower, while the X-ray diffraction patterns revealed that the clinker phase formation was quite normal (Figure 10.5). The iron oxide in the gray cement raw meal and the amorphous silica in the white cement raw meal seemed to be responsible for the enhanced reactivity in the microwave heating conditions.

Some recent studies on the microwave clinker making process, with different types of susceptors, have been reported in (6). The microwave heating patterns showed that the heating time was much shorter when compared with other modes of heating, namely, 1/8 of the electric furnace time, 1/4 for petcoke, and 1/3 for natural coal. The crystal sizes of the clinker phases were smaller in microwave heating. However, the conversion of β to Υ-C_2S was observed quite frequently during the cooling operation.

Electron beam or radiation process

A systematic experimental approach to apply radiation energy for clinker making was reported from the former Soviet Union in the 1970s and 1980s (7). In the radiation synthesis of clinker, the heating was accomplished by a high-energy beam of accelerated electrons. After conducting a series of experiments, it was established that at an absorbed power dose of 20–40 Mrad/s, the clinker making process could be completed in a few seconds, or in case of a less-powerful flux of electrons, within a few tens of seconds. It was also shown that the process was completed at a temperature that was 200°C lower than the normal temperatures required for clinker making in a rotary kiln. A block diagram of the experimental set-up is presented in Figure 10.6.

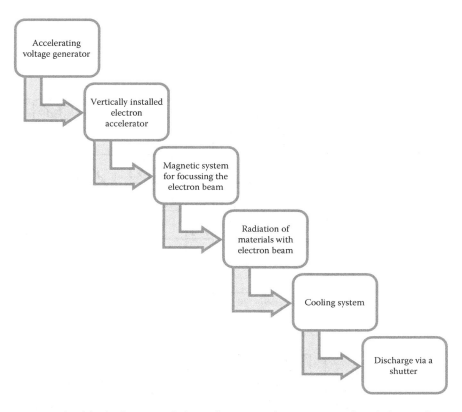

FIGURE 10.6 The block diagram of the radiation synthesis process for clinker making. (From S. K. Handoo and K. Goswami, Application of electron beam process technology for production of cement, Proceedings of Indo-USSR Seminar on Industrial Applications of Electron Accelerators, Vol. II, BARC, November 3, 1988.)

Normal Portland cement clinker synthesized by radiation was observed to be black and porous. The alite crystals were mostly acicular in shape and only a few crystals showed platy habits. The belite phase displayed both round and anhedral morphology. Liquid formation was observed to be insignificant. As normally expected, there was a linear relationship between the proportion of alite and compressive strength at all ages (Figure 10.7). The figure also shows the relation between the absorbed dose of radiation and the compressive strength of irradiated clinkers. The technology obviously had some merits of being pollution-free and possibly less energy-consuming, although overcoming the problems inherent with the use of radiation process would remain as the main challenge. In this context it may be interesting to note that during the same period there were snippets about the possibilities of using small nuclear reactors to act as the fuel source for clinker production. Plans were apparently drawn up for small test rigs somewhat in the style of a vertical shaft kiln layout, with the nuclear source in the form of a collar. The clinker formation process with nuclear bombardment was technically feasible but the radioactivity issues remained unresolved and deterrent to the scheme.

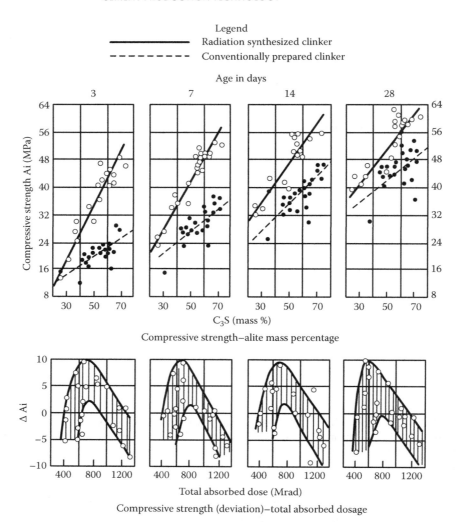

FIGURE 10.7 Relation of compressive strength, C$_3$S content, and absorbed radiation in irradiated clinkers. (From S. K. Handoo and K. Goswami, Application of electron beam process technology for production of cement, Proceedings of Indo-USSR Seminar on Industrial Applications of Electron Accelerators, Vol. II, BARC, November 3, 1988.)

10.4 Portland cement derivatives with niche application potential

It may be relevant to mention here that in a non-technical manner the word "cement" has been defined as "a powdery substance made by strongly heating lime and clay, used in making mortar and concrete" (see Compact Oxford Dictionary, Thesaurus & Wordpower Guide, 2001). This definition, in a way, comes from the manufacture of Portland cement and, hence, any treatise on new cements cannot ignore the new developments in the family of normal Portland cements (NPC), which has seen a colossal global volume growth in production and use all over the world. But it must also be recognized that the growth of the Portland cement industry has happened despite the following inadequacies in the application properties of this cement:

- It is not a ready-to-use finished product.
- Being weak in tension, plain concrete made of Portland cement is used only in compression.
- The use of RCC with improved flexural properties is confined to fairly large elements as adequate cover is essential.
- It suffers from poor toughness or crack resistance on impact.
- It has a tendency to shrink and crack on drying or cooling.
- Even well-made normal concrete is permeable to fluids, affecting its durability.
- It needs a long time to mature.
- It has problems of usage at low temperatures.

The above shortcomings have not been detected or experienced today. Back in 1846, Louis Vicat was the first to understand that calcium hydroxide formation in Portland cement concrete was the culprit causing the lack of durability. He developed the following empirical relation of oxides, which, he expected, would produce a highly durable cement:

$$(SiO_2 + Al_2O_3)/(CaO + MgO) > 1 \tag{10.1}$$

In search of a more durable hydraulic cement phase, based on the above compositional hint of Vicat, the "calcium monoaluminate" phase was discovered in 1856 by Saint-Claire Deville. This new discovery ultimately led to John Bied's patenting of the fusion process of manufacturing "Aluminous Cement" by a water-jacket furnace in 1908 in France. Thus, second-generation hydraulic cement, resistant even to marine construction, was born within 50 to 60 years of the creation of the increasingly popular Portland cement. From the 1930s to the 1940s, while there was wide use of aluminous cement in postwar reconstruction, it was also realized that the cement suffered from a serious defect of strength retrogression in certain environmental conditions. As a result, between the 1970s and 1990s there were several cases of severe building damages in UK and Europe, which led to a ban on the use of aluminous cement for structural use. Thus, the new discovery did not result in a proper replacement of Portland cement and aluminous cement became a special-purpose refractory cement, rechristened as "High Alumina Cement (HAC)" or "Calcium Aluminate Cement (CAC)" (8).

Therefore, the search for a novel generic hydraulic cement continued in the twentieth century as vigorously as before. The calcium sulfoaluminate phase ($C_4A_3\bar{S}$ or more commonly abbreviated as $C\bar{S}A$) has been known to be hydraulically active and has been in use as Kleine's compound in making expansive Portland cement. Based on this phase, a new variety of cement was produced by Zakharov in Russia by burning (at about 1300°C) a mix intermediate in composition between Portland and calcium aluminate cements with some gypsum. This cement, called "aluminous belite" cement by Zakharov, contained typically C_2S (64%), CA (12%), $C_{12}A_7$ (11%), and $C_4A_3\bar{S}$ (13%). No free lime or C_2AS phases were observed. Subsequently, in the 1980s and 1990s there were fairly extensive investigations on compositional manipulation, tailor-making of properties, and modes of application of CŚA-based cements in several countries including India. It was observed that the CŚA cements could be made and used in a similar way to normal Portland cement and they demonstrated notable versatility in application, newer dimensions of durability, and appreciable reductions in the emission of greenhouse gases during manufacture. The potential of this third-generation cement has been reviewed in (9). Yet, the global production of this variety of cement is only a few million tons and is limited to a few countries like China.

From this historical background one may conclude that some of the ancient binding materials like lime, lime-pozzolana mixes, and gypsum reigned in the construction field for thousands of years and still retain their relevance, while in the last 200 years three generations of cements, viz., NPC, HAC/CAC, and CŚA have been invented but so far the success and viability of NPC could not be matched by other types, notwithstanding the multifarious deficiencies in NPC's application properties. It is therefore not surprising that the search for new cements is still focused on modifying the Portland cement compositions.

Advances in the modified Portland cements

R&D trends in the OPC group of cements over the last several decades may be summarized as shown in Figure 10.8. It is evident from the figure that several avenues are available to improve and modify the cement properties, the illustrations of which can be taken from (10). Without

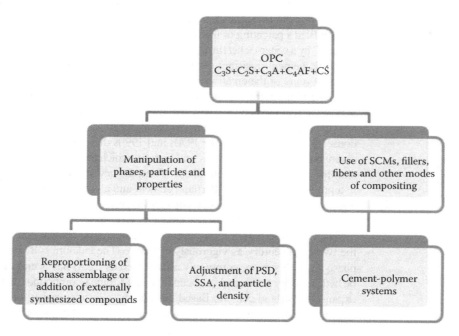

FIGURE 10.8 Different ways of improving and modifying normal Portland cement.

repeating what is either widely known or is in practice, the following derivatives of Portland cement compositions display novelties in their production and application along with substantial utilitarian values:

- high-belite or reactive-belite cement (RBC)
- calcium sulfoaluminate–belite cement (CŚA-BC)
- alinite-based eco-cement (AEC)
- Portland limestone cement (PLC)
- Multiblend Portland cement (MPC)

The status of development of the above varieties of cements is described below.

High-belite/Reactive-belite Cement High-belite cement (HBC)/reactive-belite cement (RBC) is a formulation in which there is a reversal in the preponderance of alite by belite without any significant sacrifice of early-age strength in mortar and concrete. This definition implies higher levels of early reactivity in Belite. In spite of prolonged research on the product, the commercialization goal continues to elude the industry for so far not having perfected:

- a practical and viable technique of improving the hydration rate of belite phase, adaptable to the existing process of manufacture
- a way to reduce the higher energy consumption in belite grinding

The technological requirements for the first step are rapid cooling of clinker and homogeneous dispersion of dopants or chemical stabilizers used in tiny dosages. It is known that, in countries like India and Poland, alkali-stabilized moderately high-belite (35–45%) cements have been in production and use with satisfactory performance. Taking cues from this, further studies were carried out under the author's guidance with raw mixes designed for two tentative levels of lime saturation factor (LSF)—0.82 and 0.78—and containing various dopants. All firings were carried out in a pilot rotary kiln provided with a specially designed cooler that could achieve a cooling rate of 500–600°C per minute. The relevant findings are given in Table 10.5 (11).

The effect of rapid cooling on strength was visible; when LSF was in the range of 0.81–0.82, the amount of a given dopant was near-optimal and the C_2S/C_3S ratio by the Bogue computation was not abnormally high (say, above 6). Rapid cooling was not as critical for the retention of boron and alkali oxides as for sulfate. The optimal incorporation of dopants certainly influences clinker formation, stabilization of belite polymorphs, fixing of alite crystal systems, and distortion of lattice structures of the stabilized phases. Based on a large number of experiments, it was found that the following levels of retention of the dopants in clinkers were beneficial in stabilizing the phase:

$B_2O_3 (0.5\%), K_2O (0.3\%), Mn_2O_3 (1.2\%),$

$Cr_2O_3 (-0.7\%), SO_3 (0.35\%), BaO (-1.0\%).$

Table 10.5 Effect of clinker cooling rate on the cement strength parameter

Type of dopant	Rapid cooling				Slow cooling			
	Dopant (%)	C_2S/C_3S (Bogue)	Clinker LSF	28-day comp. Str. (N/mm^2)	28-day comp. Str. (N/mm^2)	Clinker LSF	C_2S/C_3S (Bogue)	Dopant (%)
None	–	4.85	0.813	54.5	46.5	0.813	3.98	–
	–	11.05	0.780	45.0	50.5	0.784	4.21	–
B_2O_3	0.68	1.84	0.823	78.6	67.0	0.814	2.89	0.65
	0.60	13.14	0.780	64.0	67.0	0.783	5.39	0.62
K_2O	0.28	2.26	0.819	78.0	61.2	0.824	2.00	0.31
	0.14	3.51	0.789	66.0	68.5	0.792	3.14	0.18
Mn_2O_3	1.24	3.59	0.820	61.0	54.0	0.820	2.04	1.19
	1.20	4.60	0.780	49.0	41.5	0.776	5.08	1.20
Cr_2O_3	0.70	3.47	0.820	61.5	56.0	0.813	2.31	0.70
	0.68	3.69	0.794	67.7	69.6	0.793	3.54	0.68
SO_3	0.39	3.07	0.820	69.0	55.0	0.820	2.14	0.21
	0.27	6.15	0.770	47.0	47.0	0.770	6.10	0.21
BaO	0.97	8.01	0.811	51.4	–	–	–	–

It may be relevant to mention here that the lattice parameters were measured for the β-belite phase in some of the doped clinkers and the results are furnished in Table 10.6.

It is evident that the incorporation of boron and manganese oxides resulted in significant dimensional changes in cell parameters. Further, the presence of different polymorphs of belite with different dopants was determined by precise determination of 2θ values, as shown in Table 10.7. From all these results the effectiveness of B_2O_3 in making the belite phase more reactive is worth noting. In the boron-doped clinker the occurrence of α'_H and α'_L polymorphs were particularly observed in addition to the β-phase. During the last two decades, nuclear magnetic resonance (NMR) spectroscopy has found wide application in cement science. Despite some resolution problems, some recent studies have shown that the quantitative spectral deconvolution of the ^{29}Si MAS-NMR spectra of Portland cements provides reliable alite:belite ratios. Based on such advances, in the present study attempted the

Table 10.6 Changes in cell parameters of the belite phase on doping

Clinker with	Cell parameters				Cell volume (nm)³
	a (nm)	b (nm)	c (nm)	β	
No dopant	0.93761	0.66569	0.55124	91.5274°	0.34406
B_2O_3	0.94395	0.67595	0.56528	100.2715°	0.36068
Mn_2O_3	0.94545	0.67175	0.56589	94.4991°	0.35940
K_2O	0.92477	0.67384	0.55687	92.4015°	0.34701
Reference values (JCPDS)	0.93100	0.67565	0.55059	94.46°	0.34633

Table 10.7 Stabilization of different polymorphs of belite with different dopants

Clinker with	Clinker LSF	2θ values in degree			
		β	α	α'_H	α'_L
No dopant	0.82	31.100	31.669	–	–
No dopant	0.78	31.096	31,809	–	–
B_2O_3	0.82	31.058	–	32.197	32.073
B_2O_3	0.78	31.092	–	32.197	32.065
K_2O	0.82	31.092	31.669	32.196	32.074
K_2O	0.78	31.105	31.801	32.461	32.085
Mn_2O_3	0.82	31.102	–	32.465	32.076
Mn_2O_3	0.78	31.054	–	32.449	32.073
Cr_2O_3	0.82	31.063	–	32.201	32.076
SO_3	0.82	3 1.096	–	32.469	–
BaO	0.82	31.067	–	32.430	–

Table 10.8 Phase quantification by NMR spectroscopy

Dopant	Belite %			Alite %		
	NMR	XRD	Bogue	NMR	XRD	Bogue
K_2O	48.2 ± 5.0	42.87	55.4	29.9 ± 6.0	32.48	20.5
B_2O_3	53.0 + 4.0	50.67	54.8	23.5 + 5.0	27.37	21.2
Mn_2O_3	46.7 ± 4.0	44.30	52.7	30.4 ± 4.0	34.4B	22.5
Cr_2O_3	41.5 + 4.5	48.13	52.0	37.7 + 5.0	30.19	23.5

quantification of alite and belite by NMR spectroscopy (Table 10.8). In the present study an attempt was also made to estimate the distribution of dopants in the silicate and non-silicate portions of the clinkers. It was found that the incorporation of B_2O_3 in the silicate fraction was highest in the range of 95–96%.

The second problem of grindability is still under study, although some indications are available from (12) that the phenomenon of "remelting reactions" might provide some clue towards making high-belite clinker amenable to easier grinding. Alternatively, the use of chemical admixtures including grinding aids is also being explored. Under these circumstances, it is high time that efforts should be made to bring HBC/RBC into the realm of implementation, using the existing plant, unit operations, and equipment.

Calcium sulfoaluminate–belite cement (CŚA–BC) As already mentioned, the CŚA phase was considered the harbinger of third-generation cement. Since the belite phase in HBC could not reach the level of hydraulic reactivity comparable to that of alite, an idea emerged to substitute alite by CŚA in high-belite Portland cement to achieve better early-age properties in the new cement. This concept led to the development of CŚA-B cement. From the last review of the subject by the author (9), the important manufacturing aspects of this variety of cement are highlighted here.

The calcium sulfoaluminate phase is a low-sulfate mineral form and occasionally appears in clinker, when gypsum is used as a raw mix component in the range of 5–8% and when the molar ratio of $CaSO_4$ and Al_2O_3 is 1:1 or higher. It is an istropic mineral having a sodalite-like structure with a refractive index of 1.566. The partial replacement of aluminum by iron has been observed in the structure, yielding the composition $Ca_4 (Al, Fe)_6 O_{13} \cdot SO_3$, although this phase is generally represented as $C_4A_3Ś$. CŚA-B cement is made from natural raw materials like limestone, clay, bauxite, and gypsum. Bauxite as a source of alumina and lime can be substituted partially or totally by a natural rock like anorthosite or by industrial byproducts like fly ash and steel slag. The limestone component can be replaced by hydrated lime, if necessary. Similarly, the mineral gypsum can be substituted by fluorogypsum, a byproduct of the chemical industry, which may provide an additional input of fluorine as a mineralizer. The clinkering temperature is in the range of 1250°C to 1300°C. Pyroprocessing is carried out in rotary kilns with low-ash fuels. The essential phases are $C_4A_3\bar{S}$ + β-C_2S + $C\bar{S}$, while the associate phases include $C_{12}A_7$, C_4AF, and free CaO. Compounds such as C_2AS and $2(C_2S) \cdot C\bar{S}$ are deleterious. CŚA-B clinker containing $C_4A_3\bar{S}$ and β-C_2S as the major phases generally retains about 6–7% SO_3. The clinker is finally ground to about 400 m²/kg Blaine surface along with additional gypsum or hemihydrate. Typical compositions and properties of the CSA cements are given in Table 10.9.

The energy requirements (calculated from standard thermo-chemical tables) and amounts of carbon dioxide released during the formation of pure anhydrous cement phases made from $CaCO_3$ and SiO_2, Al_2O_3 or $CaSO_4$, as appropriate, are given in Table 10.10.

These data suggest that cements based on aluminate or sulfoaluminate systems are likely to be advantageous. A reduction of the limestone requirement of about 40% and an energy saving of 25% appear to be possible in cements with high $C_4A_3\bar{S}$. Coming to the application potential of these cements, one may recall that they generally form the base for expansive and shrinkage-compensating cements in many countries, but China has the longest experience of production and use of two classes of CŚA cements, e.g., the sulfoaluminate cement (SAC) and ferroaluminate cement (FAC) for general purpose construction. These two classes of CŚA cements were standardized in China in the 1980s and were used in the following types of construction: small and medium precast concrete

Table 10.9 Typical illustration of the CŚA-B clinker compositions and strength properties with 15% SO_3 in ground cement

Clinker oxide composition (%)	CŚA-1		CŚA-2	
SiO_2	11.6		13.7	
Al_2O_3	26.8		23.8	
Fe_2O_3	1.8		1.7	
CaO	49.6		51.7	
SO_3	7.1		6.1	
Clinker phases (% by XRD)	**CŚA-1**		**CŚA-2**	
$C_4A_3\bar{S}$	56		47	
β-C_2S	35		45	
C_4AF	6		6	
Free CaO	1		2	
Free $CaSO_4$	Nil		Nil	
Compressive strength (MPa) of 1:3 mortar samples	**CŚA-1**		**CŚA-2**	
1-day de-moulding	46.8	–	38.1	–
3-day de-moulding	–	50.7	–	43.0
3-day curing	45.1	–	41.1	–
7-day curing	46.4	51.1	44.6	50.2
28-day curing	58.0	63.7	62.4	60.9

Table 10.10 Energy requirement and CO_2 released during the formation of different cement phases

Cement compounds	Enthalpy of formation (kJ/kg clinker)	Carbon dioxide released (kg/kg clinker)
C_3S	1848.1	0.578
β-C_2S	1336.8	0.511
CA	1030.2	0.278
$C_4A_3\acute{S}$	~ 800	0.216

shapes including pipes, heavy prestressed elements, cold-weather concreting, and glass-reinforced concrete products. In India, the CŚA-B cement is manufactured and used in making grouting products, but the need was felt to assess its durability in a high-monsoon tropical environment before the cement could be recommended for building construction. The findings of an experimental structure exposed to such an environment for several years are shown in Figure 10.9 and the trials re-confirmed that there was no strength retrogression or any other forms of deterioration in the CŚA-B concrete used in this construction. Apart from its long-term durability, experience showed that the CŚA-B cement can be manufactured with the same facilities that are used for producing normal Portland cement and since it has the potential of low-temperature applications as well as a lower emission of greenhouse gases, CŚA cement continues to receive attention in the construction industry with a promise of success.

This treatise on the subject of CŚA cement would remain incomplete without a brief narration of the progress made by the erstwhile Lafarge group towards commercializing the belite-rich calcium sulfoaluminate cement containing calcium aluminoferrite as another major phase (13). Trade-named as Aether, the Lafarge team focused on this product in the context of exploring new avenues for CO_2 mitigation. In this development attempt, the novelty has been in combining the process of chemical stabilization of reactive belite and optimization of the phase assemblage. Between 2010 and 2013, pilot tests were conducted at the Institute of Ceramics and Building Materials in Poland, two plant-scale trials at Lafarge production units, and independent testing of Aether mortar and concrete at BRE in UK. In the final trial at the Le Teil Dry Process plant in France, about 10,000 t of clinker was produced with high levels of process control. The following findings of this developmental project are significant:

- Compared to the OPC process the carbon dioxide reduction was estimated at 25–30%.
- Because of the easier grindability characteristics of Aether clinker, the total energy saving was observed to reach up to 40%.

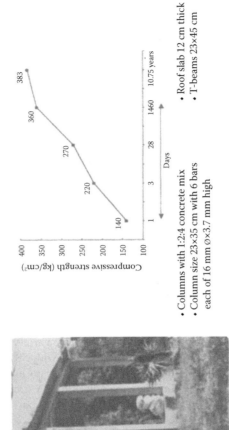

FIGURE 10.9 An experimental structure built with calcium sulfoaluminate–belite cement and its long-term exposure. (From A. K. Chatterjee, *Indian Concrete Journal*, Vol. 84, No. 11, 2010, pp. 7–19.)

- The Aether concrete demonstrated high early strength properties to the tune of 20 MPa in 6 h.
- The tentative cement contents in C20/25, C25/30, and C35/45 grades of concrete were reported to be 240, 300, and 360 kg/m^3 with corresponding w/c ratios of 0.6, 0.5, and 0.45.
- The present manufacturing plant and process are suitable for industrial production but the QC facilities should include X-ray diffractometer.
- The precast industry appears to an obvious choice for application of this cement.

Alinite based eco-cement (AEC)   Alinite is a stable silicate phase with mixed O-Cl anions having the formula $Ca_{21}Mg[(Si_{0.75}Al_{0.25})O_4]_8O_4Cl_2$, which corresponds to chemical composition of CaO (64.62%), SiO_2 (20.77%), Al_2O_3 (5.55%), MgO (2.33%), and $CaCl_2$ (6.64%). The production process of the alinite cement and its fairly large-scale use in actual construction has been narrated in (14). The wide adoption of this technology has been affected, as we know, due to the controversy regarding the durability of the alinite cement because of the likelihood of chloride corrosion of steel reinforcement. However, there is a renewed interest in cement in Japan in the context of recycling incinerated urban wastes containing chlorine ranging up to about 10% (15). The typical compositions of eco-cements obtained then had Cl$^-$ ranging from 4.0% to 0.5%. The high-chloride variety was designated as ordinary type with 70.0% alinite; the medium-chloride variety with about 2.0% Cl$^-$ was designated as rapid-hardening type A with 58% alinite and the low-Cl variety as rapid-hardening type B with no alinite but 60% alite instead. Over a period of time the non-alinitic variety was found to be more universally acceptable and its system flow chart is given in Figure 10.10.

The claimed advantages of this system are the following:

- It uses more than 500 kg of MSW incineration ash per ton of cement produced.
- The cement quality is stable and is suitable for a wide range of applications, similar to OPC.
- Dioxins are decomposed in the rotary kiln process at 1350°C and they are not reformed as the kiln exhaust gases are cooled rapidly from 800–200°C.
- The effective chloride content in clinker is below 0.1%.
- Heavy metals in the incineration ash are collected and recycled.
- Severe dissolution tests have established that there are no leaching problems of heavy metals.
- The JIS specification of eco-cement is as follows: LOI (≤ 3.0%), MgO (≤ 5.0%), SO_3 (≤4.5%). R_2O (≤ 0.75%), and Cl$^-$ (≤ 0.1%).

It may be relevant to mention here that the world's first eco-cement plant went on stream at Ichihara in Chiba Prefecture in Japan in April 2001

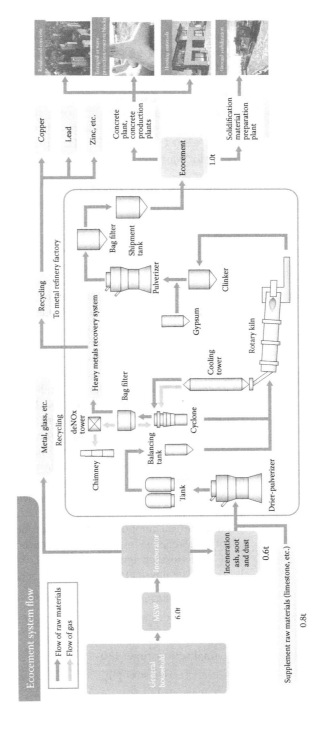

FIGURE 10.10 The system flow for manufacturing alinite-based ecocement, developed by Taiheiyo Cement Corporation, Japan. (Reprinted from the company's trade information brochure.)

with a capacity of recycling 90,000 tons of waste per year. Subsequently, in July 2006 Tokyo Tama Eco-cement facility was established with almost similar capacity to process the wastes of the Tokyo Tama region of Hinode-cho county. Observing the success of waste recycling through the alinite cement process route in Japan, many other countries have tried to explore such possibilities. One such example is the exploratory work reported from Turkey to produce alinite cement clinker from the waste material of the Solvay process of the soda industry (16). It was seen that the soda sludge with some clay and iron ore could yield the alinitic clinker by burning the raw mix at a temperature ranging from 1050 to 1200°C for 60 min. Thus, the rejuvenation of alinite cement process for waste recycling is worth noting.

Portland Limestone Cement (PLC) Historically speaking, Guyot and Ranc (17) in France had shown probably for the first time that Portland cement containing up to 25% raw limestone can perform as a satisfactory cement in terms of its workability and durability. Subsequently, of course, this variety of cement has been standardized in Europe under EN 197-1 with two classes—one with addition of 6–20% (II/A-L) and another with 21–35% (II/B-L) limestone. PLC with less than 20% limestone has been one of the largest-consumed blended cements in Europe. However, this variety of blended cement has not been standardized and commercially produced in many other countries, including a major cement-producing country like India. Hence, PLC still continues to be a subject of regional development. Against this backdrop, a fairly long-term developmental project was undertaken in the laboratories of a major cement producer under the guidance of the present author in the year 2006. After three years of lab and bench scale studies, a plant scale trial was carried out in 2009 in order to produce this variety of cement in tonnage quantity for the testing of the cement in concrete and in actual construction. An experimental building (Figure 10.11) was constructed in the same year, which is under continuous monitoring even now. For the various experimental results of this project one may refer to a previous publication of the author (18). Some selected findings are mentioned below.

The chemical composition of plant-produced PLC with 26–27% addition of limestone having 70–75% total carbonates showed 2.2–2.7% SO_3, 8–9% LOI (loss on ignition), 7–8% IR (insoluble residue), and 0.01% chlorides. The physical properties obtained were as follows:

- Blaine's surface: 330–380 m^2/kg
- normal consistency: 23.5–24.5%
- initial setting time: 115–120 min
- final setting time: 130–190 min
- compressive strength
  1-day: 12–14 MPa

FIGURE 10.11 An experimental structure built entirely with Portland limestone cement.

> 3-day: 24–27 MPa
>
> 7-day: 36–39 MPa
>
> 28-day: 50–55 MPa

- LC expansion: 0.5 mm
- autoclave expansion: 0.01–0.03%
- drying shrinkage: 0.048–0.055%
- heat of hydration (7-day): 58 cal/g

The long-duration strength test of the concrete cubes had been planned for ten years and the cubes were preserved for this purpose under appropriate curing conditions. The compressive strength results of concrete cubes for four years are presented in Table 10.11.

Table 10.11 Gain in compressive strength with age of the preserved PLC concrete cubes

Sample designation	Average compressive strength at different ages (MPa)					
	1.5 years	2.0 years	2.5 years	3.0 years	3.5 years	4.0 years
PLC-2	52.95	53.05	54.42	53.85	54.47	58.00
PLC-3	50.18	49.86	46.08	47.00	47.25	50.29
PLC-4	50.34	50.81	53.35	54.10	56.84	63.54

The long-term investigation of PLC mortars and concrete helped to arrive at the following conclusions:

- PLC can stand on its own merit even in tropical environments as a general-purpose blended Portland cement.
- Although the quality parameters would vary, depending on the raw materials used and the manufacturing process adopted, it seems that normal Portland cement clinker interground with about 20–22% limestone having 70–72% total carbonate and resulting in a cement having about 2.0–2.3% SO_3 and about 340 m²/kg Blaine's surface would perform quite satisfactorily in mortar and concrete.
- At the manufacturing stage, if the limestone component is separately ground and blended, it would obviously offer better opportunity for control of the particle size distribution of the cement and its resultant properties.
- It is obvious that there would be dilution of application properties of cement with dilution of limestone quality.
- In the Indian context, the specification of TOC in the limestone as maintained in the European standard may not be relevant and critical as this parameter turns to be significant for freeze-thaw resistance of cement.
- PLC may not give specific advantage in terms of reduction in the clinker factor, but it is certain to offer significant benefits for resource conservation and production economy.

Multi-blend Portland Cements (MPC) The concept of multicomponent blended cements has emanated from the concrete property benefits that can be realized from the packing density of cements. The denser the packing of cement, the lower the demand for mixing water into the cement paste and the denser the resulting paste, which ultimately contributes to the durability of concrete. It has also been seen that packing density is increased if one of the two materials to be blended has a significantly wider particle size distribution than the other. It should also be borne in mind that with increasing fineness of particles (< 1 µm), interparticulate adhesive forces occur that can lead to agglomeratation. The effectiveness of combining different mineral admixtures, such as silica fume and blast furnace slag, or fine limestone powder and fly ashes of different median particle size, such as 5, 10, or 20 µm, with NPC have been reported in (10).

In this context, another research endeavor on developing a ternary blended Portland cement containing calcined clay pozzolana and ground limestone, named limestone calcined clay cement (LC3), is worth mentioning. The technology was developed by a cement research team led by Aalborg Cement Company in Sweden and the innovation was submitted for patenting in different countries in 2010–2013 (19). The technology was also explored by other research groups in other countries (20). The fundamental principle behind the effectiveness of such a ternary composition is shown in Figure 10.12.

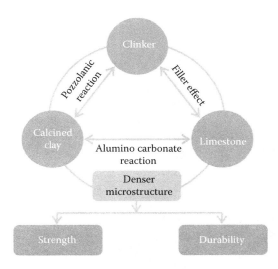

FIGURE 10.12 Synergy of the blending components in the Portland calcined clay limestone cement.

Generally speaking, the cement formulation consists of 50% Portland cement clinker, 30% calcined clay having about 60% kaolinite, 15% impure low-grade limestone, and 5% gypsum. The cement has been produced by intergrinding of all components in ball mills at a level of a few hundred tons. The basic quality traits of the cement are indicated briefly in Table 10.12.

Good-quality concrete has been produced from this cement and trial manufacture has been carried out for pre-cast products like dense and hollow blocks, roofing tiles, etc. A two-story office building (Figure 10.13) has been built in central India for evaluating the constructability of LC3

Table 10.12 Basic characteristics of LC3 cement

Parameters	Trends in comparison to other blended cements
Fineness	High, due to the intergrinding of limestone and calcined clay
Water percentage for normal consistency	High, due to the presence of the clay component
Initial setting time	Short and often close to the threshold value
Early-age compressive strength of 1:3 mortar	High after one day curing
Late-age compressive strength of 1:3 mortar	Moderate after 28 days curing
Hydration	Synergy of calcined clay and limestone to produce monocarboaluminate ($C_3A \cdot CaCO_3 \cdot 11H_2O$) or hemicarboaluminate
Chloride ion penetration	Very high resistance due to pore refinement

FIGURE 10.13 An experimental structure constructed with limestone-calcined clay Portland cement.

concrete. The current studies in various laboratories include durability characteristics including transport properties, corrosion propensity due to carbonation and chloride ingress, alkali silica reaction, sulfate attack, and leaching. Further, economic and ecological parameters including energy, emissions, resource position, etc. are under investigation. The cement has a strong potential to bring down the clinker factor in cement, thereby reducing the emission of CO_2.

10.5 Complex building products formulation with CAC

It has already been mentioned in this chapter that calcium aluminate cement (CAC) lost the status of being a structural cement for its strength retrogression property. Although today it is regarded primarily as a refractory cement, there have been renewed efforts in recent times to manufacture a special variety of CAC that can be used to formulate niche building products of high utilitarian value in the construction sector.

This specially manufactured calcium aluminate of consistent composition of high reactivity can be used as an ingredient of dry-mix mortar, flooring formulations, tile adhesives, tile grouts, repair, and anchoring mortars. Generally, the formulations are done in combination with several other binding materials, fillers, polymers, and organic additives. Most of the formulations fall in the compositional triangle shown in Figure 10.14, in which Zone 1 refers to the binary composition for rapid setting mortars, Zone 2 demarcates compositions that have high

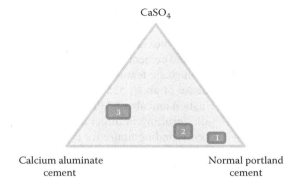

FIGURE 10.14 Typical composition domains in a CAC-OPC-CaSO4 system for building products (adapted from the trade literature of Kerneos, France and printed with the company's permission).

dimensional stability, and Zone 3 shows compositions suitable for dry-mix mortars (21). The ternary binders showed a different hydration path than products based purely on Portland cement or CAC. The microstructure of mortars of CAC-rich blends after three years of outdoor exposure was studied and it was found that the mortars contained ettringite and AH_3 as the major hydrates. Addition of Portland cement lowered the content of AH_3 but did not modify the hydration mechanism. Another interesting observation was that the presence of ettringite did not detrimentally affect the mechanical performance of the mortars, although ettringite is known to be sensitive to CO_2 (22). In other words, the ternary blends, if optimally prepared, are expected to be dense and durable

In such formulations CAC imparts special properties such as quick setting, rapid hardening, color, non-efflorescence, mechanical strength, abrasion resistance, etc. The typical characteristics of the special CAC, known as Ternal RG and manufactured by Kerneos Aluminate Technologies, are illustrated below (Table 10.13) (21):

The color and bulk density are also important parameters for characterizing this product. The product plays an important role in formulations for a wide range of applications.

Table 10.13 Properties of the special CAC

Oxide composition	Phase composition and particle size
Al_2O_3: 38.0–41.0%	Calcium monoaluminate with $12CaO.7Al_2O_3/CaO.Al_2O_3$ ratio < 0.06
CaO: 35.3–37.9%	Tetracalcium alumino-ferrite as the secondary phase
Fe_2O_3: 14.5–17.5%	PSD: below 500 μm
SiO_2: 3.5–5.0%	74–83% particles below 65 μm
	31–40% particles below 10 μm

10.6 Research thrusts towards low-carbon cement industry

Carbon dioxide emission continues to remain a major concern for the Portland cement industry all over the world. The concern emanates from the intrinsic features of the present manufacturing process, such as the release of about 535 kg CO_2 per ton of clinker from the limestone calcination and about 330 kg CO_2 per ton of clinker from the fuel combustion, resulting in direct emission of 835 kg CO_2 per ton of clinker. The corresponding figure for cement would vary, depending on the quantity of clinker used in making a ton of cement and the grinding technology adopted. In this context, the industry at large has adopted the following key levers to reduce the CO_2 emission level:

- minimizing the clinker factor
- enhancing the use of alternative fuels
- making unit operations more energy-efficient
- generating electricity with waste heat
- utilizing renewable forms of energy, to the extent possible, in place of fossil fuel

From the projection of the growth of the cement industry in the next decade and thereafter it is estimated that with the above measures the required extent of reduction of specific CO_2 emission would not be achieved unless recourse is taken to the technology of capture and storage of CO_2 for sequestration. The CCS strategy by itself is not considered viable for the cement industry unless there is value addition and utilization of captured CO_2. Hence, there is a renewed interest in developing and perfecting technologies for the capture of CO_2, its transformation into fuels and chemicals, and also for recycling to form new binding materials. The status of development in these three directions has been elaborated by the author in his earlier publications (23–25), which may be referred to for original references for topics described in the following sub-sections. The main features of these developments are highlighted below.

Capture and recovery of carbon dioxide

The process of separating a pure stream of CO_2 from the flue gas of a fuel combustion process is referred to as carbon capture. There are two stages of capture: pre-combustion and post-combustion. Pre-combustion technologies are not applicable to cement manufacture as the major proportion of the CO_2 emission comes from the calcination of limestone rather than from the fuel burnt. Hence, post-combustion technologies are more suitable for the clinker making process. Most commonly, post-combustion capture is done by using a solvent (including MEA or amine scrubbing), which reversibly reacts with CO_2 in the flue gas with subsequent application of heat to isolate the CO_2 gas into a pure stream. Apart from sorbent technologies, membrane and cryogenic technologies are also applicable. For improving post-combustion CO_2 recovery, new technologies involving oxygen enrichment and oxy-combustion are being explored. A simplified illustration of oxygen-enriched air combustion is shown in Figure 10.15. The idea is to add highly concentrated oxygen to the process with the aim of increasing the production of cement clinker, enhancing the use of low-grade fuels or improving the energy efficiency. The next step is to replace all air with pure oxygen. This technology is called oxy-fuel combustion and is schematically shown in Figure 10.16. An air-separation unit (ASU) is used to produce highly concentrated oxygen. If such pure oxygen is used as the burning medium, the temperature in the burning zone will obviously shoot up to an unacceptable level. Balancing becomes unavoidable and is done by recirculating the exhaust gas back to the burning zone.

Oxygen-enriched combustion and oxy-fuel combustion are likely to be best-suited for new cement plants that could incorporate these design features. There is also the technological option of indirect calcination, which implies calcining the limestone or raw meal without any direct mixing with fuel combustion gases (see Figure 10.17). The carbonate calcination occurs by indirect counter-flow heating, and consequently the flue gases are not mixed with the CO_2 emitted from the carbonate materials. This technology is already operating at a commercial scale for magnesite calcination. The above three additional processes for improved recovery of CO_2 are compared in Table 10.14.

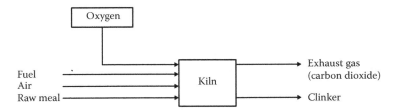

FIGURE 10.15 Oxygen-enriched air combustion process.

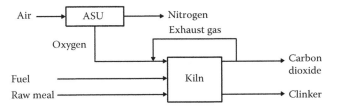

FIGURE 10.16 Oxy-fuel combustion process.

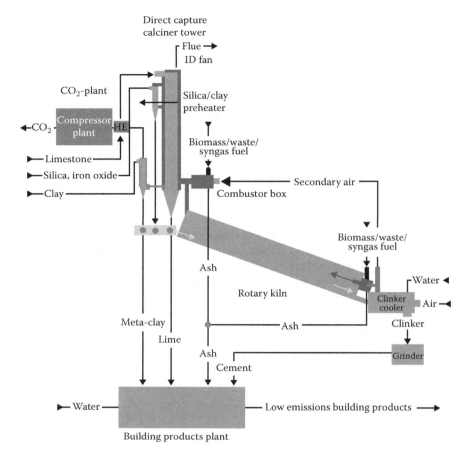

FIGURE 10.17 Limestone calcination system for uncontaminated CO_2 recovery (designed and developed by Calix Limited, Australia and reprinted with the company's permission).

Table 10.14 Comparison of the processes for improved recovery of CO_2

Comparative features	O_2 to Kiln	O_2 to Calciner	Total oxy-combustion	Indirect calcination
CO_2 capture	15%	85%	100%	65%
t-CO_2/t-O_2	1.3	5.0	3.5	N/A
Leakage	Kiln seals	Tower only	Kiln and tower	Precalciner
Quality impact	Potential	None	Potential	None
Cooler issues	High	N/A	High	N/A
Efficiency impact	High	Moderate	High	
Refractory impact	Potential	None	Potential	None

Algae cultivation with kiln exhaust gases

A novel technology has been explored in the last few years: the use of kiln exhaust gases to grow algae in bioreactors. Algae are single-celled organisms that, like plants, can use photosynthesis to grow in the presence of sunlight, water, and CO_2. They have several advantages over conventional biofuel crops, the most significant of which is that the productivity per area is between two to five times greater than that seen with normal crops and the other important aspect is that algae can be harvested continuously. Coupling the kiln exhaust gases to an algae bioreactor would result in reducing effective CO_2 and NO_x emissions with simultaneous production of valuable biomass. This technology has already been tried with some success in the energy industry. It is reported that Arizona Public Service had attached an algae bioreactor to their 1.04 MW Redhawk power plant and it had been able to produce biomass of adequate quality to be converted into biofuels, while reducing greenhouse gas emissions by approximately 80% during daylight hours. The biomass produced can be treated to yield ingredients that are commonly found in food supplements as well as in pharmaceuticals. After these valuable compounds are extracted, the remaining biomass can then be used as a kiln fuel or for the production of bio-ethanol. Several large cement companies in association with algal firms ventured into undertaking pilot trials of biosequestration of CO_2 with algae. Studies have been carried out with five micro-algae species: fresh water, sea water, calcareous, siliceous, and micro-algae with a high amount of lipids. A semi-continuous cultivation was carried out at a cement plant site in a 5.5-L photobioreactor using the gas from the stack after dedusting and scrubbing. Later, a prolonged trial was carried out in France at a cement plant without using a scrubber on two species of micro-algae (sea water and fresh water species). Both were grown separately over a period of five months in a 1 m³ tubular photobioreactor. The reactor was installed at the foot of the exhaust stack and under a greenhouse to protect from weathering effects. The reactor had two main parts: the tube part, which allowed circulation of the culture medium to an exchange tank where it was mixed with the industrial gas and the harvesting tank to recover the culture medium. The experiments in the industrial environment confirmed the following points:

- A de-dusting step was needed to prevent dust accumulation in the reactor.
- The exhaust gas could be used without damaging the micro-algae.
- The growth rates of algae with the stack gas were comparable to those with pure CO_2 (0.22 g/L.d for the seawater species and 0.24 g/L.d for the fresh-water species).
- The quality of biomass could vary with the change of limestone sources and fuel mixes.
- 1.8 t of CO_2 consumed was estimated to produce 1.0 t of biomass on certain assumptions.
- The conversion of biomass into biomethane via anaerobic digestion also appeared as an important downstream step to prove the process viability.

Thus, it is evident that it is possible to grow micro-algae by using exhaust gases from cement plants. However, the process viability has to be examined over the entire process from production of micro-algae to the use of biomethane in kilns. The available experimental data, when seen with practical assumptions, tend to indicate that the overall CO_2 mass balance is still not ready for commercialization. A technological breakthrough is awaited in the reduction of energy required for algae cultivation and the subsequent transformation processes.

Biohydrogen production from algae It might be quite relevant to mention here that algae technology has the intrinsic potential for being extended to the production of hydrogen—the clean, renewable, and eco-friendly fuel of the future. At present, the electrolysis of water and thermocatalytic reformation of hydrogen-rich compounds require high input of energy. It is therefore natural that the option of biohydrogen production has started receiving some serious attention. The possible options for biological production of hydrogen are as follows:

- direct biophotolysis (green algae)
- indirect biophotolysis (blue green algae)
- photofermentation (purple non-sulfur bacteria)
- dark fermentation (anaerobic bacteria)

The studies are still in the laboratories, but it seems that the organism *Chlamydomonas reinhardtii* has become a model for hydrogen production. More research and development efforts are necessary for metabolic engineering, optimization of the photobioreactor culture conditions, and genetic engineering of algae to enhance the production of hydrogen in a viable manner.

10.7 Technology options for converting CO_2 into fuel products

It is well known that CO_2 has low chemical activity, but it is possible to activate it towards chemical reactions by the application of temperature and pressure or by use of catalysts. It is possible to produce a variety of chemicals such as

- synthetic fuels in the form of energy vectors or energy source
- hydrogenation of CO_2 over a wide range of catalysts
- synthesis gas through reforming reactions, sometimes in multiple steps
- further treatment of synthesis gas for Fischer-Tropsch process

The F-T process is a well-known and well-characterized route that has been used in industry to produce chemicals and synthetic fuels from syngas ($CO + H_2$). By coupling the catalytic reduction of CO_2 to CO with the F-T process to produce synthetic fuels and industrial chemicals,

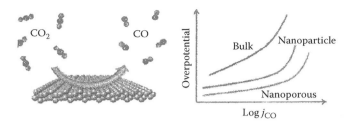

FIGURE 10.18 Dependence of CO_2 to CO conversion on the microstructural characteristics of the catalyst.

the estimated maximum reduction of atmospheric CO_2 gas emissions is 40%. However, there are technological challenges in developing the F-T process conditions as well as in designing new effective and selective catalysts. Use of the enzyme carbonic anhydrase is suggested by some researchers as the most efficient catalyst in CO_2 reactions with water, although the metallic catalysts are more extensively researched.

Electrochemical reduction of CO_2 to CO

Electrochemical CO_2 reduction using metallic catalysts has shown great promise. Gold and silver are of particular interest since both metals exhibit extremely good CO selectivity under moderate overpotentials in comparison to other metallic catalysts. However, gold is not suitable for large-scale applications owing to its high cost and low abundance and consequently there is more attention on silver as a catalyst. It is interesting to note that the conversion efficiency of the silver catalysts also varies, depending on their microstructural characteristics. Silver in general is promising because it can reduce CO_2 to CO with a good selectivity (~ 81%) and it is expected to be more stable under harsh catalytic environments than homogeneous catalysts because of their inorganic nature. Since the development of CO_2 reduction systems based on aqueous electrolytes with much higher activity is desired for large-scale processes, further studies have been continuing with silver catalyst. A notable development has taken place with a nano-porous silver catalyst, which is able to reduce CO_2 electrochemically to CO in an efficient and selective manner. CO_2 electro-reduction activity with a CO Faradaic efficiency of 92% has been achieved (Figure 10.18). The long-term performance of nanoporous calalysts in a continuous process is being explored using a flow reactor configuration. The mechanical strength of the catalyst is also under further investigation before the technology is put into practice.

10.8 Low-carbon cements and concretes

There are several low-energy low-carbon cements, a sub-set of which are the cement-like materials that are being developed by using recycled plant-emitted carbon dioxide. Out of all the diverse developments, a few relatively promising technologies are dealt with in this section, based on the information and data available in the public domain. They all adopt

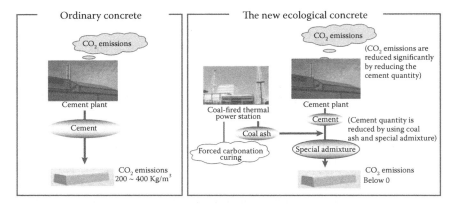

FIGURE 10.19 Typical flow chart for recycling CO₂ from exhaust gases for producing new building products.

non-traditional processes of manufacture and are expected to reduce both waste and emissions including carbon dioxide. Whether these cements would prove to be realistic alternatives to Portland cement is uncertain at this stage. However, from the available data and information it appears that, within the next five to ten years, some could well occupy niche positions in special applications and thus partially replace Portland cement, thereby reducing carbon dioxide emission. Scaling-up of the processes along with their techno-economic validation is yet to be accomplished. The present status of the technologies is briefly described below and a typical approach for recycling CO₂ from the exhaust gases of a power plant is shown in Figure 10.19.

Calera process for calcium carbonate cement

The Calera Corporation, Los Gatos, CA, USA has been engaged in the development of CO_2 capture and sequestration technologies for more than a decade. In this endeavour Calera has been specifically focusing on innovating a technology to convert carbon dioxide to a binding material by exploring the adaptability of the natural processes, such as the formation of coral reefs, ocean sediments, limestone deposits, and strong and tough shells of marine organisms.

So far as the formation of calcium carbonate is concerned, the following reactions are well known:

$$CO_2 + Ca(OH)_2 \rightarrow CaCO_3$$

$$CO_2 + 2NaOH + CaCl_2 \rightarrow CaCO_3 + 2NaCl + H_2O$$

It is, however, understandable that the $CaCO_3$ formed through the above reactions may not be a binding material, whereas the natural occurrences show that the calcium carbonate under certain circumstances can imbibe the binding property. This observation called for a fair understanding of the mechanism of precipitation of calcium carbonate, on one hand, and its polymorphic transformations, on the other.

There are three known polymorphs of $CaCO_3$: calcite, aragonite, and vaterite. It has generally been observed that from an inorganic solution first an unstable amorphous calcium carbonate phase is formed, which transforms to metastable phase vaterite or aragonite before finally forming the stable calcite phase. Precipitation of the polymorphs is affected by different factors like temperature, pH of the medium, concentration ratios of individual components, supersaturation, ionic strength, and impurities. It has also been reported that aragonite formation is favoured at higher temperatures (> 40°C) and at pH of 11 in aqueous solution, while at 24°C in the pH range of 8.5 to 10 vaterite turns out to be the major product. When the pH is more than 12, calcite is the dominant product. The precipitation of different polymorphs can be modified by the impurities in the solution. When the ionic radius of the impurities is smaller than that of Ca^{2+} cations, as in the case of Zn^{2+}, aragonite is deposited. Organic additives also show an effect on carbonate mineralization. Amino acids like glycine, aspartic acid and glutanic acid, or polysaccharides like cellulose influence the precipitation and morphology of calcite and vaterite. Living organisms like photosynthetic microalgae can induce the precipitation of $CaCO_3$ through bio-mineralization. The combination of organisms and metal ions may provide a way to form either aragonite or calcite.

The conversion of carbon dioxide to calcium carbonate requires a source of alkalinity and the availability of calcium ions. Calera technology involves the use of industrial waste streams that meet both the above requirements either in the form of $Ca(OH)_2$ or combination of caustic soda and calcium chloride occurring in the brine as starting materials. These chemicals react with carbon dioxide present in the plant exhaust gases in a reactor in an aqueous medium to precipitate calcium carbonate of desired quality. The calcium carbonate, thus formed, is dried to a free-flowing powder, which can then be used for various applications. A pilot plant, based on the above process, has been in operation for about two years in Moss Landing, CA with a capacity of producing up to 2 tonnes of calcium carbonate. The technology uses flue gas directly from the neighbouring gas power plant without any concentration of carbon dioxide. The Calera calcium carbonate can function as a supplementary cementitious material (SCM) in traditional concrete mixes. It can be used as an independent binding material in concrete products. Calera is in the process of developing wallboard and cement board products.

The Calera Corporation has estimated that their technologies may lead to sequestering of 0.5 ton of CO_2 for every tonne of cement made.

TecEco cements based on reactive magnesia

In John Harrison's original patent for TecEco Pty Ltd, an Australian Company based in Tasmania, an attempt was made to define the term "reactive magnesia" as a special type of magnesia calcined at low temperature (< 750°C) and ground to a fineness with more than 95% passing 120 μm. The general concept was that the lower the calcination temperature and finer the grinding, the more hydraulically reactive the magnesia. In this context it has been claimed that magnesia calcined at 650°C and completely ground to 45 μm or less hydrated even faster than Portland cement. The objective here is to match the hydration rate of magnesia with that of Portland cement, as the absence of this parity of reactivity could lead to a high probability of dimensional distress in magnesia-blended cements. This parity is more important for cements that are blends of Portland cement varieties and reactive magnesia in varying proportions.

The process of manufacture of TecEco cements consists of the following unit operations:

- carbonation of raw feed material
- calcination with CO_2 recycling
- grinding as required
- agglomeration or blending step

In the carbonation step the waste magnesium cations as found in process water and bitterns are carbonated to produce large quantities of nesquihonite ($MgCO_3 \cdot 3H_2O$), which is then calcined at 75°C, or preferably at lower temperatures, in a specially designed kiln to produce reactive magnesia. The carbon dioxide generated in this step is fed back into the process of carbonation. After obtaining the required fineness of low-temperature-calcined magnesia, it is used as a binder to agglomerate large amounts of nesquihonite to produce synthetic carbonate aggregate or to be blended with other materials to form different varieties of cements.

Calix and novacem processes for magnesia-based products

Like TecEco Pty Ltd, two other companies—Calix Limited (Pymble, NSW, Australia) and Novacem Ltd. (an Imperial College start-up)—had focused on a magnesium oxide and carbonate system to develop carbon-negative cements by utilizing CO_2 from the kiln exhaust gases. The Calix product was composed of 30–80% magnesia-bearing components in both oxide and carbonate phases and 20–70% of another silica-alumina component. Such formulations were tried out in the production of building materials such as cements, mortars, grouts, etc. with a lower carbon footprint than Portland cement. The company holds several patents on such processes.

The Novacem development attempted to utilize natural magnesium silicate ores containing iron oxide, such as ultrabasic rocks like serpentine, as the starting raw material, instead of magnesite. Precursor material like olivine (($Mg,Fe)SiO_4$) is carbonated in an autoclave process at a temperature of 180°C and pressure of 150 bars. The magnesium carbonate obtained from the silicate phase is decarbonated at 700°C and the carbon dioxide released during this part of the process is returned back to the process for carbonation of magnesium silicate. The Novacem cement composition is a blend of MgO that absorbs CO_2 during the process of hardening and at least one magnesium carbonate (either hydrated or unhydrated) having the formula $xMgCO_3 \cdot yMg(OH)_2 \cdot zH_2O$, wherein x is at least 1 and at least one of y or z is greater than 0. The binder composition may optionally comprise a hygroscopic material like NaCl. The magnesia component, when mixed with water in the presence of the magnesium carbonate produces a rosette-like morphology.

The product underwent extensive evaluation both in paste and concrete forms. According to the inventors, the strength development in Novacem cement happened through the formation of magnesium silicate hydrate phase, similar to the calcium silicate hydrate phase in Portland cements.

The significantly lower embodied carbon in Novacem cement was due to the following factors:

- The feedstock was non-carbonate.
- Emission of CO_2 was low due to low-temperature processing (0–150kg CO_2/t cement).
- The CO_2 available is used in forming a carbonate phase.
- The total typical emissions are estimated at -50 kg to +100 kg CO_2/t cement.

The Novacem Company has already operated an experimental batch pilot plant. A semi-commercial plant of 25,000 t capacity was planned but could not be set up. For commercial reasons, the company's technology and intellectual property ultimately came to Calix Limited, which had more or less an identical objective of technology development.

Solidia cement and concrete

Solidia cement, a trade-marked product developed by Solidia Technologies USA, is a low-lime non-hydraulic binder, the setting and hardening characteristics of which are derived from a reaction between carbon dioxide and calcium silicates such as wollastonite and pseudo-wollastonite ($CaO.SiO_2$), rankinite ($3CaO.2SiO_2$), and an amorphous meliolitic phase (Ca-Al-Si-O). During the carbonation process, calcite ($CaCO_3$) and silica gel (SiO_2) form and impart binding properties to the product (11).

The technology was first demonstrated in May–July 2012, at IBU-TEC, Weimar, Germany with the help of a natural gas-fired lab rotary kiln (0.3 × 7.0 m) by producing the clinker at 1200°C. After concluding an agreement with Solidia Technologies, further development of cement applications has been taken up by Lafarge. The process of making Solidia clinker is the same as that of Portland cement clinker. The basic raw materials are the same and the plant and machinery normally used in the Portland cement plants can be used to produce the Solidia cement. The total lime content of Solidia clinker is in the range of 45–50%, which is 30% lower than what is required for the Portland clinker. This difference results in an almost equivalent reduction in the energy required for the calcination of limestone in Portland clinker making. The burning temperature of Solidia clinker is about 250°C lower than that of Portland cement clinker, which is expected to give a saving in the fuel requirement. The absence of alite or tricalcium silicate and belite or dicalcium silicate in the Solidia clinker makes it less sensitive to the cooling rate after discharge from the kiln. Hence, the cooler heat losses should be markedly reduced. The grindability of Solidia clinker is expected to be softer, demanding less energy for grinding. The non-hydraulic nature of Solidia cement eliminates the need for gypsum addition as a set controller. For the same reasons, no special storage arrangements are needed for Solidia clinker/cement. All the above process advantages in making Solidia clinker and cement should definitely reflect in reduced emission of CO_2.

The processes of making concrete and concrete products with Solidia cement and Portland cement do not differ, except for the curing process that leads to setting and hardening of Solidia concrete. The Solidia concrete does not set, harden, and cure until it is exposed simultaneously to water and gaseous CO_2. The process is mildly exothermic and the reactions occur in an aqueous environment as follows:

$$CaO.SiO_2 + CO_2 \rightarrow CaCO_3 + SiO_2$$

$$3CaO.2SiO_2 + 3CO_2 \rightarrow 3CaCO_3 + SiO_2$$

The above reactions demand a CO_2-rich atmosphere, but can be conducted at ambient gas pressures and at fairly low temperature, viz., 20–50°C. These parameters are well within the facilities of most of the precast concrete manufacturers. It may be borne in mind that compared to the hydration reactions in Portland cement concrete, the carbonation reactions in Solidia concrete are speedier but the completion of the curing process depends on the ability of CO_2 gas to diffuse through the entire volume of a concrete component. Experiments have shown that thin roof tiles (~ 10-mm thick) can be cured in less than 10 h, while large parts like railway sleepers (~ 250-mm thick) take about 24 h. Concrete with different strengths can be designed by using Solidia cement. Compressive strengths up to 70 MPa with corresponding flexural strength of up to 8 MPa have been observed in test specimens.

One cubic meter of Solidia concrete requires 127 kg of water to flow but no water gets chemically bound, as the curing process is dependent on CO_2 gas. Most of the water (~ 80%) evaporates from the concrete and can be condensed and recycled. Solidia cement contains about 250–300 kg of CO_2 per ton. In addition, the total reduction of CO_2 emission at the clinkering stage is about 30%. In totality the CO_2 footprint associated with the manufacturing and use of cement can be reduced by 70%. Interestingly, this level of achievement can be accomplished from the present and familiar plant, equipment, process, raw materials, and supply chains. The problems of water scarcity and resource conservation can also be addressed through this technology. However, the durability issues of Solidia cement and concrete are yet to be fully comprehended. One cannot overlook the drop in pH to 9.5–10 in Solidia concrete from a level of 12 in Portland cement concrete. Hence, it is all the more necessary to study the reinforcement corrosion as well as chloride and sulphate resistance of Solidia concrete, which are reportedly being taken up for investigation.

CO_2-SUICOM technology

The technology of CO_2-SUICOM was developed by the Chugoku Electric Power Co. Inc., Denki Kagaku Kogyo Kabushiki Kaisha, and Kajima Corporation of Japan for a new ecological concrete that could bring down the CO_2 emissions levels below zero by capturing CO_2 emitted from thermal power stations. The technology makes use of a special additive in the form of $\gamma\text{-}C_2S$ and coal ash as well as a special carbon dioxide curing chamber. The special admixture $\gamma\text{-}C_2S$ reacts with CO_2 to produce $CaCO_3$ in the form of calcite and vaterite as well as SiO_2 gel. Other materials for making eco-concrete include coal ash, blast furnace slag, and other SCMs with low carbon footprints in addition to the traditional concrete-making materials. The carbonation curing is done in a special facility having a flue gas feeding unit, a system for contacting the flue gas with water or water vapor, a carbonation tank with shielded space for the object to be cured, a flue gas circulator, and a temperature and humidity-regulating device. The flue gas is fed to the curing tank without regulating its concentration and flow rate. The volume of the objects to be carbonated may be such that it can absorb the total amount of carbon dioxide available, taking into account the supply of CO_2 from the emitting source including the emissions attributable to the power consumption of the carbonation curing equipment. The temperature and humidity control system is designed with industrial steam or water fed to the heat exchanger without additional electricity consumption. The technology is suitable for precast products of various strength and quality specifications. The products have been used in both road and building construction and the benefits realized are as follows:

- The amount of captured CO_2 surpassed the CO_2 emissions of the materials.
- The efflorescence of the carbonated concrete was less than the normal concrete in the interlocking blocks.
- The abrasion resistance of the carbonated concrete was higher than that of normal concrete.
- The carbonated mortar was denser than the normal mortar.

SWOT analysis of the carbonation technologies in development

All six of the technologies narrated here are in the innovation stage, in which the concept proving phase has been crossed over with convincing laboratory evidence. It is, however, observed that even for the technologies for which the pilot-scale trials have been reported, not enough details are in the public knowledge. The technologies are still far away from standardization, industrial adoption, and market penetration. They are disruptive in nature but at the bottom rungs of application. Hence, what is important is to admit them as potential competitors of chemical cements, modified Portland cement such as calcium sulpho-aluminate-belite cement, and alkali-activated cements such as the geopolymer type. A very tentative SWOT analysis of the above technologies is given in Table 10.15 for a general appreciation.

Table 10.15 Comparison of the technologies in development at a glance

Criteria	Calera	TecEco	Calix	Novacem	Solidia	Kajima
Raw materials base	H	M	M	L	H	H
Scientific soundness of the concept	H	H	H	H	H	H
Ease of process adoption	H	M	M	M	H	H
Stage of development	Pilot plant	Not known	Not known	Pilot plant	Pilot plant	Pilot plant
Potential viability	H	M	M	M	H	H
Product versatility	H	H	M	M	H	H
Up-scaling potential	H	L	L	L	H	H
Preventive barriers	Some	Many	Many	Some	Less	Less
Time to industrial adoption	Medium	Long	Long	Medium	Short	Short
Overall feasibility	H	M	M	M	H	H

Note: H = High; L = Low; M = Moderate.

10.9 Nanotechnology in cement research

Nanotechnology has been defined as the ability to work at the nanometer level (< 100 nm) to achieve purposefully engineered structures with novel and improved size-dependent properties and functions. The key element is the use of materials with nano-dimensions, at least in one direction, in small but optimum proportions. A reinforced concrete building is a nano-enabled finished product, although seldom recognized as such. It is being progressively observed that the application of nano-structured materials or nano-modification of bulk materials as relevant to cement and concrete technology has the potential of minimizing matter consumed during the whole lifecycle, minimizing negative environmental impacts, and finally ensuring a definite comfort of use. The author has reviewed the subject recently (26) and has shown that there has been a specific emphasis on research on the application of nanosilica, nano-graphene oxide, polymer-clay nano-composites, nano-carbonate, nano-titania, nanoalumina, nano-zinc oxide, and nanocellulose in cement and concrete. The position of nanomaterials vis-à-vis other SCMs and concrete components is shown in Figure 10.20 (27). Several mono-phase and multi-phase nano-enabled cement products are available in the market or industrially applied in construction, although the volume of use is still limited and often experimental. The effect of even the same nano material in paste, mortar, and concrete is not the same or linear. Hence, it is essential to carry out thorough trials before actual application. Further, improvement of certain concrete characteristics with incorporation of nano-additives is noteworthy. Self-compacting and self-cleaning concretes are prominent examples in this regard. Incorporation of nano-silica in the SCC mixtures improves the properties of both the fresh and hardened concrete. The self-cleaning property

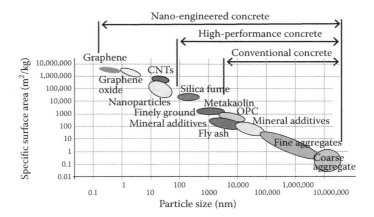

FIGURE 10.20 Comparison of nanofillers with conventional SCMs and other concrete components. (From S. Chuah et al., *Concrete in Australia*, Vol. 41, No. 1, 48–53, 2015. Adapted by Chuah et al. from K. Sobolev and M. F. Gutierrez, 2005, as communicated by *Concrete in Australia*.)

on the concrete surface is imparted by nano-titania particles under the influence of ultraviolet radiation. Studies have shown that nano particles and nanotechnology can make the steel reinforcement bars remarkably corrosion-resistant. Graphene is a material consisting of single layers of carbon atoms arranged in a hexagonal lattice. The graphene layers can be formed into carbon nanotubes and carbon fibers. A carbon nanotube has a tensile strength of 150 times that of steel and at the same time it is approximately six times lighter. The flexural and compressive strength properties of mortar and concrete can be significantly increased with very small additions (< 0.1%) of carbon nanotubes by refining the microstructure. It may be relevant to mention here that through a collaborative research program of MIT (USA) and CNRS (France) a nanotechnology-based binding material is in the process of development. The core of the technology lies in topological functionalization of the basic building blocks of the cement hydrate phase CSH and precipitation of such particles in multiple sizes (28). All in all, nanotechnology has the potential of ushering in an era of industrial revolution and ensuring promising advancements in the construction industry but it must be understood that the long-term environmental impacts and human safety aspects of manufacturing and using nano materials are yet to be studied properly.

10.10 Non-hydraulic cements

Non-hydraulic cements are a class of binding materials that set and harden with chemicals and not water. These cements have high potential for niche applications. Since the present discourse is on hydraulic cements and concretes, non-hydraulic materials are not discussed here, except a brief mention of geopolymers, as they hold a high promise for reduced CO_2 emission. Introduced in the 1970s, these are inorganic alumina-silicate polymers, synthesized from materials of geological origin or byproducts such as fly ash, slag, etc. They are formed by polycondensation of monomers (sialate) under highly alkaline conditions that yield polymeric chains, which are cross-linked with the aid of temperature, forming a gel-like structure with a short-range order. Geopolymer technology shows considerable promise for application in the concrete industry. In terms of reducing global warming, geopolymer technology could reduce CO_2 emission by up to 80% (29).

10.11 Summary

Although "chemistry" and "engineering" are the main props on which rests the basic process of clinker making and cement production, over the years the advances in engineering have overtaken the endeavours of unravelling the intricacies of fundamental chemical reactions and phase transformations in the manufacturing process. As a result, kilns of gigantic capacities of 12,000 t/d have been set up. A large number of engineering innovations have been seen in the cement production process. They pervade in the design and operation of large pyroprocessing systems consisting of five- or six-stage preheaters with low-pressure-drop cyclones; precalciners performing carbonate decomposition reactions in seconds with 60% heat input directly into the precalciner burners; three- or five-channel burners capable of burning multiple fuels including alternative refuse-derived fuels; short two-support rotary kilns with length:diameter ratios of 10–14; efficient grate coolers with heat recuperation; and several other developments.

The raw and cement grinding systems have been made more efficient and new milling systems, such as vertical roller mills, hydraulic roll presses, horizontal roller mills, combined systems of ball mills and hydraulic roll presses, etc. have been put into operation.

It is important to note that there have been global efforts to find newer pyroprocessing technologies that are not dependent on the conventional rotary kiln process. The fluidized-bed system, microwave heating, sol-gel process, electron beam radiation, etc. were investigated decades back but the studies have not been taken to the logical end of developing viable technologies. A review of the status of the technologies worked upon, and the search for newer opportunities, continues.

So far as the modified Portland cements are concerned, high-belite Portland cement and calcium sulfoaluminate–belite cement have good potential for niche applications. Portland limestone cement, which is a regular product in the European Union, can be proliferated in other countries after due investigations. Alinite cement, revived in Japan as an eco-cement, can be considered for effective disposal of high-chloride municipal solid wastes.

The newer opportunities presented by multi-ingredient formulations based on calcium aluminate cement for niche applications are important for modern construction, and necessary emphasis is being accorded to their development and application. New cements in various stages of development, specifically with a view to reducing the carbon dioxide emission, are on the threshold of technological breakthroughs. Most of these technologies are patent-protected. In the context of climate change issues, a strategy to get into the club of low-carbon cement development is crucially important. For similar objectives, accompanied with resource conservation, the broad target is to reduce the proportion of clinker in finished cements. More use of mineral admixtures with low carbon footprints, production of multi-component cements, the application of nano materials, and an examination of the viability of non-hydraulic binding materials like geopolymers are the future trends of research in cement application.

References

1. A. K. CHATTERJEE, Modernisation of cement plants for productivity and energy conservation, in *Cement and Concrete Science & Technology* (Ed. S. N. Ghosh), Vol.1, Part 1, ABI Books Pvt Ltd, New Delhi (1991).
2. HIROSHI UCHIKAWA, Present problems in cement manufacturing, *Journal of Research of the Onoda Cement Company*, Vol. 45, No. 128, 1994.
3. Kawasaki Heavy Industries Ltd, Catalogue No. 3F2428 on Fluidized bed advanced cement kiln systems (FAKS), Japan (2000).
4. C. H. PAGE, C. H. THOMBARE, R. D. KAMAT, AND A. K. CHATTERJEE, Development of sol-gel technology for cement manufacture, in *Advances in Cementitious Materials* (Ed. S. Mindess), *American Ceramic Society*, Ohio, USA (1991).
5. YI FANG, DELLA M. ROY, AND RUSTUM ROY, Microwave clinkering of ordinary and colored Portland cements, *Cement And Concrete Research*, Vol. 26, No. 1, 1996.
6. C. K. PARK, J. P. LEE, AND D. M. ROY, Clinkerization of Portland cement raw meal in the microwave processing system, Proceedings of the 12th ICCC, Montreal, Canada, July, 2007.
7. S. K. HANDOO AND K. GOSWAMI, Application of electron beam process technology for production of cement, Proceedings of Indo-USSR Seminar on Industrial Applications of Electron Accelerators, Vol. II, BARC, November 3, 1988.
8. A. K. CHATTERJEE, Calcium aluminate cement, National Seminar on Refractory Raw Materials, Central Glass & Ceramic Research Institute, Kolkata, 2012.
9. A. K. CHATTERJEE, Re-examining the potential of calcium sulphoaluminate cements from the perspective of versatility. Durability and GHG emission, *Indian Concrete Journal*, Vol. 84, No. 11, 2010, pp. 7–19.
10. A. K. CHATTERJEE, Special cements, in *Structure and Performance of Cements* (Eds J. Bensted and P. Barnes), Spon Press, London (2002).
11. A. K. CHATTERJEE, High belite Portland cement—An update on development, characterization and applications, Proc. 11th International Congress on Cement Chemistry. Vol. 1, Durban, South Africa, 2003.
12. K. FUKUDA AND S. ITO, Improvement in reactivity and grindability of belite-rich cement by remelting reaction, *Journal of the American Ceramic Society*, Vol. 82, No. 8, 1999.
13. Lafarge Project Aether®, www.aether-cement.eu.
14. A. K. CHATTERJEE, Special and new cements in a historical perspective, Proceedings of 3rd Congresso Brasiliero de Cement, Vol. 2, ABCP, Sao Paulo, 1993.

15. Hiroshi Uchikawa and Hiroshi Obana, Ecocement – the frontier of recycling of urban composite wastes, *World Cement*, November, 1995.
16. A. G. Kesim, M. Tokyay, I. O. Yaman, and A. Ozturk, Properties of alinite cement produced by using soda sludge, *Advances in Cement Research*, Vol. 25, No. 2, 2012, 104–111.
17. R. Guyot and R. Ranc, Controlling the properties of concrete through the choice and quality of cements with limestone additions, Int. Conference on the Utilization of Fly ash, Silica Fume, Slags and other By-products in Concrete, Montebello, Canada, 1983.
18. A. K. Chatterjee, New cements and binding materials, Proceedings of the 14th NCB International Seminar on Cement and Building Materials, New Delhi, December, 2015.
19. Aalborg Cement, Patent on "Portland limestone calcined clay cement," US Patent No. 20120055376.
20. K. Scrivener, Aurelie Favier, and Francois Avet, Finding a plentiful alternative, *International Cement Review*, February, 2015.
21. www.kerneos.com, brochure_BC_TER.
22. S. Lamberet, L. Amethieu, and K. Scrivener, Microstructural development of ternary binders based on calcium aluminate cement, calcium sulfate and Portland cement, in Proceedings of the Centenary Conference on Calcium Aluminate Cement, Avignon, France, June–July, 2008.
23. A. K. Chatterjee, Green cement: towards a sustainable future, *Indian Cement Review*, May, 2016.
24. A. K. Chatterjee, Veering towards carbon capture & transformation— An emerging technological need for carbon dioxide abatement strategy, 3rd International Conference on Alternative fuels and Raw Materials in Cement Industry, CMA, New Delhi, March, 2017.
25. A. K. Chatterjee, Alternative cements and concretes in development with recycled plant-emitted carbon dioxide, *Cement Wapno Beton*, No. 2, 2017.
26. A. K. Chatterjee, Nanoparticulate matters in cement, concrete and construction – Performance and EHS perspectives, *Indian Concrete Journal*, Vol. 91, No. 2, February, 2017.
27. S. Chuah, Xiangyu Li, Wen Hui Duan, Jay G. Sanjayan, and Chien Ming Wang, Graphene oxide nanosheets for enhancing concrete durability, *Concrete in Australia*, Vol. 41, No. 1, 48–53, 2015.
28. http://newoffice.mit.edu/2014/stronger-greener-cement-0925.
29. B. Vijaya Rangan, Studies on low-calcium fly ash based geopolymer concrete, *ICI Journal*, Ocober–December 2006.

CHAPTER ELEVEN

Global and regional growth trends in cement production

11.1 Preamble

In less than two centuries the global cement industry has undergone phenomenal growth to reach a production level of over 4 billion tons per year. It has positioned itself as an inexpensive, reliable, durable, and almost indispensable basic material of building construction. Combined with water and mineral aggregates it forms concrete, mortar, and other cement-based products, which, taken together, turn out to be the second-most consumed substance in the world after water. At the same time it is also evident that the growth of the cement production base is strongly dependent on the growth of the construction industry that provides the built environment to mankind for working, living, and leisure.

The volume of cement entering world trade is low relative to overall production and consumption, typically accounting for about 5% in aggregate terms. This is because of the low unit value of cement, the widespread availability of raw materials for cement production, and the link between economic growth and cement consumption. All these factors favor domestic production on a large scale. Consequently, the growth strategy of the cement industry has been essentially zonal or regional and future growth also seems to be dictated by the same zonal or regional patterns of consumption. The cement demand pattern will obviously be related to the economic development of those regions that will focus on housing and infrastructure build-up.

However, since future building requirements will be more compliant with new norms of construction technology, environmental demands, and social needs, the growth features of the cement industry may not be an extension of the past and will definitely display certain discontinuities. The emerging trends of ultra-high-rise buildings, deep subsurface structures, very-long-span bridges, marine airports, etc. will call for a new breed of cements in consonance with state-of-the-art construction technologies. Further, the process of normal cement production has an intrinsic problem of carbon dioxide emission, which will have to be tackled more innovatively in step with the expansion of the production base. The industry will also have to be better equipped to meet the social and environmental challenge of providing the most effective process to dispose of wastes from other industries.

This chapter deals with the near-term growth trend of the cement industry in the context of the above imperatives.

11.2 Capacity and production growth perspectives

Global cement production has increased from a meager level of 32 million tons per year in 1920 to 4200 million tons per year in 2016. However, if the data are viewed and compared for the nine separate decades in this time span, one would find that the growth rate has declined in almost every decade. Notwithstanding the declining growth rate, the expansion of the production base has been phenomenal.

Looking at the regional share of world cement production over seven decades of the previous century (see Figure 11.1), it is interesting to note that the production share of the Americas came below that of Europe in the pre-war years and continued to remain so until the 1990s. From 1930 to 1980, Europe held the predominant share of cement production. Asia overtook the Americas in 1965 and Europe in 1980. Although the former USSR had shown a rising trend from the 1950s to the 1970s, the fragmentation of the union thereafter made the picture anomalous. The African continent had shown a rising trend but of small magnitude. Oceania maintained a very small and gently declining production share through every decade (1).

In recent times the global cement industry continued to grow from 2.60 billion tons per year in 2007 to 4.20 billion tons per year in 2016.

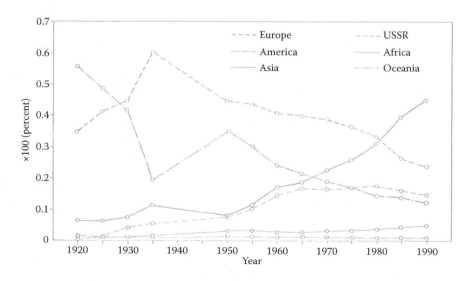

FIGURE 11.1 Regional share of world cement production in the twentieth century.

This trend was attributed to the massive expansion of the Chinese market with an increase of more than 800 million tons of annual cement production since 2007. Modest cement production growth rates have taken place in Brazil, Egypt, India, Iran, Saudi Arabia, and Vietnam, while production volumes remained flat or decreased in the USA, Japan, and Western Europe. The only European countries to show signs of recovery in the intervening period were Russia, Turkey, and Spain. The impact of the 2008 recession took some time to hit European countries and has subsequently been severe on the construction sector and consequently on the cement industry.

The regional share of world production of cement from 2001 to 2013 is shown in Figure 11.2, which shows more or less the same trends as observed in Figure 11.1 with the exception that the Asian share had grown more steeply than the other regions. World production of cement in 2013 was estimated at about 4 billion tons and the country-wise distribution is presented in Figure 11.3. The slowed-down trends of the member countries of CEMBUREAU and the EU 28 taken together during the same period are shown in Figure 11.4 (2).

World production of cement in 2016 was 4.2 billion tons, which showed an increase of 2.4% over the previous year but about 4% over 2013. The production pattern of the top 15 cement-producing countries in 2016 has been compared with that of 2013 in Table 11.1. It is important to note that, barring China, India, and the USA—the top three

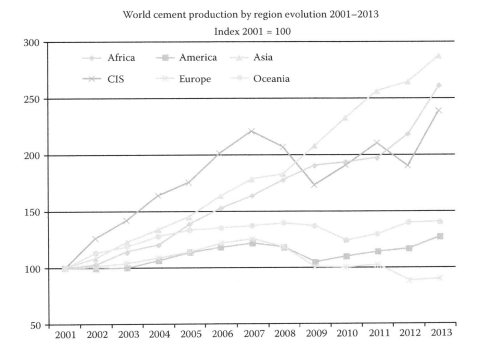

FIGURE 11.2 Regional growth trends in cement production during 2001–2013.

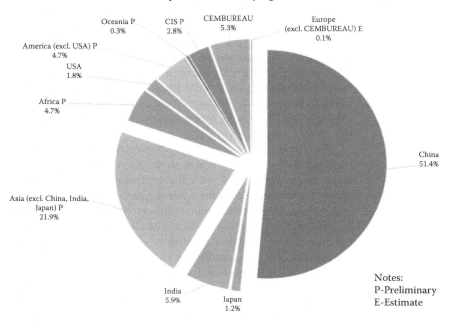

FIGURE 11.3 Country-wise distribution of cement production. (From CEMBUREAU, available at http://cembureau.eu/cement-101/key.facts-figures. Brussels.)

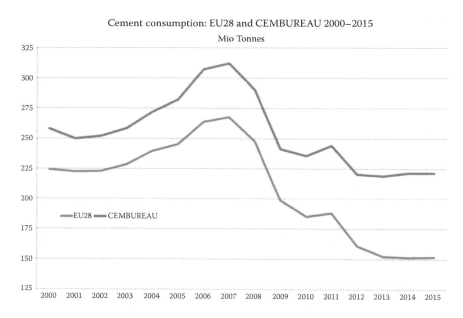

FIGURE 11.4 Cement production trends of the member countries of CEMBUREAU and all EU countries. (From CEMBUREAU, available at http://cembureau.eu/cement-101/key.facts-figures. Brussels.)

Table 11.1 Comparison of the top cement-producing countries in 2013 and 2016

Rank	Country	Production in 2013 (million tons)	Production in 2016 (million tons)
1	China	2300	2410
2	India	280	290
3	USA	78	85.9
4	Iran	75	53.0
5	Brazil	70	60.0
6	Turkey	70	77.0
7	Russia	65	56,0
8	Vietnam	65	70.0
9	Japan	53	58.0
10	Saudi Arabia	50	63.0
11	South Korea	49	48.0
12	Egypt	46	50,0
13	Mexico	36	36.0
14	Indonesia	58	63.0
15	Thailand	35	42.1
	World total	4043	4200

producers—the ranking of other countries keeps changing due to economic changes.

However, the Freedonia Group, an industry research organization, has projected that the growth rate in the immediate future will be higher at 4.5%, taking the production to 5.2 billion tons in 2019 (3). Growth will be driven by healthy increases in construction activity in emerging economies throughout the Asia-Pacific and Africa-Mideast regions. India is expected to post the fastest growth in cement demand of any major national market, advancing 8% per year through 2019. Some other countries in the Asia-Pacific region, such as Vietnam, Indonesia, and Pakistan, will also display high growth rates, although China will continue to be the largest driver of growth but with a somewhat slower pace. A consolidated region-wise growth pattern is presented in Figure 11.5. The growth of cement demand in North America is projected to improve significantly, while Western and Eastern Europe will show signs of recovery during 2014–2019 as compared to 2009–2014.

Top cement-producing companies and support infrastructure

To counteract the problems of growth deceleration due to the economic crisis of 2008, the major cement-producing companies took initiatives to consolidate the industry and improve operational efficiency. The most notable initiative in this context is the merger of two very large multi-national companies, Lafarge and Holcim, and the planned acquisition of Italcementi by HeidelbergCements. In Asia, Chinese and Indian companies also took steps towards consolidation of the industry. The top ten companies with their capacities are tentatively listed in Table 11.2, broadly on the basis of their websites. The relative ranking of these

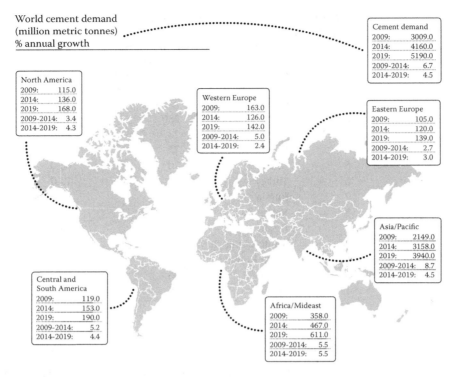

FIGURE 11.5 Region-wise projection of cement demand. (From General News, Global demand for cement to reach 5.2 billion tonnes, *Concrete in Australia*, Vol. 41, No. 4, 2016. Reproduced with the permission of the Concrete Institute of Australia, courtesy of David Miller, CEO.)

Table 11.2 Top ten cement-producing companies of the world

Tentative ranking	Country	Company	Approximate capacity (million tons/year)
1	China	CNBM	400
2	Switzerland/France	LafargeHolcim	386
3	China	Anhui Conch	285
4	China	Jidong Development	130
5	Germany	HeidelbergCements	129
6	China	Shansui (Sunnsy)	100
7	Mexico	Cemex	94
8	India	Aditya Birla Group	91
9	China	China Resources	78
10	Taiwan/China	Taiwan Cement	69

companies cannot be strictly ensured due to frequent changes of capacity and ownership (4).

Apart from the companies listed above, Eurocement in Russia and Votorantim in Brazil, with capacities in the range of 45 million tons per year, are worth mentioning. According to the Global Cement Directory 2016, there are 2273 active integrated cement plants around the world. It is also interesting to note that for improving the resource productivity and energy efficiency of the cement industry, significant support is extended by several international organizations, such as the World Business Council for Sustainable Development, International Energy Agency, United Nations Industrial Development Organization, The World Bank, etc. In addition, all the major cement-producing countries have their national federations as well as research establishments having their own organizational structures and programs of technical activities.

11.3 National economy versus cement consumption

It is an axiomatic truth that the volume of cement consumed varies directly with the performance of the construction industry. As the construction industry plays a critical role in the national economies of all countries through the provision of basic infrastructure, residential and commercial buildings, and maintenance and retrofitting of existing facilities, it is logical to presume that as the national economy progresses, the quantum of construction activity increases. The volume of cement consumed, therefore, bears a strong relation with the performance of economy. Based on this hypothesis, the per capita consumption of cement has been related to per capita gross national product (GNP). Interestingly, the relationship has been perceived to be almost parabolic (see Figure 11.6) in the sense that the per capita consumption of cement increases with an increase in per capita GNP up to a point, after which the relationship

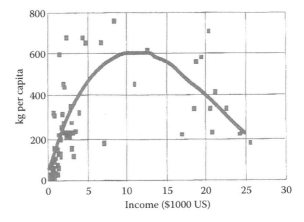

FIGURE 11.6 Relationship of cement consumption and GNP per person.

reverses. However, from the plots in the figure it is evident that the scatter is too high and the pattern is only tentative.

A more detailed study (5) was carried out by the author, involving a total population of 167 countries, which showed that the countries studied could be divided into four categories: low-, middle-, upper-middle-, and high-income economies (Figure 11.7). Although in each subgroup the per capita consumption of cement showed considerable scatter, in the low- and middle-income economies there was a discernible linearity between per capita GNP and per capita cement consumption. Economic growth in the higher echelons of economies did not appear to bear any strong relation between the two parameters. No direct relation could be established with other factors like area, population density, urbanization, etc. The demand for cement in different sectors of construction and the per capita cement consumption of a few selected countries are illustrated in Table 11.3 (6).

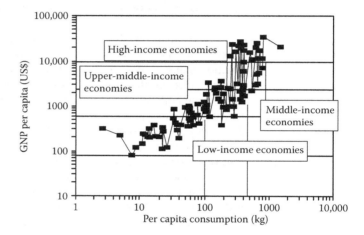

FIGURE 11.7 GNP in different tiers of economy versus per capita consumption.

Table 11.3 Comparison of sector-wise cement demand and per capita cement consumption of a few selected countries

Country	Residential (%)	Commercial (%)	Infrastructure (%)	Per capita (kg)
India	63	16	21	185
Indonesia	72	16	12	225
Brazil	56	19	25	346
China	25	30	45	1555
Italy	36	35	29	402
USA	22	22	56	290

11.4 Change drivers of production and application of cement

Since the cement industry has grown essentially on the basis of sustaining technologies rather than disruptive developments, the past trends may serve as strong indicators of the future. Up until the 1980s, growth was driven by the scale and cost of production and product quality was evaluated on the basis of paste and mortar properties at the time of dispatch. After the 1980s, energy conservation and pollution control became the prime drivers. Product quality became more intimately linked up with concrete applications. From the beginning of this century it was realized that sustainability would have to be of paramount importance to increase further capacity of cement production, which had already crossed 2 billion tons of production per year in 2003, 3 billion tons in 2009, and 4 billion tons in 2013. As already stated, the projection for 2019 is 5.2 billon tons per year.

The most redeeming feature of the global demand projection is that blended cements will account for 75% of the quantity. Presuming this to be the likely scenario of the industry at the end of 2019 and presuming most optimistically that the clinker factor of the blended cements will be as low as 0.5, one may reasonably estimate that the industry would require about 5000 million tons of limestone, more than 100,000 MW of power, about 600 million tons of coal or its equivalent fuel, and about 2000 million tons of blending materials of different types. The resource requirement for the industry is envisaged to be very large, even in the near future, and if the business proceeds on an "as-usual" basis resource consumption by, say, 2050 will be colossal. Resource conservation and sustainability will, therefore, be the prime driver of change now and in the future, integrating with other industries to utilize their byproducts and wastes to substitute the natural raw materials and non-renewable fuel.

It is important to note that the quantities of industrial wastes and municipal disposal have rapidly increased in recent years with the advances of the industry and the rise in the standard of living. The social cost for health, hygiene, and social welfare has been steadily increasing. Hence, the environmental and social pressure of waste utilization can be effectively interfaced with the problems of natural resource depletion by the cement industry. This would call for appropriate development of technology on one side and social systems on the other, so as to imbibe "environment-cleaning capability" in the production process.

Another unavoidable push of increasing intensity for the cement industry is to move to a less-polluting regime, particularly in terms of CO_2 and other gaseous emissions. Hence, the development of "environmental load-reducing cement" is a cardinal direction of growth for the industry. Combining the current and emerging trends, the change drivers for the manufacturing industry can be summarized as follows:

- energy efficiency
- man and machine productivity

- compliance with pollution regulations
- natural resource conservation
- maximization of alternative fuels and raw material use
- high plant availability with all operational efficiency and environment compatibility

With all the above change drivers the industry will restructure itself towards energy efficiency, resource efficiency, and sustainability.

Cement performance requirements

The cement industry has to recognize the fact that there is no market for cement itself but rather a market for concrete- and cement-based products. It is essential that these end products are durable, user-friendly, environmentally compatible, and competitive in the market. Cement quality has long been governed by national and international standards, which are essentially prescriptive. In recent times there has been progressive introduction of performance standards that are found to be more pertinent for multi-component blended cements, which are likely to pervade more and more into the world of cement application. These standards would meet the needs of both users and producers more effectively for efficient application of cement without imposing severe restrictions on ingredients to be used in manufacture. In this changing context, the performance of cement deserves to be reviewed and modified from the following perspectives:

- adjusting workability to application: controlling the handling time; maintaining workability for a specified duration; self-leveling of surfaces; concrete pumping; self-compaction of concrete without vibration, and so on
- inherent suitability for producing high-strength and high-performance concretes meeting the requirements of easy placement, attaining high early-age strength, long-term abrasion or impact loading, low permeability and chemical resistance, long-term durability, and other application-specific attributes
- fulfilling the service-life requirement of structures built with cement-based materials by predictive and preventive maintenance, which may include, if technologically feasible, such intelligent functions of cement as self-controlling by pH, temperature, atmosphere, and light

The above performance requirements can be met only if the hydration reactions of cement and the resultant micro-structure of concrete and cement-based products are controlled by the compositing of compatible materials but possessing diverse properties:

  i. high and low hydration
  ii. varying reactivity at different temperature ranges
  iii. SCMs and fillers having distinctly different particle size distribution
  iv. endothermic and exothermic substances
  v. low-dose property-modifying chemical admixtures

The above measures cannot be taken routinely with traditional constituent materials and normal mixing, placing, and curing methods. New materials, new equipment, and new construction methods are often needed along with training programs to help achieve the desired results. Significant improvements in concrete construction and its performance in recent times would not have been possible without the use of supplementary cementitious materials and chemical admixtures. While the use of individual materials like fly ash, blast furnace slag, silica fume, and other pozzolanas is well established, the emerging trend is to adopt blending of multiple components and also mixing of ultrafine materials. This has also led to a rapid rise in the consumption of chemical admixtures throughout the world during the past two decades. Interesting developments have been occurring in the field of super-plasticizers, air-entraining agents, anti-washout admixtures, and waterproofing materials. Some new super-plasticizers that can be independent of the addition procedure but are able to maintain slump levels for at least 1–2 hours are replacing the traditional products. Anti-washout admixtures, water-soluble cellulose-type, or polyacrylamide-type polymers have been used for the underwater placement of concrete, the need for which is bound to grow in the future. Polymer-modified paste or slurry with a very high polymer:cement ratio of 50% or more is applied on millions of square meters of reinforced concrete surfaces each year as a liquid-applied waterproofing membrane with a thickness ranging between 2–4 mm. An interesting waterproofing material that solidifies in water is in use as a shock-absorbing, waterproof backfill material for tunnels and drains.

The performance of concrete and other cement-based products can be improved with the addition of fibers and polymers. Fiber-reinforced concrete is generally produced by adding steel, polymer, glass, and carbon fibers, and such addition leads to improvements in flexural strength, toughness, and impact resistance. Slurry-infiltrated fiber concrete with micro-fiber content up to 20% and compact reinforced concrete with both fiber and particle reinforcement are potential products of future niche applications. Polymers have already made a considerable headway in the field of concrete and cement-based products. They are used for varied purposes such as improved bonding, pre-sealing, reducing permeability to water and aggressive chemicals, self-leveling, dust-proofing, and a host of other functions. Polymer-modified mortar and concrete are the preferred materials for repair and restoration of deteriorated concrete structures.

While the above-mentioned advances in cement-based materials have indeed been impressive and beneficial, an almost entirely new set of exotic materials is waiting in the pipeline for wider commercial exploitation in viable situations. These are essentially composite materials formed by densification through pressure and heat with significant manipulation of microstructure, ultimately attaining properties approximating those of either fired or chemically bonded ceramics. Examples of such developments include DSP (densified systems containing homogeneously arranged ultra-fine particles), MDF (macro defect-free cement), and RPC (reactive powder concrete).

Objectives for high-performance building materials

Although there is no unanimity of views on different directions of the development of new cement-based products and their relevance, viability, and purpose, the objectives can either be to fill up a known gap of application or to create a new opportunity for application. However, it has been clearly understood that the costs of non-durable infrastructure are enormous and natural disasters like earthquakes or tornados can at any time prove how fragile our built environment is. Hence, the objectives for developing high-performance concrete and other cement-based products are primarily to achieve the following technical goals for the constructed facilities:

- 100% increase in service life with commensurate reduction in maintenance costs
- 50% reduction in time to construct a given facility
- 100% increase in energy-absorbing capacity without any compromise in strength and stability
- 50% increase in service life of infrastructure after repair, retrofit, and renovation

The above targets are tentative and for guidance. They are not set by any formally constituted national or international organization but they do indicate the enormous benefits that may flow from such objectives in the nation's economy and to people's quality of life. In practice, they should act as the change drivers for formulating cement composition to meet the challenges of high-duty application as well as improving performance in conventional use. The specific attributes of high flexural strength, rapid-hardening, low heat, shrinkage compensation, crack-healing, wash-out resistance, etc. can make up for deficiencies in the normal range of products.

11.5 Future design of cement plants

From the foregoing discourse on the principal change drivers for manufacturing, it is evident that the scale of clinker production will be the most important parameter. Single kilns with throughput ranging from 10,000 to 12,500 tons/day are operating in various countries. It appears that kilns with capacities of 15,000 tons/day and above may appear in the fold of the industry. But, despite the scale advantage, such large-capacity kilns may not be the most frequently installed systems all over the world because of potential difficulties of logistics and cement evacuation. The present trend of kiln installations indicates preference of capacity in the range of 4000–6000 tons/day. It is important to consider that the product management capability of existing and new installations will have to be engineered in such a manner that multiple blended products manufactured from the same clinker may be stored, packed, and dispatched simultaneously. Apart from the compulsion of economy of scale for the bulk production of clinker and market demand for different types of blended cement, the need for separate manufacturing facilities for niche products with specialized processing equipment as satellite plants seems to be emerging. More than the economy of scale, the operating excellence of such plants will lie in delivering products with unique application properties. Fluidized-bed sintering plants of low volume cannot be ruled out for such specialized processing plants.

It is imperative that bulk clinker plants are designed to be as energy-efficient as possible. In all probability, clinkering plants would consist of five or six stages of cyclones in preheater, low NO_x precalciners, ultra-short rotary kilns, multi-channel burners, and advanced heat recuperation technology, which will offer specific heat consumption below 3 MJ/kg clinker. Apart from high-performance grate coolers and tertiary air utilization, which are currently in practice, further opportunities that still exist for waste heat utilization will be harnessed to produce electrical power, industrial steam, or hot water for local heating purposes. The concept of waste burning is now an accepted practice. Looking to the future, cement kilns are likely to take up the increasing role of environment cleaning by utilizing alternative fuels and raw materials without creating additional emission problems.

Away from the actual kiln, further advances will be seen in the introduction of new-generation separators, the substitution of ball mills by roller mills and hydraulic roll presses both in raw milling and finish grinding, improved design of process fans, a shift from pneumatic to mechanical transport of materials, and the use of high-efficiency electric motors and drives. The application of ultrasound techniques to separation and milling technology seems to be nearer to commercial adoption. Noise pollution reduction will be emphasized more in coming years. Future plants with their associated mining and crushing plants will face more stringent regulatory controls, leading to much quieter and cleaner clinker and cement manufacture.

The worldwide focus and debate on CO_2 emissions is expected to intensify over the coming years. In addition to the ongoing practices of clinker substitution, use of alternative fuels and raw materials, and cogenerating electricity and energy conservation, the capture and recycling of CO_2 will move towards being viable through significant research programs currently being carried out. Transformation of CO_2 into value-added binders or fuel or both are the directions of research and development, which the industry worldwide will continue to track and support.

Operational controls including quality in cement works of the future are likely to become more heavily based on real-time information. Since the end-properties of cement are dependent on chemistry and mineralogy, even more sophisticated methods for monitoring raw material mix, the homogenizing process, clinker making, and finish cement grinding operations will be adopted with on-line analyzers leading the way forward. With such real-time data process parameters will be programmed to optimize the performance of the entire plant. Further, the operator response to process fluctuations, elimination of partial or no-load running of equipment, and higher levels of process automation will enhance the performance of plants.

Cement plants are monotonous and not very aesthetically pleasing, either from a distance or from the vicinity. Harmonization of plant layout and architecture with a plant's surroundings is a social requirement that has hardly been recognized. Some plants have adopted color screening in order to camouflage the starkness of the kiln line against the background of the surrounding countryside. Many others are trying to create visually appealing cement plant landscapes. Future plant designs should unquestionably develop such ideas further.

11.6 Growth of the Indian cement industry: a case study

The Indian cement industry has from its inception in 1910–1914 been playing the role of a key core sector player, but with fluctuating growth profiles. With two huge tasks of creating modern infrastructure and providing "housing to all," the cement industry always continues to be a commodity of great importance. Like in other countries, the cement industry in India suffers from the cyclic nature of demand and supply. Consequently, the industry passes through periodic slow-down and upswing phases. The growth trends of the Indian cement industry were recently reviewed by the author (7) and the salient aspects are presented in the following sections.

National policies influencing growth trends

At the time of Indian independence in 1947, annual cement production capacity was only 1.87 million tons, while the population was 469 million, which meant only 4 kg per capita consumption of cement. The national government of the time took cognizance of the importance of cement in the economic development of the country and decided, in the interest of consumers, to protect the industry with controls on pricing, distribution, and freight. With these controls, the profitability of cement companies was quite low and uncertain. Consequently, investment in this sector was poor and capacity additions were meager and incommensurate with the demands and market shortages. This disparity between the demand and supply of cement continued for several decades. The first attempt to rectify the situation came in 1977, when there was an announcement by the government to bring the industry on a sound footing. Between 1977 and 1991 there were several changes in national policies pertaining to the cement industry. The important milestones were:

- September, 1977: Provision of guaranteed 12% post-tax return on the net worth on new investments in cement.
- February, 1982: Introduction of partial decontrol on the price and distribution of cement.
- March, 1989: Announcement of total decontrol of price, distribution, and freight equalization of cement.
- July, 1991: Abolition of the licensing system for setting up cement plants.

Cement consumption, population growth, and national economy

Thereafter, one could observe a visible movement towards privatization and globalization of the Indian cement industry. Multinational companies started displaying their interest in the acquisition of the cement units in India. The net effect was the gigantic growth of the industry, as shown in Figure 11.8, which specifically relates to the years with upswing of production and capacity creation.

A strange anomaly in the Indian cement industry is the disproportionately low per capita consumption of cement as compared to other emerging economies (see Table 11.4). Although it is difficult to explain such an anomaly, it is generally observed that the huge rural sector in India depends essentially on local natural building materials and not on factory-produced cement. Nevertheless, the consumption of cement per person has continued to rise, as shown in Table 11.4.

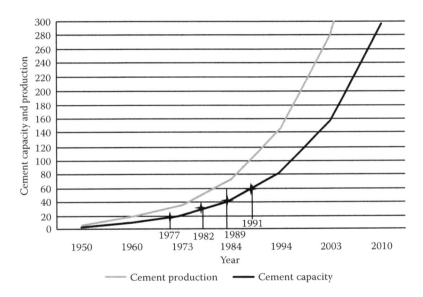

FIGURE 11.8 Growth trends of cement production and capacity in India.

Table 11.4 Growth in per capita cement consumption

Year	Cement consumption (million tons)	Population (million)	Per capita consumption (kg)
1947	1.87	469	4
1950	2.65	361	7
1960	7.83	439	18
1970	13.99	548	26
1982	20.41	683	30
1990	47.34	835	57
2000	100.45	970	97
2010	230.00	1210	178

Table 11.5 Comparison of GNP and cement consumption per person in India

	Per capita	
Year	GNP (Rs.)	Cement consumption (kg)
1950–1951	280	7
1960–1961	372	18
1970–1971	793	26
1980–1981	2008	30
1990–1991	6293	57
2000–2001	21,990	113
2010–2011	25,000	178

Although the per capita consumption of cement has increased further to 185 kg in the last couple of years, it is much lower than the world average of 285 kg and far below the levels obtained in many other developed and developing nations (say, 500–1000 kg). In parallel, it is interesting to compare the cement consumption per person with gross national product (GNP) per person. Table 11.5 summarizes the data for GNP per capita and cement consumption per capita in India over a span of six decades.

Since India stands at the middle-income block (see Figure 11.7), the cement consumption and the industry have a large window of opportunity to grow as the economic performance of the country continues to improve. Against this background the Indian cement industry is targeting a production capacity of 550 million tons by 2020, when further uplift of the national economy will promote significant growth of the construction industry and cement consumption.

Current technological status

Cement plants in India have grown in clusters in proximity of mega limestone deposits. New capacities will continue to be built in the same clusters either in the form of brownfield expansions or greenfield sites. This growth pattern has led to certain interesting techno-commercial features of the industry, such as the following:

a. The concept of "principal markets" has developed for each plant along with a strategy for retaining the hold on principal markets.
b. Most of the large plants set up satellite grinding units closer to the market place with clinker fed from the "mother" unit.
c. Techno-logistic approaches are being developed to utilize the far-flung hilly deposits, on one side, and beneficiation technologies for using marginal grade limestone deposits, on the other.

Other technological features are briefly highlighted below.

Limestone resources: in accordance with the available statistics of limestone inventory in the country, the gross reserves are 97,430 million tons, while the proven reserves are 22,476 million tons or 23% of the gross reserves. Sizable reserves are available in inaccessible areas, difficult terrains, reserved forests, bio-zones, coastal regulated zones, etc. Hence, one of the major challenges before the industry is to support the growth rate with an expanded raw materials base of proven quality and quantity.

Evolution of process profiles: while the cement plants that came into existence prior to the 1980s were mostly small kilns of 600–1500 tons/day capacity based either on wet or dry processes, the later plants were built essentially with 3000 tons/day single kilns exclusively of dry process. Economies of scale dictated the setting of large plants and the trend continues for still larger capacities in the range of 7000–8000 tons/day. At present, the largest kiln in India operates with a production level of 12,500 tons/day, which is regarded as the largest operating kiln in the world (6 m diameter × 96 m length; six-stage, two-stream preheater with SLC). Over the decades the process profiles of the industry have changed, as illustrated in Table 11.6. The basic feature is that 96% of clinker production comes from preheater-precalciner kilns. Another process feature worth mentioning is the operation of a set of precalciner kilns fed with froth floatation-based beneficiated limestone slurry passing through vacuum filters and crusher dryer systems.

Fuel usage and thermal energy consumption: so far as the fuel usage pattern is concerned, the industry is primarily dependent on bituminous coal with an average ash content of about 35–38%. Based on economic viability, a few plants have switched over fully or partly to imported coal, and high-sulfur, low-ash petroleum coke sourced from within or outside the country. The significant increase in the requirement of the above fuels in the last five years is shown in Table 11.7.

Table 11.6 Changing process profile of the Indian cement industry

Item	1950	1960	1970	1983	1995	2001	2006
Wet process							
Number of kilns	32	70	93	95	61	32	26
Capacity (t/d)	9151	25,011	38,441	39,641	25,746	13,910	11,420
% of total	97.3	94.4	69.5	41.1	12	5	3
Dry process							
Number of kilns		1	18	50	97	117	128
Capacity (t/d)		300	11,865	51,265	188,435	282,486	375,968
% of total		1.1	21.5	53.2	86	93	96
Semi-dry process							
Number of kilns	1	3	8	9	8	8	8
Capacity (t/d)	250	1200	5000	5500	5244	5260	4195
% of total	2.7	4.5	9	5.7	2	2	1
Total kilns	33	74	119	154	166	157	162
Capacity (t/d)	9401	26,511	55,306	96,406	219,425	310,706	391,583
Average kiln capacity(t/d)	285	358	465	626	1322	1921	2417

Table 11.7 Trend of increase in fuel demand by the cement industry

Sr. no	Type of fuel	2006–2007 (million tons)	2012–2013 (million tons)
1.	Indigenous bituminous coal	13.68	39.72
2.	Imported bituminous coal	3.50	
3.	Petroleum coke	2.50	
4.	Lignite	1.00	
5.	Coal for captive power plants	5.10	18.25
	Total	26.68	59.97

The break-up of fuel requirements in 2012–13 has not been attempted as the fuels are interchangeable and their relative proportions would depend on price and availability. The overall increase in fuel demand by the cement industry during the last decade has been almost fourfold. The availability and cost of fuel are major concerns for the industry. Serious attempts are being made by the industry to use alternative fuels such as used car tires, municipal solid wastes, organic solvents, etc. along with some wood chippings and agricultural shells and husks. As stated in a previous chapter, India generates over 6 million tons of hazardous waste, about 50 million tons of municipal waste, and approximately 400 million tons of non-hazardous industrial and agricultural waste. In different parts of the country different types of agricultural wastes are generated, although their collection and use pose practical difficulties. On a country-wide basis, present thermal substitution by secondary and alternative fuels is estimated at about 5% of the total heat consumption.

So far as the specific heat consumption is concerned, the national average is estimated at 725 kCal/kg of clinker, taking into account all kilns of different vintages, capacities, and processes as well as the quality of fuel used. The trend of reduction in specific heat consumption in kilns with higher capacities and five or six stages of preheaters and precalciners is shown in Figure 11.9. It may be relevant to mention that the best value of thermal energy consumption is 667 kCal/kg of clinker.

Electricity consumption: nationally speaking, the average specific power consumption of modern plants is reportedly 82 kwh/t of cement, while the overall range is 70 to 90 units. For old plants it varies from 100 to 120 units. Specific power consumption is observed to be variable even for the same type and grade of cement. A typical set of data is furnished in Table 11.8. The best value achieved in the country is 68 kWh/t of cement.

Captive power generation: the cement industry in India has preferred the route of self-sufficiency in captive power generation so as to minimize its dependence on the national power grid. At the end of 2009 the captive power generation in the cement industry was as follows:

- diesel-based: 958.06 MW
- coal/gas-based: 1177.49 MW
- wind farms: 84.90 MW

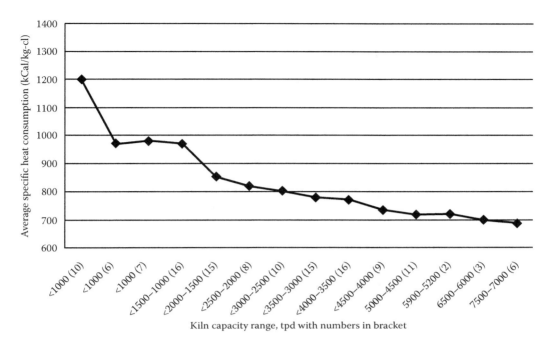

FIGURE 11.9 Relationship between kiln capacity and heat consumption in India.

Table 11.8 Specific power consumption vis-à-vis type of cement

Type of cement	Production × 10^6 t	Specific power consumption (kWh/t cement) Range	Average
Ordinary Portland cement (grades 43 and 53)	55.97	85–110	90
Portland pozzolana cement	60.23	75–105	85
Portland slag cement	10.73	75–105	80

There is a visible emphasis in the industry on increasing the installation of wind farms and moving towards renewable energy sources. In the context of captive power generation it may be worthwhile to mention the adoption of waste heat recovery systems in the Indian cement industry. It is generally estimated that 4.4 MW of electricity can be generated from 1 million tons per annum-capacity plants by waste heat recovery. As of now there is a tentative estimate that the industry has a potential of generating about 400 MW through this technology but only 13.5 MW is in operation and 71.5 MW is under installation. Although the capital cost of cogeneration of power is two times that of coal-based thermal power plants, the electricity generation cost is one-twentieth that of thermal

390 CEMENT PRODUCTION TECHNOLOGY

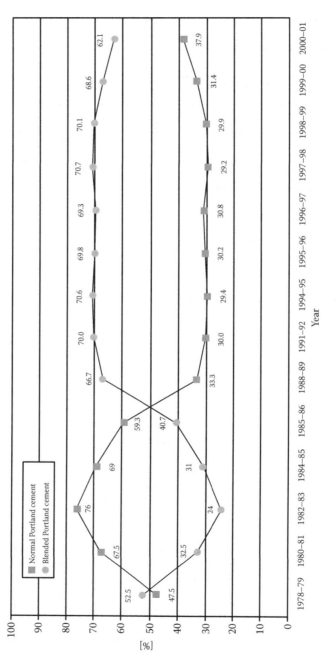

FIGURE 11.10 Transformation of product mix in India.

power. This cogeneration approach also has the benefit of reducing carbon dioxide emission.

Beneficiation of phosphogypsum: the Indian cement industry is primarily dependent on phosphogypsum coming from the fertilizer industry, as the occurrence and reserves of mineral gypsum are scarce. In fact, the industry has often taken recourse to importing mineral gypsum from other countries, as the presence of impurities like phosphorus and fluorine in phosphogypsum often makes it unsuitable for use particularly in the manufacture of blended cements. Hence, the industry is in the process of trying out technologies for improving the pH of phosphogypsum, removing impurities, enhancing its flow properties, and so on.

Product mixing: the Indian cement industry has shown a major shift in product mixing during the last decade (Figure 11.10). More and more blended cements are being produced. The same trend is likely to continue in the future with more intensity to meet the sustainability and environmental compulsions.

Future direction: the cement industry in India has seen phenomenal growth during its century-long history. However, this growth has been accompanied by fluctuating fortunes as the industry moved from control to decontrol, from shortage to surplus, from a seller's to a buyer's market. The present trends of mergers and acquisitions by the domestic and multinational players, the changing profile of the concrete and construction industry, low per capita consumption of cement, a complex raw materials base, strong environmental and sustainability compulsions, and a highly competitive market have created an environment in which the cement industry will have to grow and grow rapidly. Notwithstanding the adversities, there is strong optimism that the industry will grow to greater heights by adopting multipronged strategies of reinforcing its raw materials base, by adopting newer technologies of fuel combustion, by reducing clinker content in composite cements, by reducing carbon dioxide emission, and by finding new avenues of using cement and concrete. The country will have to achieve at least twice the present per capita consumption of cement as early as possible to support the economic progress.

References

1. A. K. CHATTERJEE, The cement industry: a global techno-commercial perspective, IOM Seminar, Mumbai, India, January, 1996.
2. CEMBUREAU website, http://cembureau.eu/cement-101/key.facts-figures.
3. General News, Global demand for cement to reach 5.2 billion tonnes, *Concrete in Australia*, Vol. 41, No. 4, 2016.
4. *Global Cement* website, www.globalcement.com/magazine/articles/964-preview-the-top-100-global-cement-companies-and-global-per-capita-capacity.trends.
5. A. K. CHATTERJEE, The future of cement: a global review, *Rock Products Cement Edition*, March, 1996.
6. ICR Research, India: gateway to a new era?, *International Cement Review*, July 2014.
7. A. K. CHATTERJEE, The Indian cement industry: its past, present and future perspectives, *Cement Wapno Beton*, No. 2, 2014.

CHAPTER TWELVE

Epilogue

The cement industry is globally a gigantic one, with a production level exceeding 4 billion tons per year, and the industry that shaped up in 1924 on the foundation of rudimentary Portland cement has survived and grown over almost two centuries to produce—on a massive scale—a large variety of well-composed Portland cements to meet the diverse applications of our built environment. Clearly, this would not have happened without innovation, research, and development.

Cement, as we all know, is a small but essential and unalterable ingredient of concrete, which is consumed in our society in quantities second only to water. Hence, its significance is universally realized. This essential construction material is produced by medium-to-large local business houses in different countries as well as quite a few large multinational companies in different parts of the world. Due to its operational experience, the modern industry is characterized by the following features:

a. High capital investment and economies of scale mark high entry barriers.

b. Fuel and freight are critical cost inputs.

c. There is hardly any threat of substitutes emerging in the foreseeable future.

d. Consumer education in the product is limited, hence the market is mostly producer-dependent.

e. In some of the major cement-producing countries, inter-firm rivalry is high and endeavors are made to position the products based on cosmetic differentiation.

f. To a large extent, the industry is held accountable for carbon dioxide emission causing climate change.

g. On the other hand, the industry is recognized as an environment savior for its utilization of wastes from other industries.

h. The demand drivers for cement vary in weightage from country to country but, by and large, they originate from sectors such as:
   - housing and real estate
   - infrastructure

- commercial buildings
- industrial structures

The characteristics and demand drivers of the industry have compelled the cement players to be innovative and research-oriented. Major innovation challenges worldwide are the following:

a. improving energy efficiency
b. reducing construction costs
c. reducing carbon footprint
d. minimizing environmental impact
e. Sustainability through resource conservation and recycling

To meet the above challenges the general management and R&D set-up of the cement industry are no more limited to improving the energy efficiency and productivity of the production units, but they are compelled to extend their efforts to efficient application of cement in order to meet customer needs. Consequently, the application of cement to concrete and other cement-based products has become intimately integrated with the cement production process in all large cement companies.

Since general-purpose cements like normal Portland cement, Portland pozzolana cement, and Portland slag cement produced by different companies cannot easily be differentiated, there is a perceptible shift towards getting engaged in value-addition to cement through the development and production of:

- niche products
- precast products
- special concretes
- low-carbon special binders
- Portland cements with low clinker factor

Since the future challenges of the cement industry are common across most of the world, industry associations with or without government support have either created their own R&D establishments or are working in conjunction with academia or publicly funded research institutions.

Coming to innovation management, all large companies have commonalities of strong top management commitment, organization design supportive of R&D, provision of adequate resources, and functional integration with production and marketing. They also display similar characteristics in the business sphere:

i. The business profiles are in all cases primarily focused on cement with vertical integration to concrete via involvement in aggregate production.

j. There is an extended diversification into other construction materials and building products that supplement the use of concrete for complete construction.

k. They have sustained and continued presence over decades in the construction industry with support services towards design, supply, and execution.
l. There is an explicit emphasis on research and development for new products, while all groups ensure through their management structure serious involvement in operational efficiency, sustainability, environmental protection, energy conservation, occupational safety and health management, and social development in different geographic regions.
m. The growth pattern in all cases has been a mix of both organic expansion and business acquisition, the focus remaining on emerging markets.

As far as R&D and innovation are concerned, the commonalities of the large cement-producing groups are as follows:

n. The groups have fairly well-established and well-equipped R&D laboratories. In some cases this may be a stand-alone facility and in others it is a semi-autonomous constituent of a technical centre.
o. The R&D laboratories of all groups have strong linkages with universities and independent research institutions, although the prime in-house emphasis of these laboratories is on applied research.
p. From a corporate management perspective, R&D activity is regarded as an independent function and is not subject to other corporate functions like production, marketing, finance, human resource development, etc.
q. R&D centers operate with an adequate manpower base and rolling budgetary support.
r. Some companies oversee the R&D activity with the help of innovation committees and project steering groups (for identified developmental activities), consisting of internal and external experts. Alternatively, corporate management keeps a high-level executive in charge of innovation alone.
s. There is a strong emphasis on "innovation" and "intellectual property rights." A system of R&D capitalization is adopted by most companies.
t. Since all companies operate globally, they must also set up regional "developmental centers" to adapt research outcomes to local needs and demands.
u. Technology tracking and forecasting is considered an essential component of innovation management.
v. Notwithstanding the prime business interests of these groups being "cement," they fix up their visions, slogans, values, and goals much beyond the production of cement. Such larger visions help to develop the chain of ultimate delivery of cement.

Index

Page numbers followed by f and t indicate figures and tables, respectively.

A

Aalborg Cement Company, 350
AAQ (ambient air quality) standards, 318–319
ABB
    Expert Optimiser, 284
    SpectraFlow CM100 raw materials analyzer, 269
Abrasive type of limestone, 55
Absorption, coal ash, 47–48, 92
ACC Limited, 333
Activation energy, dissociation rate vs., 145, 146f
Admixtures, mineral, grinding, 203–204
Adoption, of computer programming, 53, 54f
Adsorption, superplasticizers, 243
Aether, 344, 346
AFR, see Alternative fuels and raw materials (AFR)
Agricultural waste, 104
Agriculture, MBM in, 117
Aids, grinding, 205–209
    dosing, 209
    flow properties, improving, 206
    on hydration, 206
    operational features of ball mill and effect, 206–208
    overview, 205–206
    for roller press systems, 208–209
    for VRMs, 208
Air
    emissions to, 108
    excess, 93, 94, 95
    primary, 92–93, 95
    secondary, 93, 94f, 95
Air classifiers, defined, 188
Air-fluidization method, 61
Air mixing, fuel and, 84–85
Air-separation unit (ASU), 355
Alabaster, 33
Algae, cultivation with kiln exhaust gases, 357–358
Alinite based eco-cement (AEC), 346–348
Alite phase, clinker, 41, 176–177, 178, 179f–180f, 181, 184, 335, 339
Alkalis, 46
Alkali sulfates, 244
Alkaloamine-based grinding aids, 205, 206

All India Tire Manufacturing Association, 114
Alternative fuels and raw materials (AFR), 103–137
    classification, 104–107
        hazardous waste, 104–107
        overview, 104, 105t
    composition and properties, 104, 105t
    co-processing, 121–124
        cement kilns, advantages of, 121–124
        defined, 121
        emission norms of cement plants, 124t
    environmental aspects, 136–137
    feasibility, 107–109
        air, emissions to, 108
        carbon footprint, 109
        noise pollution, 108
        odor issues, 108
        overview, 107–108
        stakeholders engagement, 109
        visual impacts, 108–109
        water and land, emissions to, 108
    gasification technology, 133–135
        overview, 133, 134f
        tire-derived and other secondary fuels, 133, 134f, 135
    hazardous waste
        classifying, biological parameters in, 106–107
        defined, 104, 105–106
        specification for, 106t
    inventory and material characteristics, 109–118
        biomass residues, 112, 113t
        hazardous waste, 114–115
        industrial plastic waste, 117
        miscellaneous, 117–118
        MSW, 110–112
        overview, 109–110
        scrap motor tires, 112, 114
        sewage sludge, 115–117
    overview, 103–104
    raw materials, 135–136
    systematic quality assessment, 118–121
        organic pollutants, measurement of, 118–119, 120t
        overview, 118
        requirements of test facilities, 119, 121

systemic requirements, 124–132
  overview, 124, 125f
    precalciner design considerations, 131–132
    storing, dosing, and conveying, 125–131
Alumina
  bricks, 163
  melt formation due to, 47f
Alumina modulus (AM), 43, 64
Aluminous belite cement, 337–338
Aluminous cement, manufacturing, 337
Aluminum silicates, 29, 31
Ambient air quality (AAQ) standards, 318–319
American Petroleum Institute (API), 224, 225t
Amino acids, 361
Ammonia, 317
Anatase, 31t
Anemometers, velocity measurement by, 282
Anhydrite, 202–203
Ankerite, 64
Anode-grade petcoke, 96
Anorthoclase, 31t
Anthracene, 119, 120t
Anthracite, 74, 75, 131, 132
Anti-washout admixtures, 381
Application(s)
  ESP, 309
  Portland cement derivatives, 336–352
    modified, 338–352
    overview, 336–338
  R&D, trends of, see Research and development (R&D)
  SRPC, 223
  of superplastcizers, 243
AQM monitors, 322
Aragonite, 361
Argillaceous materials, 29–32
Arizona Public Service, 357
Ash content
  absorption on clinker, 92
  in coal, 76
  combustion of coal, 91–92
  rice straw and husk, 112
Asia-Pacific Region, coal reserves in, 81, 82t
Aspartic acid, 361
Assessment
  burnability, raw mix, see Raw mix
  limestone deposits, 8–18
    classification and exploration intensity, 9–10
    dimension, quantity, and preparation of samples, 17–18
    industrial implications of categorization of reserves, 14–15
    overview, 8–9
    reliability of different categories of reserves, 14
    resource, reserve, and exploitability, 10, 11–13
    sampling, for evaluation, 15, 16–17
  systematic quality, AFR, 118–121
    organic pollutants, measurement of, 118–119, 120t
    overview, 118
    requirements of test facilities, 119, 121

ASTM
  C 114, 271
  C 595, 227
  C 618, 228
  C 989, 227
  C 1017, 241, 242
  C 183 and C 917, 254
  classification, Portland cements, 215–216
  D 396, 77
  D 2234, D 4915, and D 2013, 254–255
Atomic absorption spectrometer (AAS), 230
Attrition, 55
Automation, storing, dosing, and conveying, alternative fuels, 131

B

Babcock E-mills, 87
Bag filters, 298–303
Bagging, of cement, 211
Ball mills
  circuit, roller press to, 208
  clinker grinding system, 186–192, 202
    closed-circuit, 187f
    features, comparison, 186t
    internals, 187f
    performance, 192
    powder filling in, 188f
    separators, types of, 188, 189–191
    Tromp curve, 191, 192f
  coal in, 87
  operational features, 206–208
  Petcoke in, 98
  raw milling system, type of, 58–60
Basalt, 2
Basic bricks, 163
Batch-type blending system, 61–63
Bauxite, 342
Bed-blending, 258–260
Belite
  CSA–BC, 342–346
    clinker compositions and strength properties, 342–343, 343t
    energy requirement and carbon dioxide, 343, 344t
    experimental structure, 344, 345f
    findings of developmental project, 344, 346
  HBC/RBC, 339–342
    cell parameters of belite phase on doping, 339, 341t
    clinker cooling rate, on cement strength parameter, 339, 340t
    NMR spectroscopy, phase quantification by, 341, 342, 342t
    stabilization of different polymorphs of, 341
  phase, clinker, 44, 177, 178, 179f–180f, 181, 182f, 185, 335
Benzene, 119, 120t
BETA mill, 198
Bioconcentration process, hazardous wastes, 106
Biohydrogen, production from algae, 358

Biological parameters, in classifying hazardous wastes, 106–107
Biological sedimentary rock, 7
Biomass residues, 112, 113t
Biotite, 32
Bis (2ethylhexyl) phthalates (DEHP), 119, 120t
Bituminous coal, 74, 75, 76
Blaine's air permeability method, 233, 264
Blast furnace slag and limestone powder, 204–205
Blasting factor, 26
Blended Portland cements, 215, 226–230
  characteristics, 227
  masonry cement, 229–230
  materials, 227–228
  merits, 226–227
  overview, 226–227
  Portland fly ash cement, 229
  PSC, 228
  supersulfated cement, 230
Blending system, batch-type, 61–63
Block variance, 256
Bogue equations, potential phase computation by, 43, 44–46
Bohemite, 33
Bragg-Brentano diffractometers, 265
British cements, 213, 234
Brownmillerite, 41
Building materials, high-performance, 382
Burnability
  assessment, raw mix, see Raw mix
  factor, 66
  raw materials, 153
  raw mixes, 141, 143
Burnability features, of raw meal, 63–68
  chemical and mineral characteristics, 64
  homogeneity, 65
  minor constituents, effect of, 64–65
  product fineness, 63
  reactivity and burnability, 65–68
    experimental approaches, 66–68
    theoretical approaches, 65–66
Burners, kiln, 155–156
Burning zone refractories, 164

C

CAC, see Calcium aluminate cement (CAC)
Calcination, limestone, 144
Calcined coke, 96
Calcite, 144, 145, 361, 365
Calcitic limestone, 19
Calcium aluminate cement (CAC)
  characteristics, 353
  complex building products formulation with, 352–353
  phase, 177, 178, 178t, 179, 181, 185, 219, 221f, 222t, 223, 237
  Portland cement, derivative, 337
  refractory materials and kiln zonation, 163–164
Calcium aluminoferrite (brownmillerite) phase, 178, 179, 181, 182f, 185, 219, 221f, 222t, 223, 237, 344

Calcium carbonate cement
  Calera process for, 360–361
  dissociation rate and temperature, 144–145
  minerals, 18, 19t
Calcium hydroxide
  crystalization, 221, 233, 237
  formation in Portland cement, 337
  hydration of silicates, 237
  injection system, 315
  slag cement, 240
Calcium langbeinite, 313
Calcium monoaluminate phase, 337
Calcium silicate hydrates
  deposition of, 221
  phase, 363
Calcium sulfate, 313
  hemihydrate, 235
Calcium sulfoaluminate–belite cement (CSA–BC), 342–346
  clinker compositions and strength properties, 342–343, 343t
  energy requirement and carbon dioxide, 343, 344t
  experimental structure, 344, 345f
  findings of developmental project, 344, 346
Calcium sulfoaluminate phase, 226, 337
Calera process, for calcium carbonate cement, 360–361
Calibration, reference materials for, 270–272
Californium 252, radioactive, 258
Calix processes for magnesia-based products, 362–363
Calorific intensity, defined, 76
Calorific value, fuel, 74, 76, 79, 92, 112, 154
Capacity and production growth perspectives, 372–377
  during 2001–2013, 373
  CEMBUREAU, countries of, 373, 374f
  country-wise distribution, 373, 374f
  top cement-producing countries, 373, 375–377, 375t
  twentieth century, 372
Captive power generation, Indian cement industry, 388, 389, 391
Capture, carbon
  defined, 354
  post-combustion, 354, 355
  pre-combustion, 354–355
Carbide sludge, 136
Carbonation technologies, SWOT analysis of, 365, 366t
Carbonatite, 7
Carbon capture
  defined, 354
  post-combustion, 354, 355
  pre-combustion, 354–355
Carbon dioxide (CO_2)
  capture and recovery of, 354–356
  concentration of, 144, 145
  CO_2-SUICOM technology, 364–365
  electrochemical CO_2 reduction, 359
  emission, 168–169, 288–289, 354, 359, 365
  energy requirement and, CSA–BC, 343, 344t
  flue gas component, 273
  fuel products, technology options for converting, 358–359
  measurement technique, 275–276
  presence and quantity, 278

Carbon footprint, feasibility of AFR project, 109
Carbon monoxide
　electrochemical CO_2 reduction, 359
　emissions, 291
　flue gas component, 273, 278–279
　measurement technique, 275–276
　presence and quantity, 278
　production, 84
Carbon nanotubes (CNTs), tensile strength, 367
Categories, of reserves, 14–15
Cell parameters, of belite phase on doping, 339, 341t
CEMBUREAU, countries of, 373, 374f
Cement rock, defined, 42
Cement(s)
　AFR, see Alternative fuels and raw materials (AFR)
　clinker grinding and cement making, see Clinker grinding
　consumption, national economy vs., 377–378
　environmental mitigation and pollution control technologies, see Environmental mitigation and pollution control technologies
　fuels, in clinker production, see Fuels
　global and regional growth trends, see Global and regional growth trends
　hydration, see Hydration
　performance requirements, 380–381
　plant-based QC practice, see Plant-based QC practice
　Portland cements, see Portland cements
　production, mineral resources for, see Mineral resources
　pyroprocessing and clinker cooling, see Pyroprocessing, clinker cooling and
　raw mix, see Raw mix
　top cement-producing countries, 373, 375–377, 375t
　trends of R&D, see Research and development (R&D)
Central inverted-cone silos, 210
Central Pollution Control Board, 106, 107, 114, 318
CEN/TS 15442, 118
Chalk, 19
Change drivers, of production and application, 379–382
　cement performance requirements, 380–381
　high-performance building materials, 382
　overview, 379–380
Characteristics
　chemical and mineral, raw meal, 64
　clinker, 176–185
　　broad characteristics, 184–185
　　granulometry, 184
　　phase composition, 176–178
　　proportions of clinker phases and their microstructure, 178–184
　dust, 293–313
　　cyclones and multicyclones, 297–298
　　ESP, 303–310
　　fabric filters/bag filters, 298–303
　　fugitive dust control methods, 311–313
　　gravel bed filter, 310–311
　　operational controls and design measure, 296
　　overview, 293–296
　　trends and progress in dust abatement technologies, 296–297
　fuels, 74–81
　　calorific value, 74
　　coal, oil, and gas, comparative behavior of, 80–81
　　coal, ranks and properties of, 74–77
　　liquid fuels, types and properties, 77–79
　　natural gas and synthetically produced gaseous fuels, properties, 79–80
　hazardous waste, 105–106
　material, AFR, 109–118
　　biomass residues, 112, 113t
　　hazardous waste, 114–115
　　industrial plastic waste, 117
　　miscellaneous, 117–118
　　MSW, 110–112
　　overview, 109–110
　　scrap motor tires, 112, 114
　　sewage sludge, 115–117
　material, energy conservation and, 200–205
　　blast furnace slag and limestone powder, 204–205
　　grinding of mineral admixtures and fillers, 203–204
　　gypsum, dehydration of, 202–203
　　overview, 200–202
　Portland cements, 219–225
　　blended, 227
　　chemical composition, 219
　　color, 219
　　hydration of pure cement compounds, 220
　　hydrophobic Portland cement, 224
　　LHPC, 223–224
　　oil well cements, 224–225
　　phase composition-property relationship, 222t
　　principal phases, 219
　　reactivity, 220, 222
　　RHPC, 222–223
　　SRPC, 223
　　stages, 221
　　strength development, 221f
　　white and colored, 224
Characterization
　minerals and rocks, 2–6
　Portland cements, 230–235
　　compositional aspects, 232
　　density and related parameters, 231
　　fineness, effect of, 233
　　heat evolution over time, 233–235
　　hydraulic cements, 230–231
　　setting behavior, 233
　　testing and, 230–231
Chemical analysis, online systems for, 268
Chemical properties
　gypsum, 34–35
　raw meal, 64
Chemical requirements, of API, 225t
Chemistry, of combustion, 83–86
　flame emissivity, 86
　fuel and air mixing, 84–85
　heat transfer, 86
　overview, 83–84
Chemoluminescence detectors, 276

Chert, 18
Chevron method, 56–57
China, 8, 229
   standard specifications for cements in, 218–219
Chlamydomonas reinhardtii, 358
Chlorine compounds, 292
Chlorine content, in raw meal, 117
Chlorite, 31, 32
Chromite bricks, 164
Chugoku Electric Power Co. Inc., 364–365
Circular stockpile, 57–58
Classification(s), 104–107
   ASTM, portland cements, 215–216
   broad, 104–107
      hazardous waste, 104, 105–106
      overview, 104, 105t
   of dust categories, 289
   hazardous waste
      biological parameters in, 106–107
      defined, 104, 105–106
   limestone deposits, 9–10, 19
   overview, 104, 105t
   petcoke, 96–97
   rocks, 4t
Classifiers, defined, 58
Clay minerals, 29–32
Clean gas, defined, 273
Clinker(s)
   cooling rate, on cement strength parameter, 339, 340t
   defined, 175
   phases, monitoring online kiln for, 268–270
Clinker grinding, cement making and, 175–211
   bagging of cement, 211
   characteristics, 176–185
      broad, 184–185
      granulometry, 184
      phase composition, 176–178
      proportions of clinker phases and microstructure, 178–184
   dispatch of, 210
   energy conservation and material characteristics, 200–205
      blast furnace slag and limestone powder, 204–205
      grinding of mineral admixtures and fillers, 203–204
      gypsum, dehydration of, 202–203
      overview, 200–202
   grinding aids, 205–209
      on cement hydration, 206
      dosing, 209
      flow properties of cement, improving, 206
      operational features of ball mill and effect, 206–208
      overview, 205–206
      for roller press systems, 208–209
      for VRMs, 208
   overview, 175
   storage of cement, 209–210
   systems, 185–200
      ball mills, 186–192

      high-pressure grinding rolls, 195–198
      horizontal roller mill (horomill), 198–200
      overview, 185–186
      VRMs, 193–195
Clinkering reactions and kiln systems, 143–146
Clinker making process
   Bogue equations, 43, 44–46
   burning process, Q-T diagram of, 147, 148f
   cooling, pyroprocessing and, *see* Pyroprocessing
   fluidized bed, 331–332
   formation process, 141–146
      overview, 141, 142f, 143
      reactions and kiln systems, 143–146
   fuels in, *see* Fuels, in clinker production
   liquid phase, calculation of, 46–47
   LSF and SM ratios in, 44f
   minor constituents, 64–65
   raw materials for, 41–43, 42t
Closed-circuit ball mill system, 187f
Closed-circuit crushing plant, 55, 56f
Coal(s)
   ash absorption, calculation of, 47–48
   combustion of, 91
   composition, 76
   gas, 79, 80t
   gasification processes, 133–135
   preparation and firing, 87–92
      ash absorption on clinker, 92
      characteristics on combustion, 91–92
      direct system, 87, 88f
      indirect system, 87, 89f
      semi-direct system, 87, 90f
   ranks and properties of, 74–77
   resources of world, 81–83
      in Asia-Pacific Region, 81, 82t
      inventory in India, 82–83
      regional distribution, 81t
      R/P ratio, 81
      status, 82
   thermal characteristics of coal, oil, and gas, comparison, 80–81
Coarse-grained compact limestone, 20f
Color, Portland cements, 219, 224
Combined grinding, 200
Combustible wastes, 104
Combustion
   chemistry and physics, 83–86
      flame emissivity, 86
      fuel and air mixing, 84–85
      heat transfer, 86
      overview, 83–84
   of coal, 91
   kiln, 155–156
   multistage, 316–317
   oxy-fuel combustion, 355
   oxygen-enriched air combustion, 355
   post-combustion capture, 354, 355
   pre-combustion capture, 354–355
   relation of process parameters with, 92–95
      excess air, 93, 94
      flame temperature, 94–95

primary air, 92–93
secondary air, 93, 94f
zone conditions in cement kiln, 122t
Compositional aspects, Portland cements, 232
Compositions
AFR, 104, 105t
clinker, CSA–BC, 342–343, 343t
coal, 76
limestone, 18–24
phases, of clinker, 176–178
Portland cements, *see* Portland cements
Compressive strength, in hydrating portland cement, 232
Computation, raw mix
adoption of computer programming, 53, 54f
overview, 48
step-wise matrix method, 49–53
stoichiometric requirements, 41, 42–48
clinker liquid phase, 46–47
coal ash absorption, 47–48
overview, 41–43, 44f
potential phase computation by Bogue equations, 43, 44–46
trial-and-error method, 49
Computer-aided run-of-mine limestone QC, 256–257
Computer programming, adoption of, 53, 54f
Concentration, LC, 107
Concretes, low-carbon, 359–366
Calera process for calcium carbonate cement, 360–361
Calix and novacem processes for magnesia-based products, 362–363
CO_2-SUICOM technology, 364–365
overview, 359–360
Solidia cement and concrete, 363–364
SWOT analysis of carbonation technologies in development, 365, 366t
TecEco cements based on reactive magnesia, 361–362
Conditional simulation, 257
Cone-bottom silos, 210
Cone shell method, 57
Confederation of Indian Industries, 247
Conservation, energy
material characteristics and, 200–205
blast furnace slag and limestone powder, 204–205
grinding of mineral admixtures and fillers, 203–204
gypsum, dehydration of, 202–203
overview, 200–202
Consistency, of cement paste, 231
Consumption, of cement, 385, 386
Control
kiln, strategies, 169–170
storing, dosing, and conveying, alternative fuels, 131
Controlled flow gate (CFG), 159
Control(s), raw mix, 260–263
homogeneity, monitoring, 262–263
hot meal quality monitoring, 263
overview, 260, 261f
XRF, 261–262

Conveying systems, alternative fuels, 125–131
automation and control, 131
to kilns, 128, 130
overview, 125–127
Coolers, clinker, 157–161
effects on clinker quality, 159–161
overview, 157–159
Cooling, clinker
pyroprocessing and, *see* Pyroprocessing
Co-processing, of alternative fuels, 121–124
cement kilns, advantages of, 121–124
defined, 121
emission norms of cement plants, 124t
Coquina, 19
Corona discharge, 305
Corona power-based method for monitoring ESP performance, 321–322
Corrective materials, 32–33
Corrosivity, hazardous waste, 105
COSMA, 269
Costs
fuel oil, 78
petcoke, 96
raw materials and fuels, 103
sweetener-grade limestone, 20
Cradle-to-gate LCA, 247
Cristobalite, 31t
Cross bar cooler, 159
Crosscut samplers, 254
Crude oil, 77
Crushing operation, 53, 55–56
Crystalline limestone, 7
Curing, temperature on, 238–239
Cut-off stripping ratio (COSR), 25
Cyanide residues, 117
Cyclone design, preheater, 151, 152
Cyclone gravel bed filter, 311
Cyclone preheater kiln, *see* Preheater-precalciner systems
Cyclones, 297–298
separators, high-efficiency, 190–191

D

Datamine, 28
Decarbonation
capacity of preheater, 147
reactions, limestone, 143–145
Degradation, of refractory lining in service, 166
Dehydration
defined, 34
gypsum, 202–203, 210
Delayed petcoke, 96
Density, Portland cements, 231
Deposition, of calcium silicate hydrates, 221
Deposits, limestone, 8–18
classification and exploration intensity, 9–10
dimension, quantity, and preparation of samples, 17–18
industrial implications of categorization of reserves, 14–15
overview, 8–9
reliability of different categories of reserves, 14

resource, reserve, and exploitability, 10, 11–13
sampling, for evaluation, 15, 16–17
Design(s)
    ESP, 303–304
    future, of cement plants, 382–384
    measure, dust, 296
    precalciner systems, 131–132
    preheater-precalciner systems
        cyclon, progress in, 151, 152
        variations, 150, 151
    quarry, 27–29
    raw mix, 48–53
        step-wise matrix method, 49–53
        trial-and-error method, 49
    silos, 210
Desulfurization mechanism, 314
Diaspore, 33
Dicalcium silicate, 41, 363
Diesel oil, 77
Differential scanning calorimeter (DSC), 230
Differential thermal analyzer (DTA), 230
Diffusion charging, 306
Dihydrate form, 202
Dimension, of samples, 17–18
Dioxins, 118
Direct system, of coal firing, 87, 88f
Discharge electrodes, ESP, 304, 305, 306
Dispatch, of cement, 210
Dispersive mode, IR radiation, 276
Disruptive innovation, 327
Dissociation rate
    activation energy *vs.*, 145, 146f
    calcium carbonate, 144–145
Distributed control system (DCS), 284
Distribution, limestone occurrence, 7–8
Docking station, AFR, 125, 127f
Dolomite, 64
    bricks, 164, 166
Dolomitic limestone, 19
Dome silos, 210
Dosing
    grinding aids, 209
    LD, 106
    systems, alternative fuels, 125–131
        automation and control, 131
        to kilns, 128, 130
        overview, 125–127
        tires and whole large packages, 130
Double roll crushers, 55
Drilling operations, 27–28
Dry gas, 79
DSC (differential scanning calorimeter), 230
Durable concrete, cements for, 245–247
Dust
    abatement technologies, trends and progress in, 296–297
    emissions, 136, 289
    generation and broad characteristics of, 293–313
        cyclones and multicyclones, 297–298
        ESP, 303–310
        fabric filters/bag filters, 298–303
        fugitive dust control methods, 311–313
        gravel bed filter, 310–311
        operational controls and design measure, 296
        overview, 293–296
        trends and progress in dust abatement technologies, 296–297
    pollution monitoring, 292–293
Dynamic impact, crushing method, 54
Dynamic pressure, defined, 281

E

E-axis, economic and social viability, 11
Economic Commission for Europe (ECE), 10
ECS/Process Expert, 284
Eisen Portland Cement Association, 228
Electricity consumption, Indian cement industry, 388, 389t
Electric resistance furnace, 334
Electrochemical CO_2 reduction, 359
Electron beam/radiation process, 334–336
Electronic waste (e-waste), 117–118
Electrostatic precipitator (ESP), 303–310
    advantages and limitations, 309–310
    application, 309
    in cement making process, 278
    components, 303–304
    design, 303–304
    disadvantage, 310
    discharge electrodes, 304, 305, 306
    dust abatement technologies, 296
    dust-related problems, 309
    efficiency, 306
    efforts, 323–324
    gas sampling, 280
    hot ESPs, 309
    moisture conditioning, 307, 308f
    negative ions, 305, 306
    operation, 307
    operational principle, 305
    performance, corona power-based method for monitoring, 321–322
    problems of design and construction, 309
    resistivity, 307–309
        temperature, 307, 308f
Emissions
    air, 108
    carbon dioxide, 168–169, 288–289, 354, 359, 365
    carbon monoxide and hydrocarbon, 291
    dust, 136, 289
    greenhouse gases, 344
    kiln, energy consumption and, 167–169
    nitrogen oxide, 169, 290, 315–319
        AAQ standards, 318–319
        overview, 315
        reduction, primary methods for, 315–317
        reduction, secondary methods for, 317–318
    norms of cement plants, 124t
    sulfur dioxide, 289–290, 313–315
        overview, 313–314
        prevention and abatement opportunities, 314–315
    water and land, 108

Emissivity, flame, 86
EN 197, 227t
EN 197-1, 216, 217t, 227, 348
EN 197-1:2000, 216, 217t
EN 197-4, 217
EN 413-1, 217–218
EN 13649:2001, 120t
EN 13725:2003, 108
EN 15359:2011, 111
Energy
    conservation and material characteristics, 200–205
        blast furnace slag and limestone powder, 204–205
        grinding of mineral admixtures and fillers, 203–204
        gypsum, dehydration of, 202–203
        overview, 200–202
    consumption and kiln emissions, 167–169
Energy-dispersive XRF (EDXRF), 261
Entrained- flow gasifier, 133
Environmental aspects, AFR, 136–137
Environmental Management Systems, 323
Environmental mitigation and pollution control technologies, 287–324
    dust, generation and broad characteristics of, 293–313
        cyclones and multicyclones, 297–298
        ESP, 303–310
        fabric filters/bag filters, 298–303
        fugitive dust control methods, 311–313
        gravel bed filter, 310–311
        operational controls and design measure, 296
        overview, 293–296
        trends and progress in dust abatement technologies, 296–297
    monitoring techniques, 321–323
        AQM monitors, 322
        corona power-based method for monitoring ESP performance, 321–322
        manual monitoring, 323
        opacity/photometric measurement, 322
    nitrogen oxide emissions, 315–319
        AAQ standards, 318–319
        overview, 315
        reduction, primary methods for, 315–317
        reduction, secondary methods for, 317–318
    noise pollution, 319–321
        level in cement plants, 319–320
        overview, 319
        reduction, 320–321
    outlook, 323–324
    overview, 287–288
    pollutants emitted into atmosphere, 288–293
        lesser concern, 290–292
        overview, 288–289
        pollution monitoring, basic concepts, 292–293
        principal pollutants, 289–290
    sulfur dioxide emissions, 313–315
        overview, 313–314
        prevention and abatement opportunities, 314–315
Environmental Protection Agency (EPA), 119
ESP, *see* Electrostatic precipitator (ESP)
Estimation variance, 256–257
Ettringite, 223, 237, 240, 353

Eurocement, 377
European Pollution Release and Transfer (E-PRTR) data, 119
European standard, types of portland cements, 216–218
European Union, 115
Evaluation, sampling of limestone deposits for, 15, 16–17
Evolution
    heat, 233–235
    process profiles, 387
Excess air, 93, 94, 95
Expansive cement, phase-modified portland cements, 225–226
Expansive grouts, 226
Experimental approaches, reactivity and burnability of raw meal, 66–68
Expert Optimiser, 284
Exploitability, limestone deposits, 10, 11–13
Exploitation, limestone occurrence, 7–8
Exploration intensity, limestone deposits, 9–10
External recirculation, 85
Extractive gas sampling system, 274–276
Extrinsic properties, minerals, 3

F

Fabric filters (FFs), 296, 298–303
False set, 202, 233
F-axis, field project status and feasibility, 11
FCB, 198
FCT ACTech COSMA, 269
Feasibility, AFR project, 107–109
    air, emissions to, 108
    carbon footprint, 109
    noise pollution, 108
    odor issues, 108
    overview, 107–108
    stakeholders engagement, 109
    visual impacts, 108–109
    water and land, emissions to, 108
Feldspar, 18
Felted fabrics, 299
Ferric oxide, melt formation due to, 47f
Ferroaluminate cement (FAC), 343
Ferruginous limestone, 19
Fertilizer sludge, 136
FFs (fabric filters), 296, 298–303
Field charging, 306
Fillers, grinding, 203–204
Filtering velocity, 299
Filtration mechanisms, in FFs, 298, 299f
Fine-grained moderately compact limestone, 20f
Fineness
    effect of, 233
    raw meal, 63
Fineness modulus (FM), 184
Firing shrinkage, test method based on, 67, 68
Firing system
    coal, 87–92
        ash absorption on clinker, 92
        characteristics on combustion, 91–92
        direct system, 87, 88f

indirect system, 87, 89f
  semi-direct system, 87, 90f
  petcoke, 98–99
Fischer-Tropsch (F-T) process, 358–359
Flame(s)
  emissivity, 86
  ionization method, 276
  momentum, 156
  photometer, 230
  temperature, 94–95
Flash point, 77
Flash set, 202, 233
Flat-bottom silos, 210
Flexi-coke, 96
Flowability, defined, 206
Flow properties, grinding aids for, 206
FLSA, 285
FLSmidth, 132, 193, 284
Flue gas analysis, 272–280
  in cement production, 278
  extractive sampling and measurement techniques, 274–276
  gas sampling points and probes, 279–280
  interlock and warning systems, 278–279
  overview, 272–274
  special considerations for gas sampling systems, 276–278
Fluid coke, 96
Fluidization, of dry raw meal, 61
Fluidized bed clinker making process, 331–332
Fluidized-bed gasifier, 133
Fluidized-bed sintering plants, 383
Fluoride-bearing sludge, 135
Fluoride residues, 117
Fluorine compounds, in feed constituents, 291
Fluorogypsum, 342–343
Fly ash cement, hydration chemistry of, 240–241, 242f
"Foraminifera" fossil, in limestone, 21f
Fossiliferous limestone, 19
Fourier-transform infrared spectroscope (FT-IR), 230
Fractional distillation, process of, 77
Freedonia Group, 375
Free lime (CaO) grains, 182, 183f
Free lime temperature relation, 67
FT-IR (Fourier-transform infrared spectroscope), 230
Fuel-grade petcoke, 96
Fuels, AFR, see Alternative fuels and raw materials (AFR)
Fuels, in clinker production, 73–100
  characteristics, 74–81
    calorific value, 74
    coal, oil, and gas, comparative behavior of, 80–81
    coal, ranks and properties of, 74–77
    liquid fuels, types and properties, 77–79
    natural gas and synthetically produced gaseous fuels, properties, 79–80
  coal preparation and firing, 87–92
    ash absorption on clinker, 92
    characteristics on combustion, 91–92
    direct system, 87, 88f
    indirect system, 87, 89f
    semi-direct system, 87, 90f
  coal resources of world, 81–83
    in Asia-Pacific Region, 81, 82t
    inventory in India, 82–83
    regional distribution, 81t
    R/P ratio, 81
    status, 82
  combustion, basic chemistry and physics, 83–86
    flame emissivity, 86
    fuel and air mixing, 84–85
    heat transfer, 86
    overview, 83–84
  combustion, relation of process parameters with, 92–95
  defined, 73
  overview, 73
  petcoke as substitute fuel, 96–100
    classification, 96–97
    firing of, 98–99
    grinding, 97–98
    merits and demerits, 99–100
    overview, 96
    production, world status, 96
    properties, 97
  types, 73
Fuel usage, Indian cement industry, 387, 388f, 389f
Fugitive dust (FD)
  control methods, 300, 311–313
  pollutants of principal concern, 289
  primary sources of generation, 294–295
Fuller Company, 331
Furans, 118
Future design, of cement plants, 382–384
F-value, 239–240

G

Gas analysis, flue, see Flue gas analysis
Gas(es)
  coal, 79, 80t
  dry, 79
  flow, measurements, 280–281
  natural gas, 2, 79–80
  oil, 79–80
  producer, 79, 80t
  R/P ratio for, 81
  thermal characteristics of coal, oil, and gas, 80–81
  water gas, 79, 80t
  wet, 79
Gasification technology, AFR, 133–135
  overview, 133, 134f
  tire-derived and other secondary fuels through gasification, 133, 134f, 135
Gas sampling systems, special considerations for, 276–278
G-axis, geological knowledge, 11
GBFs (gravel bed filters), 296, 310–311
GCV (gross calorific value), 74, 79, 82–83
Generation, of dust, 293–313
  cyclones and multicyclones, 297–298

ESP, 303–310
   advantages and limitations, 309–310
   application, 309
   components, 303–304
   design, 303–304
   disadvantage, 310
   discharge electrodes, 304, 305, 306
   dust-related problems, 309
   efficiency, 306
   hot ESPs, 309
   moisture conditioning, 307, 308f
   negative ions, 305, 306
   operation, 307
   operational principle, 305
   problems of design and construction, 309
   resistivity, 307–309
   temperature resistivity, 307, 308f
fabric filters/bag filters, 298–303
fugitive dust control methods, 311–313
gravel bed filter, 310–311
operational controls and design measure, 296
overview, 293–296
trends and progress in dust abatement technologies, 296–297
GEOVIASurpac, 28
Gibbsite, 33
Glauconite, 32
Global and regional growth trends, 371–391
   capacity and production growth perspectives, 372–377
     during 2001–2013, 373
     CEMBUREAU, countries of, 373, 374f
     country-wise distribution, 373, 374f
     top cement-producing countries, 373, 375–377, 375t
     twentieth century, 372
   change drivers of production and application, 379–382
     cement performance requirements, 380–381
     high-performance building materials, 382
     overview, 379–380
   future design of cement plants, 382–384
   Indian cement industry, case study, 384–391
     consumption, population growth, and national economy, 385, 386
     current technological status, 386–391
     national policies, 384–385
     overview, 384
   national economy *vs.* cement consumption, 377–378
   overview, 371–372
Global Cement Directory, 377
Global warming potential (GWP), 247
Glutanic acid, 361
Glycine, 361
GNP (gross national product), 377–378, 385, 386
Goethite, 31t, 32
Grades, Portland cements, 214–219
   ASTM classification, 215–216
   European standard, 216–218
   Indian standards, 218
   overview, 214–215
   standard specifications for cements in China, 218–219

Granite, 2
Granulated blast furnace slag, defined, 227
Granulometry, clinker, 184
Graphene, 367
Grate coolers, 157–159
Gravel bed filters (GBFs), 296, 310–311
Gravity coolers, 157
Gravity impact, crushing method, 54
Green coke, 96
Greenhouse gases, emissions, 344
Grinding
   aids, in cement manufacture, 205–209
     on cement hydration, 206
     dosing, 209
     flow properties of cement, improving, 206
     operational features of ball mill and effect, 206–208
     overview, 205–206
     for roller press systems, 208–209
     for VRMs, 208
   clinker, *see* Clinker grinding
   index, HGI, 87, 97
   mineral admixtures and fillers, 203–204
   petcoke, 97–98
   process, cement, 264–265
     RRSB plot of PSD, 265
   rolls, high-pressure, 195–198, 201, 202
Grit separators, 188, 189
Gross calorific value (GCV), 74, 79, 82–83
Gross national product (GNP), 377–378, 385, 386
Growth
   global and regional, *see* Global and regional growth trends
   sustaining technologies, 328–330
     advanced engineering features, 328, 329
     changes in technology and scale of operation, 328, 329f
     Portland cement chemistry, critical milestones in, 328t
     pyroprocessing, course-changing developments in, 330t
     raw and finish grinding processes, tentative technological milestones, 330t
Gypsum
   dehydration of, 202–203, 210
   fluorogypsum, 342–343
   natural, 33–35
     chemical properties, 34–35
     producing countries, 33–34
   phosphogypsum, beneficiation of, 391
   transportation, 247
Gyratory crushers, 55

H

Hammer crushers, 55
Hardened cement paste, microstructure, 237–238
Hardening, defined, 220
Hardgrove grindability tester, 77, 121
Hardgrove grinding index (HGI), 87, 97
Hardness, Mohs scale of, 4, 33

Hard type of limestone, 55
Hauling factor, 26
Hazardous waste
　AFR inventory and material characteristics, 114–115
　classifying, biological parameters in, 106–107
　defined, 104, 105–106
　incinerators, cement kilns *vs.*, 122t
Heat balance, in rotary kiln system, 153, 154–155
Heat evolution over time, portland cement, 233–235
Heat transfer, 86
HeidelbergCements, 375
Hematite, 31t, 32
Hemihydrate, 34, 35, 202–203
High Alumina Cement (HAC), 337
High-belite cement (HBC), 339–342
　cell parameters of belite phase on doping, 339, 341t
　clinker cooling rate, on cement strength parameter, 339, 340t
　NMR spectroscopy, phase quantification by, 341, 342, 342t
　stabilization of different polymorphs of, 341
High-efficiency cyclone separators, 190–191
High-performance building materials, 382
High-pressure grinding rolls, 195–198, 201, 202
High-range water reducers (HRWR), 241, 242–243
Holcim, 375
Homogeneity
　raw meal, 65
　raw mix, 262–263
Homogenization process, post-milling, 60–63
Homogenizing efficiency, chemical fluctuations using, 255–256
Horizontal roller mills (horomills), clinker grinding system, 185, 186, 198–200
　industrial installation, 200f
　typical illustration, 199t
　working principle, 198f
Horomills, *see* Horizontal roller mills (horomills)
Hotdisc combustion, for AFR, 132
Hot ESPs, 309
Hot meal quality monitoring, 263
HRWR (high-range water reducers), 241, 242–243
Human machine interfaces (HMIs), 284
Hybrid grinding, 200
Hybrid milling systems, for raw materials, 58–60
Hydration
　defined, 220
　delayed, 222
　grinding aids on, 206
　pure cement compounds, 220
　reactions, Portland cements, 235–245
　　cement, 236–237
　　fly ash cement, chemistry of, 240–241, 242f
　　microstructure of hardened cement paste, 237–238
　　overview, 235–236
　　slag cement, chemistry of, 239–240
　　superplasticizers, 241, 242–245
　　temperature on, 238–239
Hydraulic cements, characterization, 230–231
Hydraulic modulus, defined, 43

Hydraulic roll press, 195–198
Hydrocarbon emissions, 291
Hydrogen, calorific value of fuel, 76
Hydrogen sulfide, 291
Hydrophobic portland cement, 224
Hydrous aluminum silicates, 29

I

iGantt, 28
Igneous rocks, 2
Ignitability, hazardous waste, 105
Ignition temperature, of petcoke, 97–98
IHI-Chichibu, 331
Illite, 30t, 31
IMACON analyzer, 268
IMA Engineering: Quarcon, 268
Impact crushers, 55
Incombustible wastes, 104
India
　AAQ standards, 318
　ACC Limited in, 333
　agricultural crop residues in, 112
　All India Tire Manufacturing Association, 114
　ash content in coal, 76
　Central Pollution Control Board, 106, 107
　clinkers, characteristics of, 184–185
　coal inventory in, 82–83
　coal reserves, 81, 82, 101
　electronic waste in, 118
　fly ash generating countries, 229
　fuel oils in, 79t
　granulated slag and fly ash, 228
　growth in cement, 375
　gypsum production in, 33
　Indian Bureau of Mines, 8, 12
　limestone reserves in, 13t
　limestone rocks of different geological ages, 24t
　mineral resources, 12
　National Council of Cement and Building Material in, 9, 20, 270
　petcoke-producing country, 96
　plastic waste in, 117
　PSC, 228
　Rajasthan gypsum, 34
　sewage sludge in, 115
　tire-derived fuel, 114
Indian cement industry, case study, 384–391
　consumption, population growth, and national economy, 385, 386
　current technological status, 386–391
　　captive power generation, 388, 389, 391
　　electricity consumption, 388, 389t
　　evolution of process profiles, 387
　　features, 386
　　fuel usage and thermal energy consumption, 387, 388, 389f
　　future direction, 391
　　limestone resources, 386–387
　　phosphogypsum, beneficiation of, 391
　　product mixing, 390f, 391

national policies, 384–385
overview, 384
Indian standards, types of portland cements, 218
Indirect system, of coal firing, 87, 89f
Indonesia, coal inventory, 83t
Induction period, defined, 145
Inductively coupled plasma spectrometer (ICP), 230
Industrial implications, of reserves, 14–15
Industrial plastic waste, 117
Industrial wastes, 104
Infrared radiation, 275–276
Inline calciners (ILC), 150
Insoluble anhydrite, 34, 35
Inter-granular pores, limestone with, 21f
Interlock systems, flue gas analysis, 278–279
Internals, ball mill, 187f
International Energy Agency, 377
Intrinsic properties, minerals, 3–4
Inventory
    AFR, 109–118
        biomass residues, 112, 113t
        hazardous waste, 114–115
        industrial plastic waste, 117
        miscellaneous, 117–118
        MSW, 110–112
        overview, 109–110
        scrap motor tires, 112, 114
        sewage sludge, 115–117
    coal, in India, 82–83
Ion-specific potentiometry, 275
Iron ore, limestone and, 50, 51–53
IS 269:2015, 218
ISO 14000, 14001 and 14004, 323–324
Isokinetic gas sampling, 119

J

Jaw crushers, 55
J-Catrel process, 110f
Jet entrainment, process of, 84–85
Juli flora, 112

K

Kajima Corporation of Japan, 364–365
Kaolinite, 30t, 31
Kawasaki Heavy Industries, 331
Kerneos Aluminate Technologies, 353
Kerosene oil, 77
KHD-Humboldt-Wedag, 190
Kiln(s)
    ampere, 170
    control strategies, 169–170
    cyclone preheater, see Preheater-precalciner systems
    dosing and conveyance to, 128, 130
    emissions, energy consumption and, 167–169
    exhaust gases, algae cultivation with, 357–358
    feed, 53
    inlet, 279
    monitoring for clinker phases, online, 268–270
    operation monitoring, 263–264

torque, 170
zonation, refractory materials and, 163–165
Kiln systems
    burners and combustion, 155–156
    clinkering reactions and, 143–146
    preheater-precalciner, volatiles cycle in, 161–162
    refractory lining materials in, 162–166
        degradation, in service, 166
        kiln zonation, 163–165
        overview, 162–163
        suspension preheater section, 166
    rotary, 152–155
        heat balance, 153, 154–155
        overview, 152–153, 154f
Kleine's compound, 337
Kriging, 257
Krupp Polysius group, 190, 193
K-type Portland cements, 226
Kuel's Index, 65

L

Lafarge group, 284, 344, 363, 375
Land
    emissions to, 108
    plaster, 33
Laterite, 32, 33
Le Chatelier method, 231
Lepidocrocite, 32
Lepol grate pre-heater, 152
Le Teil Dry Process plant, 344
Lethal concentration (LC), 107
Lethal dose (LD), 106
LHPC (low-heat Portland cement), 215, 223–224
Life cycle assessment (LCA), for cement, 246–247, 323
Lignite, 74, 75
Lignosulfonate, 243
Lime combinability temperature, 67
Lime-rich sludge, 135, 136
Lime saturation factor (LSF), 43, 44f, 64, 66, 339
Limestone
    calcination, 144
    calcination system, for uncontaminated CO_2 recovery, 355, 356f
    calcined clay cement (LC3), 350–352
    cement-grade, specification of, 20, 22
    cement production, 1–2
    classification, 19
    decarbonation reactions, 143–145
    deposits, assessment of, 8–18
        classification and exploration intensity, 9–10
        dimension, quantity, and preparation of samples, 17–18
        industrial implications of categorization of reserves, 14–15
        overview, 8–9
        reliability of different categories of reserves, 14
        resource, reserve, and exploitability, 10, 11–13
        sampling, for evaluation, 15, 16–17
    iron ore, 50, 51–53
    microstructures, 20f, 21f

mining, 24–26
occurrences
  distribution and exploitation, 7–8
  nature of, 6–8
  typical physicochemical properties of, 22–24
powder, blast furnace slag and, 204–205
production, 8
QC, computer-aided run-of-mine, 256–257
resources, Indian cement industry, 386–387
sand, 50, 51–53
shale, 50, 51–53
types, 55
Limitations
  cyclone/multicyclones, 297–298
  ESP, 310
  fabric filters, 303
Limonite, 32
Liquid fuels, types and properties, 77–79
Liquid phase, clinker, calculation of, 46–47
Lithium metaborate, 262
Lithium tetraborate, 262
Loading rate, 26
Local exhaust ventilation, 311
Loesche, 193, 194
Longitudinal stockpile, 58
Long-term integrated mine planning, 27
Lots, sampling units in, 253
Low-carbon cement(s), 359–366
  Calera process for calcium carbonate cement, 360–361
  Calix and novacem processes for magnesia-based products, 362–363
  CO_2-SUICOM technology, 364–365
  industry, research, 354–358
    algae cultivation with kiln exhaust gases, 357–358
    carbon dioxide, capture and recovery of, 354–356
    overview, 354
  overview, 359–360
  Solidia cement and concrete, 363–364
  SWOT analysis of carbonation technologies in development, 365, 366t
  TecEco cements based on reactive magnesia, 361–362
Low-heat Portland cement (LHPC), 215, 223–224
LSF (lime saturation factor), 43, 44f, 64, 66, 339
LUCIE, 284

M

Macropores, 238
Magma, defined, 2
Magnesia, 46, 64
  based products, Calix and novacem processes for, 362–363
  bricks, 163, 164, 166
  limestone, 19
  reactive, 361–362
Magnesium carbonate, 362
Magnesium silicate hydrate phase, 363
Magnetite, 32
Manual monitoring, 323

Maptek Vulcan, 28
Marble limestone, 7
Masonry cement, 215, 229–230
Master rollers, 194
Material(s)
  argillaceous/clay, 29–32
  blended portland cements, 227–228
  building, high-performance, 382
  characteristics, AFR, 109–118
    biomass residues, 112, 113t
    hazardous waste, 114–115
    industrial plastic waste, 117
    miscellaneous, 117–118
    MSW, 110–112
    overview, 109–110
    scrap motor tires, 112, 114
    sewage sludge, 115–117
  characteristics, energy conservation and, 200–205
    blast furnace slag and limestone powder, 204–205
    grinding of mineral admixtures and fillers, 203–204
    gypsum, dehydration of, 202–203
    overview, 200–202
  corrective, 32–33
  raw
    alternative fuels and, see Alternative fuels and raw materials (AFR)
    for clinker making, 42t
  refractory lining, in kiln system, 162–166
    degradation, in service, 166
    kiln zonation, 163–165
    overview, 162–163
    suspension preheater section, 166
Matrix effect, 262
Matrix method, step-wise, 49–53
Measurements
  organic pollutants, 118–119, 120t
  photometric, 322
  process, plant-based QC practice, 280–284
    gas flow and velocity, 280–281
    overview, 280
    temperature, 282–284
    velocity measurement by anemometers, 282
  techniques, for different gases, 274–276
Meat and bone meal (MBM), 117
Mechanical separators, 189–190, 190f
Mechanical shaking methods, 300, 301f
Mercury, mineral in liquid form, 2
Mesopores, 238
Metamorphic rocks, 2
Micromine, 28
Microphones, decreasing, 188
Micropores, 238
Microstructures
  cement hydration products, 245
  clinker phases, 178–184
    alite and belite crystals, 179f–180f
  hardened cement paste, 237–238
  limestone, 20f, 21f
  oolitic limestone, 21f

Microwave heating, 333–334
Migration velocity, 306–307
Milling operation, raw, 58–60
Mineral admixtures, grinding, 203–204
Mineral characteristics, raw meal, 64
Mineralizers, use of, 68, 69
Mineral reserves, defined, 12
Mineral resources, for cement production, 1–37
    argillaceous/clay materials, 29–32
    characterization, 2–6
    corrective materials, 32–33
    defined, 12
    limestone
        cement-grade, specification of, 20, 22
        classification, 19
        composition and quality of, 18–24
        microstructures, 20f, 21f
        mining, 24–26
        occurrences, typical physicochemical properties of, 22–24
    limestone deposits, assessment of, 8–18
        classification and exploration intensity, 9–10
        dimension, quantity, and preparation of samples, 17–18
        industrial implications of categorization of reserves, 14–15
        overview, 8–9
        reliability of different categories of reserves, 14
        resource, reserve, and exploitability, 10, 11–13
        sampling, for evaluation, 15, 16–17
    limestone occurrence
        distribution and exploitation, 7–8
        nature, 6–8
    natural gypsum, 33–35
        chemical properties, 34–35
        producing countries, 33–34
    overview, 1–2
    properties, 3–4
    quarry design and operational optimization, 27–29
    raw materials on unit operations, 36–37
    rocks
        characterization, 2–6
        classification, 4t
        physical and technological properties of, 4–6
    thermo-chemical reactivity of clay minerals, 31–32
Mining
    limestone, 24–26
    opencast, 24, 25–26
    operational issues, 26
    terrace, 25
Mitsubishi fluidized-bed calciner (MFC), 147, 148f
Model–based predictive techniques, 284
Modified Portland cements, advances in, 338–352
    AEC, 346–348
    CSA-BC, 342–346
        clinker compositions and strength properties, 342–343, 343t
        energy requirement and carbon dioxide, 343, 344t
        experimental structure, 344, 345f
        findings of developmental project, 344, 346
    HBC/RBC, 339–342
        cell parameters of belite phase on doping, 339, 341t
        clinker cooling rate, on cement strength parameter, 339, 340t
        NMR spectroscopy, phase quantification by, 341, 342, 342t
        stabilization of different polymorphs of, 341
    MPC, 350–352
    overview, 338–339
    PLC, 348–350
Mohs scale of hardness, 4, 33
Moisture conditioning, of dust, 307, 308f
Moisture content
    in biomass residues, 112
    combustion of coal, 91
Momentum, flame, 156
Monitoring
    hot meal quality, 263
    kiln operation, 263–264
    online kiln, for clinker phases, 268–270
    pollution, basic concepts, 292–293
    raw mix homogeneity, 262–263
    techniques, pollutants, 321–323
        AQM monitors, 322
        corona power-based method for monitoring ESP performance, 321–322
        manual monitoring, 323
        opacity/photometric measurement, 322
Monolithic products, 163
Montmorillonite, 30t, 31, 32
Moving-bed gasifier, 133
MPC (multi-blend Portland cements), 350–352
MSW (municipal solid waste), 110–112
M-type Portland cements, 226
Multi-blend Portland cements (MPC), 350–352
Multicyclones, 297–298
Multiple-compartment silos, 210
Multistage combustion, 316–317
Municipal solid waste (MSW), 110–112
Muscovite, 31t, 32

N

Nanotechnology, in cement research, 366–367
Naphthalene, 119, 120t
National Council for Cement and Building Materials, 9, 20, 270
National economy, cement consumption vs., 377–378
National Institute for Occupational Safety and Health, 295
National Institute for Standards and Technology (NIST), 270, 271
National policies, influencing growth trends, 384–385
Natural gas, 2
    properties, 79–80
Natural gypsum, 33–35
    chemical properties, 34–35
    producing countries, 33–34
Negative corona, 305
Nesquihonite, 362

Net calorific value (NCV), 74
Neural Net technology, 284
Nitrogen
   calorific value, 76
   in dried sludge, 117
   flue gas component, 273
Nitrogen oxide, 137, 169
   emissions, 315–319
      AAQ standards, 318–319
      overview, 315
      reduction, primary methods for, 315–317
      reduction, secondary methods for, 317–318
   flue gas component, 273
   measurement technique, 275–276
   pollutant, 290
   pollution monitoring, 293
   presence and quantity, 278
NMR (nuclear magnetic resonance) spectroscopy, phase quantification, 341, 342, 342t
NMVOC (non-methane volatile organic compounds), 118, 119, 120t
Noise pollution, 319–321
   AFR project, 108
   level in cement plants, 319–320
   overview, 319
   reduction, 320–321, 383
Non-dispersive mode, IR radiation, 276
Non-hydraulic cements, 367–368
Non-methane volatile organic compounds (NMVOC), 118, 119, 120t
Normal Portland cement (NPC), 214–215, 218, 219, 222–223, 335, 336
Novacem processes for magnesia-based products, 362–363
NPC (Normal Portland cement), 214–215, 218, 219, 222–223, 335, 336
Nuclear magnetic resonance (NMR) spectroscopy, phase quantification, 341, 342, 342t

O

Odor issues, 108
Oil(s)
   gas, 79–80
   R/P ratio for, 81
   thermal characteristics of coal, oil, and gas, 80–81
   well cements, 224–225
Oil well cement (OWC), 215
Olivine, 362
Online quality control, in cement plants, 267–272
   cement quality control, 270
   chemical analysis, 268
   kiln monitoring for clinker phases, 268–270
   overview, 267
   reference materials for calibration, 270–272
Oolitic limestone, 7, 19, 21f
Opacity/photometric measurement, 322
OPC (ordinary Portland cement), 218, 219, 222–223
Opencast mining, 24, 25–26
Operation, of ESP, 307
Operational controls, dust, 296
Operational features, of ball mill, 206–208
Operational issues in mining, 26
Operational optimization, 27–29
Operational principle, ESP, 305
Optical microscopy, 45
Optical polarized microscope, 230
Optical pyrometers, 282, 284
Ordinary Portland cement (OPC), 218, 219, 222–223
Organic pollutants, measurement of, 118–119, 120t
Organic sedimentary rock, 7
Organo-chlorine compounds, 121
O-Sepa model, 190, 191f
Overburden ratio, 26
Oxy-fuel combustion, 355
Oxygen
   in flue gas, 278
   measurement, 275
Oxygen-enriched air combustion, 355

P

Pack-set, 206, 233
PAHs (polycyclic aromatic hydrocarbons), 119, 120t
Palygorskite–Attapulgite–Sepiolite, 30t
Parameters, biological, in classifying hazardous wastes, 106–107
Particle density, of portland cement, 231
Particle size, combustion of coal, 91
Particle size determination (PSD), 121
Particle size distribution (PSD)
   cement, 201–202, 205
   determining, 231, 264
   RRSB plot of cement, 265
Pavillion8 MPC, 284
PCBs (polychlorinated biphenyls), 119, 120t, 121, 123
PCDDs (polychlorinated dibenzodioxins), 118, 119, 120t, 137
PCDFs (polychlorinated dibenzofurans), 118, 119, 120t, 137
PCPs (polychlorinated phenols), 121
Peat, 74, 75
Peiffer (MPS), 193
Performance(s)
   ball mill, 192
   cement, 380–381
   cement milling systems, comparison, 185t
   corona power-based method for monitoring ESP, 321–322
   high-performance building materials, 382
   improvers, 228
   requirements, of API, 225t
Periclase, 45, 64, 159, 163, 164, 176, 182, 184
Petcoke, 73
   ignition properties, 132
   precalciner systems, 131
   as substitute fuel, 96–100
      classification, 96–97
      firing of, 98–99
      grinding, 97–98
      overview, 96
      production, world status, 96
      properties, 97

Petrol, 77
Petroleum
　coke, see Petcoke
　derivatives, 77
　mineral in liquid form, 2
PGNAA (prompt gamma neutron activation analysis), 258–260, 269–270
Phase-modified portland cements
　expansive cement, 225–226
　ultra-rapid-hardening cement, 226
Phases, of clinker
　composition, 176–178
　proportions and microstructure, 178–184
Phosphogypsum, beneficiation of, 391
Photometric measurement, of opacity, 322
Physical properties, of rocks, 4–6
Physical requirements, of API, 225t
Physicochemical properties, of limestone occurrences, 22–24
Physics, of combustion, 83–86
　flame emissivity, 86
　fuel and air mixing, 84–85
　heat transfer, 86
　overview, 83–84
Phytotoxicity, 107
Pit Navigator, 284
Pitot tube, basic construction, 281
Planetary coolers, 157–158
Planning, quarry, 27–29
Plant-based QC practice, 251–285
　cement grinding process, 264–265
　computer-aided run-of-mine limestone, 256–257
　flue gas analysis, 272–280
　　in cement production, 278
　　extractive sampling and measurement techniques, 274–276
　　gas sampling points and probes, 279–280
　　interlock and warning systems, 278–279
　　overview, 272–274
　　special considerations for gas sampling systems, 276–278
　kiln operation monitoring, 263–264
　online QC in cement plants, 267–272
　　cement quality control, 270
　　chemical analysis, 268
　　kiln monitoring for clinker phases, 268–270
　　overview, 267
　　reference materials for calibration, 270–272
　overview, 251–252
　preblending operation, 258–260
　process measurements, 280–284
　　gas flow and velocity, 280–281
　　overview, 280
　　temperature, 282–284
　　velocity measurement by anemometers, 282
　raw mix control, 260–263
　　homogeneity, monitoring, 262–263
　　hot meal quality monitoring, 263
　　overview, 260, 261f
　　XRF, 261–262

RRSB plot of PSD, 265
sampling guidelines, 252–254
　principles, 252
　purpose, 252
　scheme, establishing, 252–254
sampling stations, 254–256
　general statistical considerations, 255–256
total process control system, 284–285
XRD, for phase analysis, 265–267
Plastic waste, industrial, 117
PLC (Portland limestone cement), 215, 348–350
Pneumatic blending systems, 61–63
Pneumatic conveyance, 128, 130
Pneumatic retractable tube samplers, 254
Polexpert KCE/MCE, 284
Pollutants
　emitted into atmosphere, 288–293
　　lesser concern, 290–292
　　overview, 288–289
　　pollution monitoring, basic concepts, 292–293
　　principal pollutants, 289–290
　organic, measurement of, 118–119, 120t
Pollution
　monitoring, basic concepts, 292–293
　noise, 319–321
　　level in cement plants, 319–320
　　overview, 319
　　reduction, 320–321, 383
Polyacrylamide-type polymers, 381
Polyacrylate superplasicizers, 244
Polychlorinated biphenyls (PCBs), 119, 120t, 121, 123
Polychlorinated dibenzodioxins (PCDDs), 118, 119, 120t, 137
Polychlorinated dibenzofurans (PCDFs), 118, 119, 120t, 137
Polychlorinated phenols (PCPs), 121
Polycyclic aromatic hydrocarbons (PAHs), 119, 120t
Polyethylene, high-density, 211
Polypropylene, high-density, 211
Polysius, 284
Portland cements
　chemistry, critical milestones in, 328
　derivatives, application, 336–352
　　modified, 338–352
　　overview, 336–338
　limestone cement, 204, 205
　mineral admixtures, grinding, 203
　slag cement, 204
Portland cements, composition and properties, 213–249
　blended, 226–230
　　characteristics, 227
　　masonry cement, 229–230
　　materials, 227–228
　　merits, 226–227
　　overview, 226–227
　　Portland fly ash cement, 229
　　PSC, 228
　　supersulfated cement, 230
　characteristics, 219–225
　　chemical composition, 219
　　color, 219
　　hydration of pure cement compounds, 220

INDEX 413

hydrophobic Portland cement, 224
LHPC, 223–224
oil well cements, 224–225
phase composition-property relationship, 222t
principal phases, 219
reactivity, 220, 222
RHPC, 222–223
SRPC, 223
stages, 221
strength development, 221f
white and colored Portland cement, 224
characterization and practical implications of
    properties, 230–235
    compositional aspects, 232
    density and related parameters, 231
    fineness, effect of, 233
    heat evolution over time, 233–235
    setting behavior, 233
    testing and characterization, 230–231
durable concrete, 245–247
grades and varieties, 214–219
    ASTM classification, 215–216
    European standard, 216–218
    Indian standards, 218
    overview, 214–215
    standard specifications for cements in China,
        218–219
hydration reactions, 235–245
    of cement, 236–237
    chemistry of fly ash cement, 240–241, 242f
    chemistry of slag cement, 239–240
    microstructure of hardened cement paste,
        237–238
    overview, 235–236
    superplasticizers, 241, 242–245
    temperature on hydration and curing, 238–239
NPC, 214–215, 218, 219, 222–223
OPC, 218, 219, 222–223
overview, 213–214
phase-modified Portland cements
    expansive cement, 225–226
    ultra-rapid-hardening cement, 226
strength grades, 222
Portland fly ash cement, 229
Portlandite, 237
Portland limestone cement (PLC), 215, 348–350
Portland pozzolana cement (PPC), 215, 218, 229
Portland slag cement (PSC), 215, 218, 228
Positive corona, 305
Post-combustion capture, 354, 355
Pour point, defined, 78
Powitec's Pit Navigator, 284, 285
PPC (Portland pozzolana cement), 215, 218, 229
Practical implications, of properties
    Portland cements, 230–235
        compositional aspects, 232
        density and related parameters, 231
        fineness, effect of, 233
        heat evolution over time, 233–235
        setting behavior, 233
        testing and characterization, 230–231

Preblending operation, plant-based QC practice,
    258–260
Precalciner systems
    basic configurations, 149–150
    design considerations, 131–132
    preheater and, 146–152
        design variations of, 150, 151
        overview, 146–151, 152f
        preheater cyclone design, progress in, 151, 152
    types, 150
Pre-combustion capture, 354–355
Preheater-precalciner systems, 146–152
    design variations of, 150, 151
    kiln
        sulphur cycle in, 160
        volatiles cycle in, 161–162
    overview, 146–151, 152f
    preheater cyclone design, progress in, 151, 152
Pre-homogenizing systems, 56–58
Preparation
    coal, 87–92
        ash absorption on clinker, 92
        characteristics on combustion, 91–92
        direct system, 87, 88f
        indirect system, 87, 89f
        semi-direct system, 87, 90f
    process for raw mix, 53, 55–63
        crushing operation, 53, 55–56
        homogenization process, 60–63
        pre-homogenizing systems, 56–58
        raw milling operation, 58–60
    of samples, 17–18
Prevention, SO_2 emissions, 314–315
Primary air, 92–93, 95
Primary fuels, 73
Primary methods, for NO_x emission reduction, 315–317
Principles
    plant-based QC practice, 252
    SNCR, 317
Probes
    gas sampling points, 279–280
    sampling, 274
Processing, raw mix, see Raw mix
Process measurements, plant-based QC practice, 280–284
    gas flow and velocity, 280–281
    overview, 280
    temperature, 282–284
    velocity measurement by anemometers, 282
Process parameters with combustion, relation of, 92–95
    excess air, 93, 94
    flame temperature, 94–95
    primary air, 92–93
    secondary air, 93, 94f
Producer gas, 79, 80t
Product fineness, raw meal, 63
Production growth perspectives, 372–377
    during 2001–2013, 373
    CEMBUREAU, countries of, 373, 374f
    country-wise distribution, 373, 374f
    top cement-producing countries, 373, 375–377, 375t
    twentieth century, 372

Product mixing, Indian cement industry, 390f, 391
Programmable logic controllers (PLCs), 284
Promine, 28
Prompt gamma neutron activation analysis (PGNAA), 258–260, 269–270
Properties
　alternative fuels, 104, 105t
　chemical
　　gypsum, 34–35
　　raw meal, 64
　clay mineral groups, 29, 30t
　clinker phases, 178–184
　coal, 74–77
　CSA, 343
　flow, of cement, grinding aids, 206
　fuels, 74
　liquid fuels, 77–79
　minerals, 3–4
　natural gas and synthetically produced gaseous fuels, 79–80
　petcoke, 97
　physical, of rocks, 4–6
　physicochemical, of limestone occurrences, 22–24
　Portland cements, see Portland cements
　practical implications of, cements, 230–235
　　compositional aspects, 232
　　density and related parameters, 231
　　fineness, effect of, 233
　　heat evolution over time, 233–235
　　setting behavior, 233
　　testing and characterization, 230–231
　rocks, physical and technological properties of, 4–6
Proportioning, raw mix, see Raw mix
Proportions, of clinker phases, 178–184
PSC (Portland slag cement), 215, 218, 228
Pseudowollastonite, 363
PTFE, 301, 303
Pulse energization system, 309
Pulse jet filters, 301, 302f
Purpose, sampling, plant-based QC practice, 252
Pyrite, 18, 87
Pyrometers, optical, 282, 284
Pyrophyllite, 32
Pyroprocessing
　course-changing developments in, 330t
　technologies, potentially disruptive, 331–336
　　electron beam/radiation process, 334–336
　　fluidized bed clinker making process, 331–332
　　microwave heating, 333–334
　　sol-gel process, 332f, 333
Pyroprocessing, clinker cooling and, 141–170
　coolers, 157–161
　　cooling effects on clinker quality, 159–161
　　overview, 157–159
　energy consumption and kiln emissions, 167–169
　formation process, 141–146
　　clinkering reactions and kiln systems, 143–146
　　overview, 141, 142f, 143
　kiln systems
　　burners and combustion, 155–156
　　control strategies, 169–170
　　preheater-precalciner, volatiles cycle in, 161–162
　　refractory lining materials in, 162–166
　overview, 141, 142f
　preheater-precalciner systems, 146–152
　　overview, 146–151, 152f
　　preheater cyclone design, progress in, 151, 152
　rotary kiln systems, 152–155
　　heat balance, 153, 154–155
　　overview, 152–153, 154f
Pyzel process, 331

Q

Q-T diagram, of clinker burning process, 147, 148f
Quality
　assessment, AFR, 118–121
　　organic pollutants, measurement of, 118–119, 120t
　　overview, 118
　　requirements of test facilities, 119, 121
　clinker, cooling conditions on, 159–161
　limestone, 18–24
Quality control (QC)
　cement, 270
　online, in cement plants, 267–272
　　chemical analysis, 268
　　overview, 267
　practice, plant-based, see Plant-based QC practice
Quantity, of samples, 17–18
Quarry design and operational optimization, 27–29
Quartz, 18, 29, 31t

R

Radiation energy, for clinker making, 334–336
Rankinite, 363
Ranks, of coal, 74–77
Rapid fall-through (RFT) cooler, 159
Rapid-hardening Portland cement (RHPC), 215, 222–223
Ratio thermometer, 170
Raw gas, defined, 273
Raw materials
　AFR, see Alternative fuels and raw materials (AFR)
　on unit operations, 36–37
Raw meals, 53
　analysis, 262
　burnability features of, 63–68
　　chemical and mineral characteristics, 64
　　minor constituents, effect of, 64–65
　　product fineness, 63
　chlorine content in, 117
　defined, 60
　homogeneity, 65
　reactivity and burnability, 65–68
　　experimental approaches, 66–68
　　theoretical approaches, 65–66
Raw milling operation, 58–60
Raw mix, 41–70
　burnability features of raw meal, 63–68
　　chemical and mineral characteristics, 64
　　homogeneity, 65

minor constituents, effect of, 64–65
product fineness, 63
reactivity and burnability, 65–68
computation, 41, 42–53
adoption of computer programming, 53, 54f
clinker liquid phase, 46–47
coal ash absorption, 47–48
overview, 41–43, 44f, 48
potential phase computation by Bogue equations, 43, 44–46
step-wise matrix method, 49–53
stoichiometric requirements, 41, 42–48
trial-and-error method, 49
control, 260–263
homogeneity, monitoring, 262–263
hot meal quality monitoring, 263
overview, 260, 261f
XRF, 261–262
design, 48–53
step-wise matrix method, 49–53
trial-and-error method, 49
materials for clinker making, 42t
mineralizers, use of, 68, 69
overview, 41, 42t
preparation process for, 53, 55–63
crushing operation, 53, 55–56
homogenization process, 60–63
pre-homogenizing systems, 56–58
raw milling operation, 58–60
reactivity and burnability of, 141, 143
Raymond, 193
Reactive-belite cement (RBC), 339–342
cell parameters of belite phase on doping, 339, 341t
clinker cooling rate, on cement strength parameter, 339, 340t
NMR spectroscopy, phase quantification by, 341, 342, 342t
stabilization of different polymorphs of, 341
Reactive magnesia, TecEco cements based on, 361–362
Reactivity
hazardous waste, 105
raw meal, 65–68
experimental approaches, 66–68
theoretical approaches, 65–66
raw mixes, 141, 143
Rebinder's effect, 206
Receiving system, AFR, 125, 126f
Reciprocating grate coolers, 157–158
Recirculation, combustion gases, 85
Reclaiming arrangement, 58
Recovery, of carbon dioxide, 354–356
Redhawk power plant, 357
Reduction
noise pollution, 320–321, 383
NO_x emission
primary methods for, 315–317
secondary methods for, 317–318
Reference materials, for calibration, 270–272
Refining, of petroleum, 77
Refractory lining materials, in kiln system, 162–166
degradation, in service, 166

kiln zonation, 163–165
overview, 162–163
suspension preheater section, 166
Refuse-derived fuel (RDF), 110–112
Regional distribution, of coal reserves, 81t
Regional growth trends in cement production, *see* Global and regional growth trends
Rehydration, 34
Rejects, 190
Reliability, of different categories of reserves, 14
Remelting reactions, 342
Remote terminal unit (RTU), 284
Requirements
cement performance, 380–381
of test facilities, 119, 121
Research and development (R&D), in cement manufacture, 327–368
CAC, complex building products formulation with, 352–353
electrochemical CO_2 reduction, 359
low-carbon cement industry, 354–358
algae cultivation with kiln exhaust gases, 357–358
carbon dioxide, capture and recovery of, 354–356
overview, 354
low-carbon cements and concretes, 359–366
Calera process for calcium carbonate cement, 360–361
Calix and novacem processes for magnesia-based products, 362–363
CO_2-SUICOM technology, 364–365
overview, 359–360
Solidia cement and concrete, 363–364
SWOT analysis of carbonation technologies in development, 365, 366t
TecEco cements based on reactive magnesia, 361–362
nanotechnology in cement research, 366–367
non-hydraulic cements, 367–368
overview, 327–328
Portland cement derivatives, application, 336–352
modified, 338–352
overview, 336–338
pyroprocessing technologies, potentially disruptive, 331–336
electron beam/radiation process, 334–336
fluidized bed clinker making process, 331–332
microwave heating, 333–334
sol-gel process, 332f, 333
sustaining technologies in growth, 328–330
advanced engineering features, 328, 329
changes in technology and scale of operation, 328, 329f
critical milestones in Portland cement chemistry, 328t
pyroprocessing, course-changing developments in, 330t
raw and finish grinding processes, tentative technological milestones, 330t
technology options for converting CO_2 into fuel products, 358–359

Reserve(s)
 industrial implications of categorization of, 14–15
 limestone deposits, 10, 11–13
 reliability of different categories of, 14
 sizable, 387
Reserves and Resources and Minerals and Commodities, 10
Reserve-to-production (R/P) ratio, for oil and gas, 81
Residues, biomass, 112, 113t
Resistance thermometers, 282
Resistivity, of dust, 307–309
Resource, limestone deposits, 10, 11–13
Reverse air bag filters, 300, 301, 302f
Revolving-disc cooler (RDC), 159
RHPC (rapid-hardening Portland cement), 215, 222–223
Rice straw and husk, 112
Rietveld refinement technique, 266
Rittinger's law, 206
Rocks
 characterization, 2–6
 classification, 2, 4t
 defined, 2
 physical and technological properties of, 4–6
Roller mills, 87
 VRMs, see Vertical roller mills (VRMs)
Roller press systems, grinding aids for, 208–209
Roll presses
 hydraulic, 195–198
 raw milling operation, 58–60
Roman cement, 213
Rosin-Rammler- Sperling-Bennet (RRSB) plot, of PSD, 265
Rotary coolers, 157–158
Rotary kiln systems, 152–155
 heat balance, 153, 154–155
 overview, 152–153, 154f
RRSB (Rosin-Rammler- Sperling-Bennet) plot, of PSD, 265
Rubber Manufacturers Association, 114

S

Sampling
 dimension, quantity, and preparation, 17–18
 gas, points and probes, 279–280
 limestone deposits for evaluation, 15, 16–17
 plant-based QC practice
  general statistical considerations, 255–256
  guidelines, 252–254
  principles, 252
  purpose, 252
  scheme, establishing, 252–254
  stations, 254–256
 probes, 274
 systems, gas, 276–278
Sand, limestone and, 50, 51–53
Satin spar, 33
SCADA (supervisory control and data acquisition), 284
Scanning electron microscopes (SEMs), 178, 230
SCF (segregated combustible fractions), 110, 111–112
Scheme, sampling, plant-based QC practice, 252–254

SCMs (supplementary cementitious materials), 215, 361
Scrap motor tires, 112, 114
Secondary air, 93, 94f, 95
Secondary fuels, 73
Secondary methods, for NO_x emission reduction, 317–318
Sedimentary rocks, 2
Segregated combustible fractions (SCF), 110, 111–112
Selective catalytic reduction (SCR) technology, 137, 317
Selective non-catalytic reduction (SNCR) technology, 137, 169, 317
Selectivity curve, 191
Selenite, 33
Self-cleaning property, 366–367
Self-compacting concretes, 366
Semi-direct system, of coal firing, 87, 90f
Semi-VOCs, 121
SEMs (scanning electron microscopes), 178, 230
Separate-line calciners (SLC), 150
Separators, types of, 188, 189–191
Sepax of FLSmidth & Co., 190
Sepmaster-SKS of KHD-Humboldt-Wedag, 190
Sepol of Krupp Polysius group, 190
Setting behavior, portland cement, 233
Sewage sludge, 115–117
Shale, limestone and, 50, 51–53
Shrinkage-compensating concrete, 226
Siderite, 18
Silica modulus (SM), 43, 44f, 64, 66
Siliceous limestone, 19
Silos
 designs, 210
 homogenizing, 61–63
 storage, 210
Silo-set, 206, 233
Silo storage systems, 128, 129f
Silver, as catalyst, 359
Sizable reserves, 387
Slag cement, hydration chemistry of, 239–240
Slates, 31
Sludge
 process-wise, 114–115
 sewage, 115–117
Slurry, 114–115
Smoldering coal, 87
SNCR (selective non-catalytic reduction) technology, 137, 169, 317
Software(s)
 packages, basic feature, 28
 RGB camera, 285
 Rietveld, 267
 spread-sheet, 53, 54f
SOLBAS, 269
Sol-gel process, of cement manufacture, 332f, 333
Solidia cement and concrete, 363–364
Solid pollutants, flue gas component, 273
Solid recovered fuel (SRF), 110, 111–112
Solid-state electrolyte sensor, 275
Soluble anhydrite, 34
Solvay process, 348
Sound level meter, defined, 323

INDEX

Soundness, of cement, 231
Soundproof cabin, 130
South Africa, coal inventory, 83t
Specifications
   cement-grade limestone, 20, 22
   fuel oils, 78, 79t
   hazardous waste fuels, 115t
Specific surface area (SSA), 230, 231
Spectralflow Online Analyzers, 270
Spent pot-liners, 117
SRMs (standard reference materials), 261, 270, 271, 272
SRPC (sulfate-resisting Portland cements), 215, 223
SRPC (sulfate-resisting portland cements), 223
Stability, of clinker burning operation, 65
Stack gas analysis, 272–280
   in cement production, 278
   extractive sampling and measurement techniques, 274–276
   gas sampling points and probes, 279–280
   interlock and warning systems, 278–279
   overview, 272–274
   special considerations for gas sampling systems, 276–278
Stacking, 57
Stacking-and-reclaiming systems, 56
Stakeholders engagement, feasibility of AFR project, 109
Standard reference materials (SRMs), 261, 270, 271, 272
Standard specifications, for cements in China, 218–219
Static separators, 188, 189
Stations, sampling, plant-based QC practice, 254–256
Statistical considerations, sampling stations, 255–256
Status
   coal production, 82
   petcoke, 96
Steel slag, 135
Stefan–Boltzman equation, 284
Step-wise matrix method, 49–53
Stockpiling
   coals, 87
   methods, 56–58
Stoichiometric requirements, raw mix computation, 41, 42–48
   clinker liquid phase, 46–47
   coal ash absorption, 47–48
   overview, 41–43, 44f
   potential phase computation by Bogue equations, 43, 44–46
Storage systems
   alternative fuels, 125–131
      automation and control, 131
      overview, 125–127
      silo storage systems, 128, 129f
      storage yard with pit, 127–128
   cement, 209–210
Strength
   hardened cement paste, 238
   OPC/NPC, 222–223
Stripping ratio, 25
S-type Portland cements, 226
Sub-bituminous coal, 74, 75
Substitute fuel, petcoke as, 96–100
   classification, 96–97
   firing of, 98–99
   grinding, 97–98
   overview, 96
   production, world status, 96
   properties, 97
Sulfate-resisting Portland cements (SRPC), 215, 223
Sulfoaluminate cement (SAC), 343
Sulfonated melamine formaldehyde, 243
Sulfonated naphthalene formaldehyde, 243
Sulfur content
   coal, 76
   fuel oil, 78
   petcoke, 98
   sludge and sludge ash, 114
Sulfur dioxide, 136
   emissions, 313–315
      overview, 313–314
      prevention and abatement opportunities, 314–315
   flue gas component, 273
   measurement technique, 275–276
   pollutant, 289–290
   pollution monitoring, 293
   presence and quantity, 278
Sulphur cycle, in preheater-precalciner kiln, 160
Sumitomo, 331
Superplasticizers
   adsorption, 243
   applications, 243
   in cement hydration, 241, 242–245
   clinker–calcium sulphate–superplasticizer, 244f
   defined, 242
   developments, 381
   incompatibility issues, 244–245
   key function, 243
   sulfonated, 243–244
Supersulfated cement, 230
Supervisory control and data acquisition (SCADA), 284
Supplementary cementitious materials (SCMs), 215, 361
Support rollers, 194
Surface miner, 25
Suspended particulate matter (SPM), 318
Suspended solid particles (SPM), 289
Suspension preheaters, *see* Preheater-precalciner systems
Suspension preheater section, lining of, 166
Sustainable industrial development, 287–288
Sustaining technologies, in growth of cement industry, 328–330
   advanced engineering features, 328, 329
   changes in technology and scale of operation, 328, 329f
   critical milestones in Portland cement chemistry, 328t
   pyroprocessing, course-changing developments in, 330t
   raw and finish grinding processes, tentative technological milestones, 330t
SWOT analysis, of carbonation technologies in development, 365, 366t

Synthetically produced gaseous fuels, properties, 79–80
Systematic quality assessment, AFR, 118–121
    organic pollutants, measurement of, 118–119, 120t
    overview, 118
    requirements of test facilities, 119, 121
Systemic requirements, alternative fuels, 124–132
    overview, 124, 125f
    precalciner design considerations, 131–132
    storing, dosing, and conveying, 125–131
        automation and control, 131
        dosing and conveyance to kilns, 128, 130
        silo storage systems, 128, 129f
        storage yard with pit, 127–128
        tires and whole large packages, 130

T

Tailings, 190
TCLP (toxicity characteristics leaching procedure), 106, 121
TecEco cements, based on reactive magnesia, 361–362
Technological properties, of rocks, 4–6
Technological status, Indian cement industry, 386–391
    captive power generation, 388, 389, 391
    electricity consumption, 388, 389t
    evolution of process profiles, 387
    features, 386
    fuel usage and thermal energy consumption, 387, 388, 389f
    future direction, 391
    limestone resources, 386–387
    phosphogypsum, beneficiation of, 391
    product mixing, 390f, 391
Technology, converting CO_2 into fuel products, 358–359
Temperature
    calcium carbonate, 144–146
    of flame, 94–95
    on hydration and curing, 238–239
    ignition, of petcoke, 97–98
    measurements, 282–284
    resistivity, of dust, 307, 308f
Ternal RG, 353
Terra alba, 33
Terrace mining, 25
Test facilities, requirements of, 119, 121
Testing, Portland cements, 230–231
Tetracalcium aluminoferrite, 41
Theoretical approaches, reactivity and burnability of raw meal, 65–66
Thermal characteristics, of coal, oil, and gas, 80–81
Thermal energy consumption, Indian cement industry, 387, 388, 389f
Thermal NO_x, 137, 169
Thermo-chemical reactivity, of clay minerals, 31–32
Thermocouples, 282, 283f
Thermogravimetric analyzer (TGA), 230
Thermometer, ratio, 170
Thermometers, resistance, 282
Tire-derived and other secondary fuels through gasification, 133, 134f, 135
Tires and whole large packages, dosing of, 130

Tokyo Tama Eco-cement facility, 348
Topas BBQ, 267
Total organic carbon (TOC), 121
Total petroleum hydrocarbons (TPH), 121
Total process control system, 284–285
Toxicity, hazardous waste, 105–106
Toxicity characteristics leaching procedure (TCLP), 106, 121
Trace elements, 290–291, 293
Transportation, of gypsum, 247
Travertine, 19
Trends, in cement production
    global and regional growth, see Global and regional growth trends
    research and development, see Research and development (R&D)
Trial-and-error method, 49
Tricalcium aluminate, 41
Tricalcium silicate, 41, 363
Triethnolamine, 206
Tromp curve, 191, 192f
Tubular photobioreactor, 357
Tufa, 19

U

Ultra-rapid-hardening cement, 226
Ultra violet–infrared spectrometer (UV-IR), 230
Ultraviolet radiation, 276
United Nations Economic and Social Council (ECOSOC), 11
United Nations Framework Classification (UNFC), 10, 11–13
United Nations Industrial Development Organization, 377
Unit operations, raw materials on, 36–37
Unstable flame, in kiln, 95
Utilization, of fuels, 288

V

Variations, in opencast mining, 25–26
Varieties, Portland cements, 214–219
    ASTM classification, 215–216
    European standard, 216–218
    Indian standards, 218
    overview, 214–215
    standard specifications for cements in China, 218–219
Variograms, 256
Vaterite, 361, 365
Velocity, measurements
    air/gas, 280–281
    by anemometers, 282
Vermiculite, 32
Vertex mill, operating principles of, 123
Vertical roller mills (VRMs), 58–60, 98
    clinker grinding system, 193–195
        grinding aids for, 208
        main requirement, 194
        schematic diagram, 193f
        types, 193

typical flow chart, 195
working principle, 194, 195
Vicat plunger, 231
Viscosity, liquid fuels, 77–78
Visual impacts, feasibility of AFR project, 108–109
Volatile matter, combustion of coal, 91
Volatile organic compounds (VOCs), 121
Volatiles cycle, in preheater-precalciner kiln systems, 161–162
Volume shrinkage temperature relation, 68
Votorantim, 377
VRMs, *see* Vertical roller mills (VRMs)
V-separator, 196–197
VSK separator, 197

W

Warning systems, flue gas analysis, 278–279
Waste(s)
    agricultural, 104
    combustible, 104
    from diverse sources, 103
    hazardous
        AFR inventory and material characteristics, 114–115
        classifying, biological parameters in, 106–107
        defined, 104, 105–106
        incinerators, cement kilns *vs.*, 122t
    incombustible, 104
    industrial, 104
    MSW, 110–112
    plastic, industrial, 117
    storage, 125
    utilization and disposal, 288
Water
    emissions to, 108
    gas, 79, 80t
    mineral in liquid form, 2
    reducers, 241, 242–243
    sprays, control of ball mill temperature, 187
    spray systems, 311
Waterproofing materials, 381

Wavelength-dispersive XRF (WDXRF), 261
WDXRF (wavelength-dispersive XRF), 261
Wet gas, 79
Wet probe, 279
Wet scrubbing, 315
White Portland cement (WPC), 215, 224
Windrow method, 56, 57
Wollastonite, 363
Working principles
    horomills, 198f
    VRMs, 194, 195
World Bank, 377
World Business Council for Sustainable Development, 377
World cement production
    during 2001–2013, 373
    CEMBUREAU, countries of, 373, 374f
    country-wise distribution, 373, 374f
    top cement-producing countries, 373, 375–377, 375t
    twentieth century, 372
World reserves
    coal, 81–83
        in Asia-Pacific Region, 81, 82t
        inventory in India, 82–83
        regional distribution, 81t
        R/P ratio, 81
        status, of production, 82
    petcoke, 96
Woven fabrics, 299–300

X

XPAC, 28
X-ray diffractometry (XRD), 45, 230, 265–267, 268, 269, 270, 271, 334
X-ray fluorescence spectrometry (XRF), 230, 261–262, 268, 269, 271, 272

Z

Zirconia, 275
Zonation, kiln, refractory materials and, 163–165